ANALYSIS TECHNIQUES FOR RACECAR DATA ACQUISITION

Other SAE titles of interest:

Hands-On Race Car Engineer
By John H. Glimmerveen
(Product Code: R-323)

Formula 1 Technology
By Peter Wright and Tony Matthews
(Product Code: R-230)

Race Car Vehicle Dynamics
By William F. Milliken and Douglas L. Milliken
(Product Code: R-146)

Fundamentals of Vehicle Dynamics
By Thomas D. Gillespie
(Product Code: R-114)

For more information or to order a book, contact SAE at

400 Commonwealth Drive
Warrendale, PA 15096-0001
phone (724) 776-4970; fax (724) 776-0790
e-mail CustomerService@sae.org;
website http://store.sae.org.

ANALYSIS TECHNIQUES FOR RACECAR DATA ACQUISITION

JÖRGE SEGERS

SAE International™
Warrendale, Pennsylvania, USA

For permission and licensing requests contact:

SAE Permissions
400 Commonwealth Drive
Warrendale, PA 15096-0001-USA
Email: permissions@sae.org
Tel: 724-772-4028
Fax: 724-772-4891

Library of Congress Cataloging-in-Publication Data

Segers, Jorge.
Analysis techniques for racecar data acquisition / Jorge Segers.
p. cm.
ISBN 978-0-7680-1655-0
1. Automobiles, Racing--Dynamics--Data processing.
2. Automobiles, Racing--Performance--Measurement.
3. Automobiles, Racing--Testing.
I. Title.
TL243.S43 2008
629.228028'7—dc22
2007047077

SAE International
400 Commonwealth Drive
Warrendale, PA 15096-0001 USA
Tel: 877-606-7323 (inside USA and Canada)
Tel: 724-776-4970 (outside USA)
Fax: 724-776-1615
Email: CustomerService@sae.org

ISBN 978-0-7680-1655-0
SAE Order No. R-367

Printed in USA

TABLE OF CONTENTS

PREFACE

A proven way for athletes to be successful in any sporting discipline is for them to record their performance, analyze what has happened, and draw conclusions from the factors that influence that performance. Marathon runners log their running speed and distance with their heart rate to optimize their training schedules. Football players record their games on video to evaluate techniques, performance, and tactics. Chess players write down every move in a game to replay and analyze it afterward. They measure something, learn from it, and try to use it to their advantage next time.

In motor racing, sophisticated recording devices are used in conjunction with numerous sensors to record what the car and its driver are doing. Engineers often are employed full-time to maintain the system, analyze the recorded data, and draw the correct conclusions from it.

Motor racing is known for high-end technology, and this technology changes every day. Ten years ago, racecar data acquisition was somewhat limited to well-funded teams in high-profile championships. Nowadays, the cost of electronics has decreased dramatically. Powerful computers are available for very little expense. Data acquisition systems are now sold for the price of a single racing tire. This means data acquisition has become accessible to everyone.

Whatever the price of the data acquisition system, it is a waste of money if the recorded data is not interpreted correctly. This book contains enough information to prevent the investment in a data acquisition system from being a waste of money.

Whether measuring the performance of a Formula One racecar or that of a road-legal street car on the local drag strip, the dynamics of the vehicles and their drivers remain the same. Identical analysis techniques apply. This book contains a collection of techniques for analyzing data recorded by any vehicle's data acquisition system. It details how to measure the performance of the vehicle and driver, what can be learned from it, and how this information can be used to your advantage the next time the vehicle hits the track.

ACKNOWLEDGMENTS

When I began working in motor racing in 1998, I soon learned that this business is a team effort. The sum of the qualities of each member determines the team's success. Eight years later, when I wrote this book, I learned this also is a team effort, very similar to running a successful racing team. That is why I would like to begin by appropriately crediting "my" team.

First, I would like to thank everyone at SAE International for guiding this project in the right direction. Special thanks go out to Martha Swiss, intellectual property manager, and Terri Kelly, administrative assistant, as well as Terry Wilson for artwork.

A big contributor to this book was Josep Fontdecaba I. Buj, engineering director at Creuat S.L., not only for writing the greater part of Chapter 10, but especially for the many discussions we had about suspension setup and data analysis. His input added immeasurable value to this book.

David Brown and Andrew Durant at Race Technology gave me detailed insight about GPS-based data acquisition techniques. I would like to thank them for providing me with the hardware that was used to create much of the data traces used throughout this book. Their company is proof that data acquisition can be affordable for all motor racing disciplines.

I am proud to have Pi Research support the creation of this book. The information and analysis files supplied by this company were invaluable. Thanks go out to Thomas Buckler and Michael de Cock.

The following people deserve credit for taking the time to evaluate the manuscript and for providing me with invaluable feedback: Peter Wright (consultant to the FIA), Dr. Wolfgang Ullrich (head of Audi Sport), John Glimmerveen (author of the book *Hands-on Racecar Engineer*), Doug and Bill Milliken (authors of the book *Racecar Vehicle Dynamics*), and William C. Mitchell (head of Mitchell Software).

This book addresses what I know about racecar data acquisition, and what I know is influenced greatly by the people I had a chance to work with. Therefore, my great respect goes out to all the engineers, mechanics, and team owners that were there to teach me. I hope I can repay these debts when they read this book.

Every graph in this book was created by a racecar driver. Many of these graphs resulted in successful track performance, pole positions, race victories, and championships. I thank all of these drivers for providing me with data to analyze.

Finally, I would like to thank Henrik Roos of the Simbin Development Team for triggering my interest in writing. He gave me the idea to write a book on this little-documented subject in the first place.

Jörge Segers

CHAPTER 1
INTRODUCTION

***Figure 1.1** Racecar data logging systems record user-defined parameters while the car is in motion. These data can be downloaded afterwards to a computer and analyzed. (Courtesy of GLPK-Carsport)*

One of the most important weapons a racecar team can employ is information. The more information it can gather (and process), the better its judgment will be in making key decisions. Data acquisition provides engineers with the information they and the team require to evaluate vehicle performance.

What Is This Book All About?

Nowadays, almost every racecar is equipped with a data logging device that can measure almost every performance parameter of the vehicle and its driver. These measurements can be used to examine the effects of setup changes or changing track conditions, driving style, and causes of performance variations or component failures.

This book covers the use of electronic data logging systems in racecars. It is not a how-to manual for installing a data logger, selecting different components, or choosing the most appropriate configuration, although these topics are discussed briefly. This book is primarily about analyzing the endless data streams produced by the system. It is also about using this information to evaluate and optimize a given racecar's setup.

The data logging system provides information about how a car-driver combination is performing at a particular location on a racetrack. This book takes the analysis a step further and tries to determine why the car/driver is performing in this particular way at this particular place on the track. Upon completing the book, the reader will have the insight to effectively use competition car data acquisition.

Useful literature addressing racecar data acquisition and data analysis already has been published. However, in this work a mathematical approach to data analysis is emphasized, with the primary intention being to show the reader how even a limited amount of data can provide useful information about racecar dynamics.

In the early days of motor racing, many of the engineer's decisions were based on intuition and experience. The stopwatch, tire pressure gauge, pyrometer, and driver's comments served as the data logging system. Nowadays, the electronic data acquisition system provides almost everything the engineer needs to know about the car's behavior.

As the degree of competition increases, costs of racing and testing increase. Because of this, there is a greater demand for understanding the racecar dynamics to increase testing efficiency, educate drivers, and provide them with the tools to educate themselves. It is also important to provide the parameters to simulate the dynamic behavior of the car. This more than justifies the use of data acquisition, and in the past this was indeed only possible at the higher echelons of motor racing. However, times have changed, and data acquisition is now a common technique from Formula One and Indy cars to clubsport championships and karting.

The configuration of the book is as follows:

Chapter 1: This chapter is an introduction to data acquisition. What is data acquisition's purpose? What should one measure? What are the hardware requirements? What are the latest developments in data acquisition?

Chapter 2: What should one require from the analysis software? What different ways can data be displayed? How can one manipulate the data channels?

Chapter 3: During race weekends or test sessions, time limitations require the engineer to be able to quickly find what he needs in the logged data. Developing the ability to read the graphs is required.

Chapter 4: In this chapter, straight-line acceleration is analyzed as well as the ability of the racecar to overcome the external resistances acting upon it.

***Chapter 5*:** Acceleration usually is followed by braking. This chapter covers the performance analysis of the car's braking system.

Chapter 6: Most racecars carry a gearbox to adapt the vehicle torque to a wide range of velocities. This chapter discusses shifting techniques and the choice of the proper gear ratios for a given racetrack.

Chapter 7: How can the car's cornering balance be evaluated? How does one diagnose oversteering and understeering?

Chapter 8: In this chapter, methods to quantify the roll stiffness and its distribution on the front and rear axles are covered.

Chapter 9: This chapter addresses how the loads at the tires' contact patches are calculated and investigates the effects of lateral and longitudinal load transfer.

Figure 1.2 ***This graph shows the signals recorded from four potentiometers mounted on the suspension as well as the damper acceleration of the left front wheel. The area indicated shows what happened to the wheels when the photograph in Figure 1.1 was taken.***

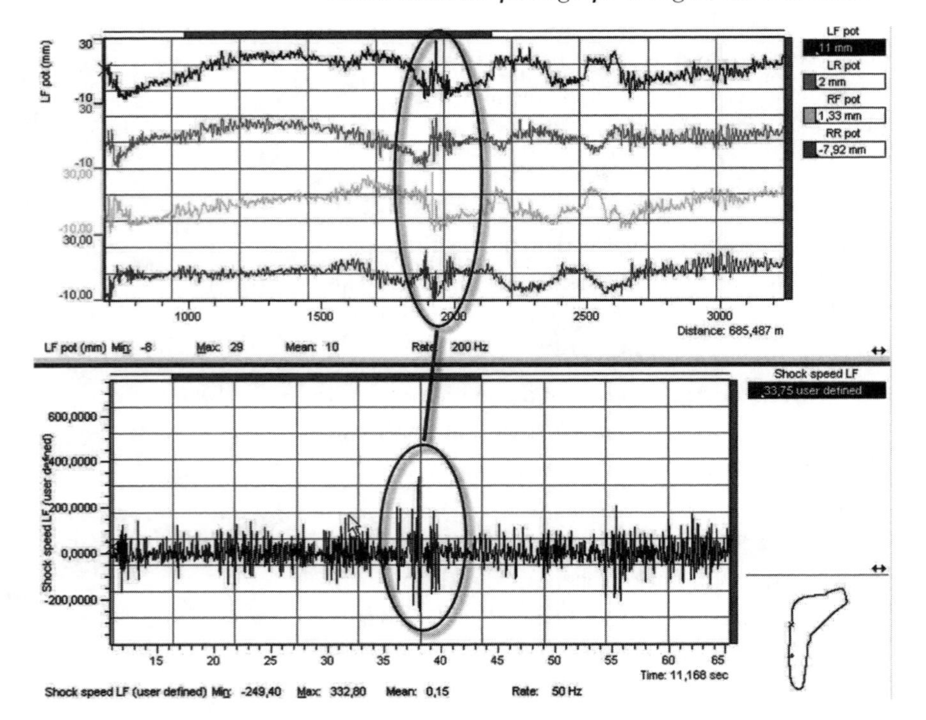

Chapter 10: The shock absorbers determine the transient behavior of the racecar. This chapter focuses on analyzing the damping characteristics of the vehicle and how the springs, shock absorbers, and antiroll bars are matched to each other.

Chapter 11: A car traveling through air is subject to aerodynamic forces. The methods used to measure these forces and an analysis of their effect on the vehicle's balance are covered in this chapter.

Chapter 12: Driving style and racing lines— how much of the traction circle is the driver using? How can data analysis help improve the driver's performance?

Chapter 13: Lap time simulation is a recent technology that has found its way into almost all levels of motor racing. With the data logging system, this is a valuable and cost-effective tool for vehicle development. This chapter illustrates how the logged data helps tune the vehicle simulation model.

Chapter 14: Data logging can be an invaluable help in selecting the right tactics for the race or practice session. This chapter shows how to use the data logging system as a predictive tool for race strategy.

Chapter 15: The last chapter introduces sensor technology. Usable data must be measured correctly. To ensure this, a basic knowledge of measurement technology is required.

What Is Data Acquisition?

Put simply, a racecar data acquisition system is an electronic memory unit that stores user-defined parameters as a function of time while the car is on the track. The stored data can be downloaded to a computer where it can be analyzed, often with specialized software packages.

Data Acquisition Categories

This analysis can be broken down into the following categories:

1. Vehicle Performance Analysis

Data analysis supports the comments of the driver. The engineer can pinpoint more easily handling problems and the locations on the racetrack where they occur. From this analysis, the engineer decides which setup changes should be made to the car for the next driving session.

2. Driver Performance Analysis

Logging the cockpit activities of drivers sheds some light on their driving style. Data acquisition makes it possible to analyze different laps by a driver or to compare the differences in style and performance between multiple drivers. This type of analysis is particularly useful in championships where more than one driver uses the same car.

3. Vehicle Development

Data logging is an invaluable tool in a racecar development program. Purpose-driven measurements aid in decision-making with regard to what direction development should be focused.

4. Reliability and Safety

By recording vital channels such as engine oil pressure and temperatures as well as battery voltages, reliability problems can be discovered before more damage to the car is done. Safety is another factor in play here (e.g., tire pressure monitoring systems).

5. Determining Vehicle Parameters

Racecar simulation software is becoming more popular. To develop a simulation model of a given racecar, all relevant parameters of the car should be known by the programmer to guarantee sufficient model accuracy. Model parameter examples include vehicle suspended and unsuspended mass, track width, wheelbase, center of gravity height, and roll center locations. Some parameters can be measured on the static vehicle or calculated; others should be measured under racing conditions.

6. Running Logs

A data logging system records the active history of a racecar. It records how long the car runs and what happens to it during this time. When this information is coupled with the vital parts of the car, a running log of component lifetimes can be created. The system keeps track of when a part should be replaced and rebuilds performed.

Data Categories

Although many of the signals often are interrelated, the data that the system measures can be divided generally into the following categories:

1. The Vital Functions of the Car

These signals include the important engine and driveline-related channels such as engine oil pressure and temperature, water temperature, fuel pressure, gearbox and differential temperature, and battery voltage. Engine revolutions per minute (RPM) also fall into this category.

2. Driver Activity

Driver activity parameters are those over which the driver has direct control. They include throttle position, steering angle, brake pedal position, and gear.

3. Chassis Parameters

These are the vehicle dynamics-related signals such as vehicle speed, lateral and longitudinal g-force (Gs), steering angle, damper position, brake line pressure, tire temperatures and pressures, ride height, and suspension loads.

Basic Data Acquisition Signals

Depending on the budget available for acquiring a data acquisition system, the possibilities are almost infinite. Data acquisition systems exist for almost any application. A traditional configuration for data acquisition starters consists of a suitable logging unit that measures the following signals for chassis and driver performance analysis:

- engine RPM,
- wheel speed,
- throttle position,
- steering angle,
- lateral acceleration, and
- longitudinal acceleration.

In addition to these channels, the vital functions of the car (e.g., fluid pressures and temperatures, battery voltage) should be logged. A beacon channel should be provided to indicate the beginning and end of a lap. Measuring the six basic signals already gives the engineer a massive amount of data to analyze.

Supplemental Data Acquisition Signals

Extended vehicle dynamics analysis can require more sensor signals to be recorded, the most important probably being suspension travel. Next to the six signals previously mentioned, the following channels are recommended (in order of priority):

- suspension (shock absorber) movement,
- brake line pressure,
- clutch pressure,
- gear position,

- speed of each wheel,
- front and rear axle lateral acceleration,
- vertical acceleration,
- tire pressures,
- ride height,
- suspension loads (strain gauges),
- tire temperatures,
- brake disc temperatures,
- yaw speed (gyroscope),
- propshaft torque,
- aerodynamic pressures (pitot tubes), and
- gear lever force.

This list can go on, and of course engineers always ask themselves what exactly they want to measure. Specific needs require specific channels to log. Most teams start data-logging the six basic channels and then extend the system step by step as they gain more experience in analyzing the data. Analysis often provides as many answers as it does new questions. However, suspension movement is usually the next logical step. When investing in a system, keep in mind the number of signals may be extended in the future, which impacts the wiring harness, available memory, and other hardware.

The engine electronic control unit (ECU) often features its own data logger that records engine-specific data. This system should be capable of communicating to the external data acquisition unit logging the chassis-related parameters. In this way, the signals from the engine ECU can be transferred and overlaid with lap-timing beacons. For engine performance analysis, the most important signals are engine RPM, throttle position, lambda, and airbox pressure.

Example of Parameters

Table 1.1 provides an overview of the parameters logged from the GLPK Racing's Dodge Viper during the 2004 Belgian GT season. The system used was a MoTeC advanced dash logger (ADL) with an internal memory of 10 Mb, communicating with the engine ECU to receive all engine-related channels.

Most engine parameters are recorded by the engine management system and sent to the data logger through a serial link. The logging unit measures and stores gearbox and differential temperatures as additional vital channels.

The six basic channels are all present: engine RPM, vehicle speed, throttle position, steering angle, and lateral and longitudinal acceleration. There are five different *g*-force channels. Lateral, longitudinal, and vertical acceleration are measured by a three-axis *g*-force sensor located near the car's center of gravity. In addition, two lateral *g*-force sensors are located on the front and rear axle. These are convenient for analyzing understeering and oversteering.

There are eight wheel-speed signals recorded by the logging unit. This might seem a bit excessive, but this particular car was equipped with an engine-controlled traction control system (TCS) and, completely separate from this, an antilock brake system (ABS). Because the team uses separate wheel-speed sensors for both systems, a failure in one does not affect the other. All eight signals are logged for analysis as well as diagnostic purposes.

Three sensors indicate what occurs in the braking system. The amount of pedal effort by the driver is recorded by a linear potentiometer. Brake line pressures are logged as well. In addition to being useful for analysis, brake line pressure readouts make it easier to adjust the brake balance.

Suspension travel is measured by four potentiometers mounted on the shock absorbers. Three locations per tire measure tire surface temperature, which accounts for another twelve sensor signals.

Table 1.1 already represents fifty-one channels that are directly logged. From these, the analysis software calculates another seventy-three channels, which brings the total to 124. Getting lost becomes a potential risk. In this case, the investment in such a system and all the sensors was justified by several reasons:

- The championship consisted of seven races on three different racetracks. To continuously improve the vehicle and driver performance, more data was required.
- During the 2004 season, the team was developing a semiactive hydraulic suspension system. Vehicle dynamic parameters were measured to compare to those measured using the conventional suspension.
- Traction control and the ABS required four wheel-speed sensors anyway, so those signals

Table 1.1 *Logged channels on GLPK's Dodge Viper GTS-R*

Channel	Description
1	Engine RPM (measured by engine ECU)
2	Engine oil temperature (measured by engine ECU)
3	Engine oil pressure (measured by engine ECU
4	Air inlet manifold pressure (measured by engine ECU)
5	Throttle position (measured by engine ECU)
6	Lambda left (measured by engine ECU)
7	Lambda right (measured by engine ECU)
8	Engine water temperature (measured by engine ECU)
9	Air temperature before throttle (measured by engine ECU)
10	Battery voltage at engine ECU (measured by engine ECU)
11	Internal temperature engine ECU (measured by engine ECU)
12	Lateral *g*-force at center of gravity
13	Longitudinal *g*-force at center of gravity
14	Vertical *g*-force at center of gravity
15	Lateral *g*-force at front axle
16	Lateral *g*-force at rear axle
17	Steered angle
18	Brake pedal position
19	Brake line pressure front
20	Brake line pressure rear
21	Damper position front left
22	Damper position front right
23	Damper position rear left
24	Damper position rear right
25	Tire temperature rear right inner
26	Tire temperature rear right middle
27	Tire temperature rear right outer
28	Tire temperature rear left inner
29	Tire temperature rear left middle
30	Tire temperature rear left outer
31	Tire temperature front right inner
32	Tire temperature front right middle
33	Tire temperature front right outer
34	Tire temperature front left inner
35	Tire temperature front left middle
36	Tire temperature front left outer
37	Battery voltage at ADL
38	Differential oil temperature
39	Gearbox oil temperature
40	Internal temperature ADL
41	Traction control wheel speed front left
42	Traction control wheel speed front right
43	Traction control wheel speed rear left
44	Traction control wheel speed rear right
45	Beacon lap time
46	ABS wheel speed front left
47	ABS wheel speed front right
48	ABS wheel speed rear left
49	ABS wheel speed rear right
50	Gear position (measured by engine ECU)
51	Gear lever force (measured by engine ECU)

(in this case, eight) were wired to the data logger as well for analysis and diagnostics.

- The team implemented lap-time simulation software and used the data acquisition system to help build a virtual model of the racecar.

More signals measured means more accurate conclusions, but it often requires more analysis skills as well. Getting the most out of the available channels is explored in this book.

Hardware

Data acquisition systems are available in various configurations, but they always have the main components in common ***(Figure 1.3)***.

Figure 1.3 General configuration of a data acquisition system

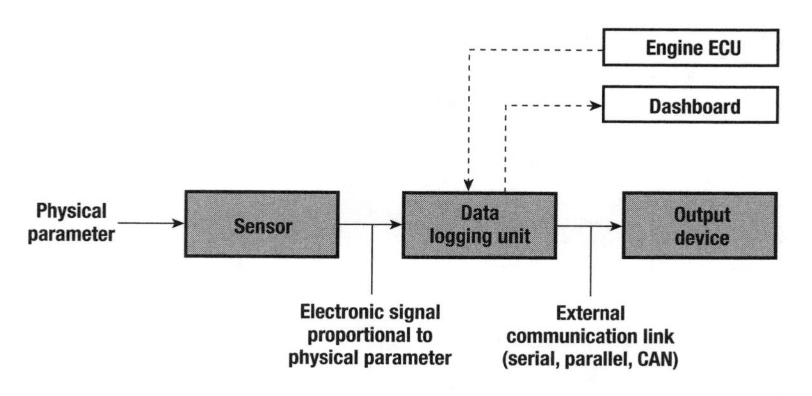

Figure 1.4 An example of a possible data acquisition hardware configuration

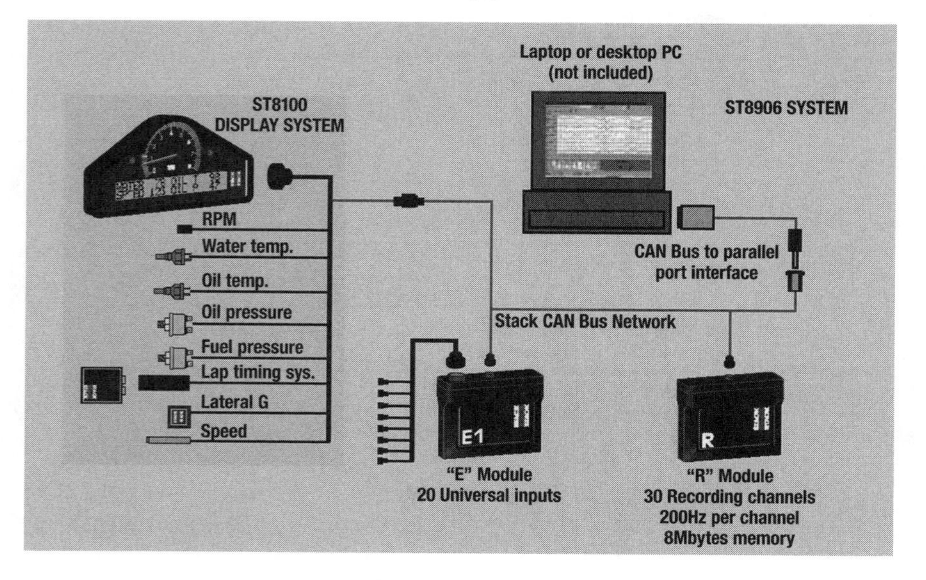

A physical parameter (e.g., pressure, temperature, speed, force) of interest is captured by a sensor that transforms the measurement into an electronic signal proportional to this parameter and understandable to the data logging unit. The most important property of the data logging unit is that it stores the measured parameters in an electronic memory. An output device (computer or laptop) can communicate with the data logger via an external link. This link is very often bidirectional because most systems offer some parameters to be configured by the user. Controller area network (CAN) communication links are becoming more popular as a replacement for serial or parallel links because of the communication (downloading and uploading) speed and the much easier addition of different devices to the system.

Via this or a separate communication link, an external display can be added to the system to visualize sensor readings to the driver. Some current systems on the market are dashboards with an integrated memory. In this case, the dashboard and data logger form one unit. Most engine ECUs offer the possibility to transfer engine-related sensor signals to an external data logger.

Figure 1.4 gives an example of a possible configuration. It concerns an engineering system from STACK Ltd. built around a CAN. The system starts from a display system with eight possible inputs (RPM, water and oil temperature, oil and fuel pressure, lap beacon, lateral G and speed). The dashboard measures these values but does not store them. Through a connection to a CAN, the measured signals are transferred to a recording module (the logger). To allow more inputs to be recorded, additional input modules can be added to the network (only one is pictured). With an interface cable, the user can link a computer to the network to download data and configure the system.

Recent Hardware Trends

During the last three decades, data acquisition systems have come a long way and basically followed the advances in microcontroller technology. The three areas of primary importance are the available memory to store data from increasing numbers of sensors, the number of possible sensor inputs, and the speed at which the data can be downloaded to an external computer.

Available logging memory is expressed in megabytes (megs) or gigabytes (gigs), and microcontroller manufacturers are able to store more available memory on an ever-decreasing microchip area. Increasing logging memory results in longer recording times or the possibility to increase the number of measured channels. A complete 24-hour race can be recorded while logging a reasonable amount of channels.

The use of external memory cards, such as Secure Digital (SD) or CompactFlash (CF) cards is becoming more popular. These come in memory capacities of up to 64 gigs and using them as logging memory makes long download times a thing of the past ***(Figure 1.5)***.

Modern data acquisition devices often are incorporated into a CAN within the vehicle. The CAN is a serial bus system suited for networking devices, sensors, and actuators within a system and was developed by Robert Bosch Gmbh. It is easier to add devices to the network (e.g., external dashboards, input expansion boxes) and make them communicate with each other. Data transfer rates between these devices are far greater, compared to parallel or serial connections.

The classic download cable plugged into a serial or parallel port has been replaced by USB cables allowing vast amounts of data to be downloaded to an external computer within seconds. The latest developments include communicating with the system in the car and downloading data from it through a wireless network. Typical transfer rates of a Wi-Fi network are 7 to 30 meg/sec, which produces acceptable download times.

A more popular feature of modern data acquisition techniques is the synchronization of video images, audio channels, and logged data. ***Figure 1.6*** shows an example in which the images (and sounds) recorded by an in-car camera are synchronized with the channels logged by the data acquisition unit. These types of systems are primarily intended to register driver action in the cockpit, but basically a camera can be aimed at almost anything, including the car's suspension and rotating shafts. An audio channel makes the system even more powerful and puts the engineer much closer to what actually happens in the car. A missed gearshift can be detected immediately from the recorded engine sounds, but also wheel spin or clutch slip can be diagnosed without filtering out the problem from different signal traces.

The accuracy of track maps can be improved greatly by adding global positioning sytem (GPS) measurements to the data logging system. Race Technology's DL1 data logger uses a 5-Hz GPS to measure position and speed. GPS position accuracy depends on various factors, but combining it with

Figure 1.5
Race Technology's DL1 data logger uses a CF memory card as logging memory. Maximum memory size depends on the card manufacturer's maximum specifications, and a download cable is not necessary.

Figure 1.6 *Video logging*

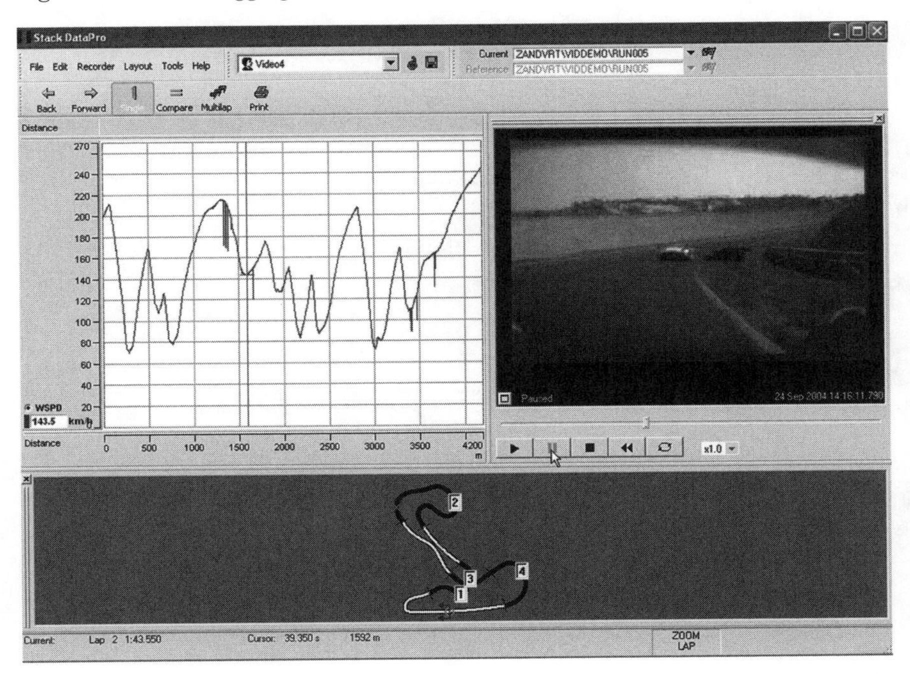

inertial corrections (integration of lateral and longitudinal acceleration) can significantly improve it.

GPS integration has two major advantages. The quality of track maps is much better than when only inertial sensor signals are used to calculate the map ***(Figure 1.7)***. A closed circuit is not required, making this technology suitable for rallying and powerboat racing. For motorcycling, it greatly facilitates the generation of a track map. Infrared timing beacons that define the beginning and end of a lap are no longer required. The second advantage of using GPS is the higher speed accuracy, which is typically within 0.1–0.2%. This is far better than that obtained with a magnetic pickup sensor measuring wheel speed. Speed accuracy is of vital importance to calculate the lap distance, and improving it increases the quality of lap segment calculations and lap overlays.

***Figure 1.7** Track map created with GPS*

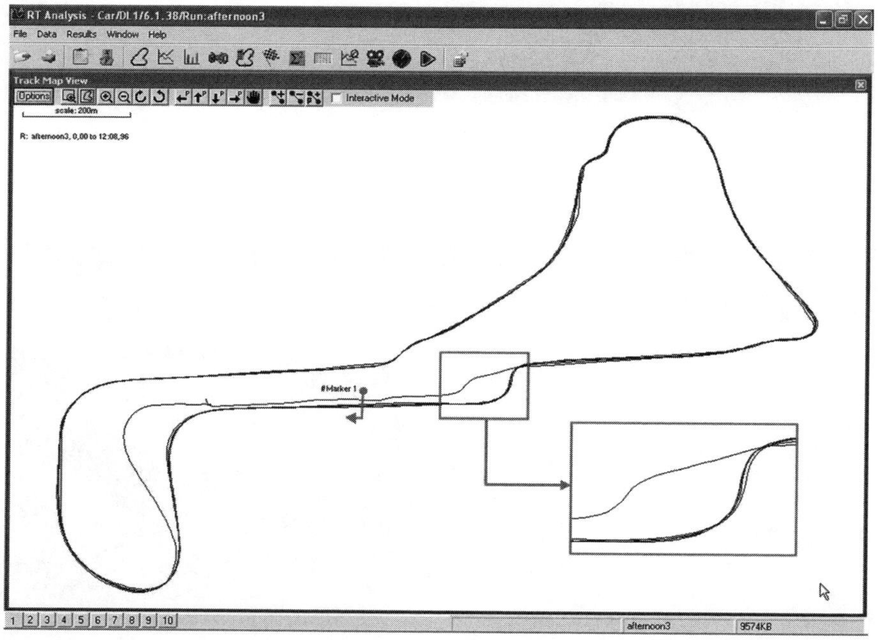

CHAPTER 2

DATA ANALYSIS SOFTWARE REQUIREMENTS

To do a good job, the right tools are required. The most important tool for data acquisition engineers is the software used to analyze data. This chapter acts as a guide for selecting a suitable software package, and tips are given on using this package effectively.

General Requirements for Data Acquisition Software

On a racetrack during a race event or test session, the time available to analyze data from the onboard logger is limited. The data acquisition engineer must provide clear answers in a very short time. Therefore, choosing the right software package for the job is absolutely essential.

Software Features

Software preferences can vary from person to person, but the most important question to be answered is, "Does the software let you customize the way the system displays the data to suit your needs?" Look for the following features:

- user-definable graph limits;
- multichannel display;
- multilap overlays and plot of time difference between compared laps;
- zooming;
- predetermined display templates;
- cursor functions (e.g., cursor data point values, set markers, distance and time location);
- plot data versus time or distance, *X-Y* graphs, histograms;
- track mapping;
- statistical data per lap and lap sections;
- data file organization;
- adding session notes to the data;
- switching between units;
- capability of creating mathematical channels;
- data filtering;
- data export to other software packages; and
- user friendliness, software support, and updates.

Different Ways of Displaying Data

Most data analysis software packages provide different ways of presenting data graphically. The most important are time and distance plots, *X-Y* graphs, and histograms.

Time and Distance Plots

Figure 2.1 shows the speed signal of a lap around Zandvoort—the upper versus time, the lower versus distance. Compared to the time plot, the distance plot expands fast sections of the track and compresses slow sections.

The distance graph indicates where an event occurred, whereas the time graph shows when an event occurred. Graphs are plotted against distance because a certain track location remains reasonably constant over different laps and correlating an event to a certain place on the racetrack is desired. Time plots are used to determine the duration of an event or the rate of change of a signal.

The engineer often wants to investigate more than one channel at a time. This can be achieved by opening multiple graphs on the computer screen or

Figure 2.1 *Speed traces from one lap around Zandvoort—the first against time, the second against distance*

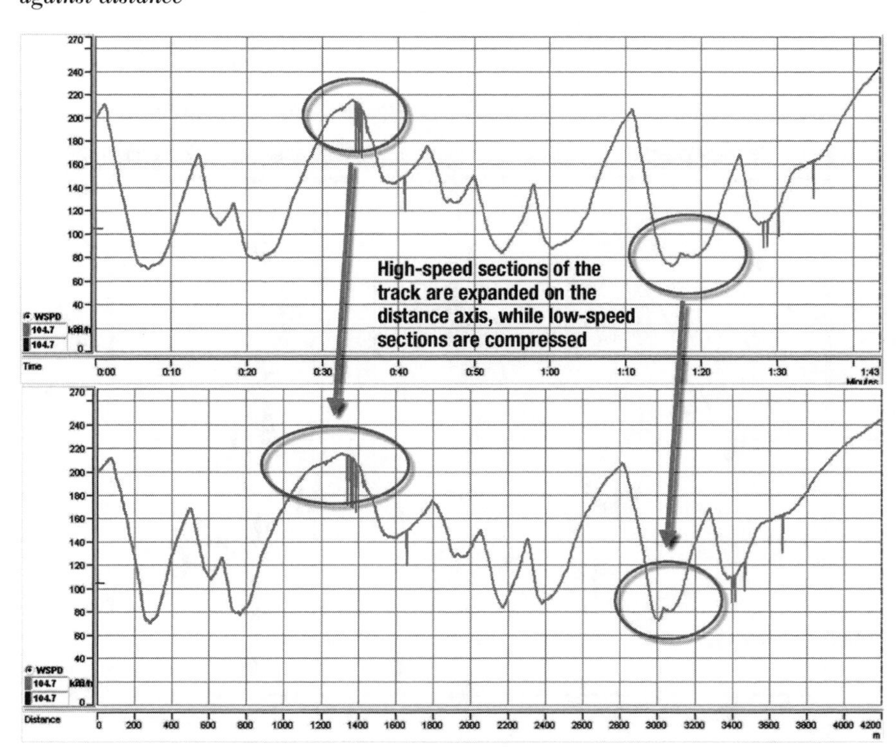

by placing the required channels in one graph. The first possibility *(**Figure 2.2**)* has the advantage of separating all signals from each other to make patterns easily recognizable. If all the channels in Figure 2.2 are placed in one graph, it becomes virtually unreadable, as illustrated in ***Figure 2.3***.

Figure 2.2 Multiple traces pictured separately

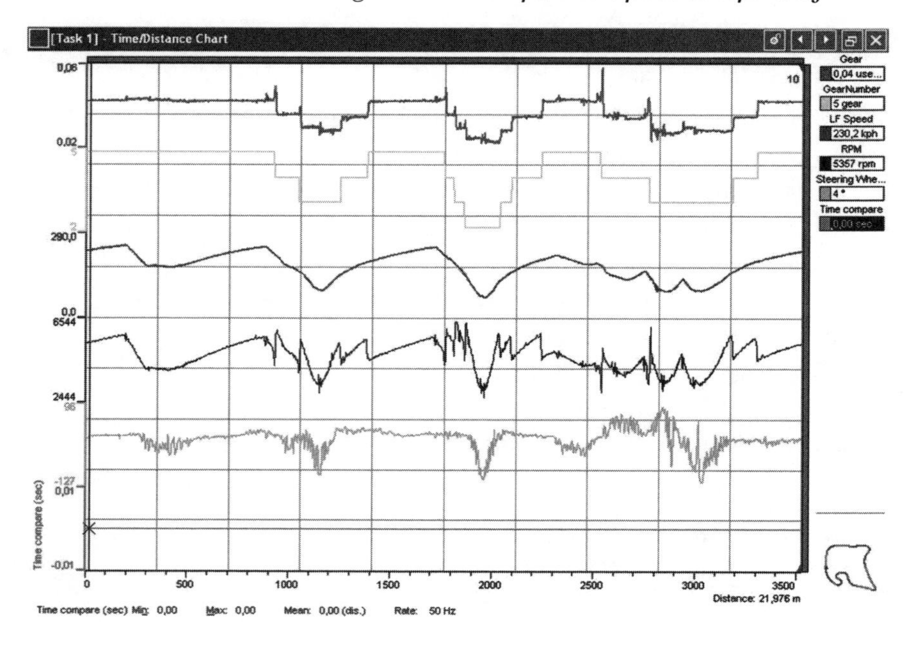

Figure 2.3 Multiple traces in one graph

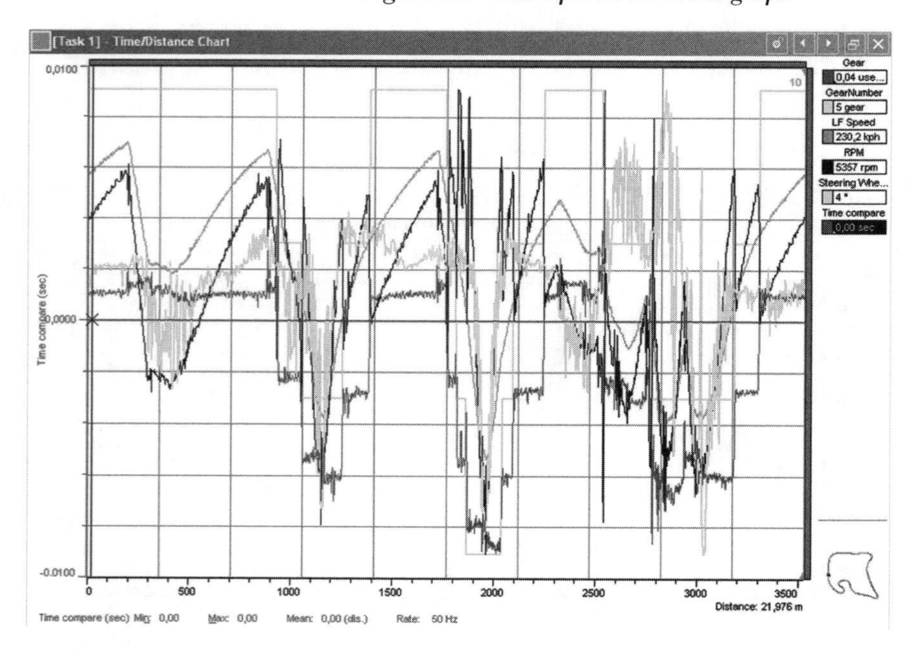

However, placing multiple channels in one graph has the advantage in that the y-axis is stretched to its maximum, while the time (or distance) axis remains the same. Variations in the signal are much more visible. ***Figure 2.4*** provides an example in which speed and throttle signals from the same lap in Figures 2.2 and 2.3 are displayed. Most software packages offer the user the capability to manually scale the signal axis to fit the data on the screen.

X-Y Graphs

When the relationship between two signals is investigated, plotting them in an *X-Y* graph can be useful. A very popular example of this feature is given in ***Figure 2.5***. This represents the vehicle's traction circle by plotting the lateral against the longitudinal g-force.

Histograms

Histograms represent the distribution of a set of data points into several ranges. Popular histograms used for racecar data analysis are RPM, throttle position, and shock velocity. Conclusions that can be drawn from them are covered in ***Chapters 3*** and ***10***. ***Figure 2.6*** shows a histogram of the vertical chassis movement of a racecar. This graph was created to analyze the effects of friction on the vertical stability of the car.

Keeping Notes with Data Files

The number of data files created during a race weekend or test session can be quite substantial. A primary contributor to this situation is the fact that the vehicle's configuration did not remain constant. Even if this were not the case, environmental conditions were probably different. To avoid confusion, an effective system of relating the recorded data to the car's configuration is required. The simplest solution is noting the file names for each session on the vehicle's setup sheet. The analysis software also may feature an editor to add notes to the respective data files.

Some software packages allow a specific setup sheet to be appended to each data file. This can be very useful, as various items on the sheet are probably necessary to perform specific mathematical operations with the data. ***Table 2.1*** is an example of a setup sheet created in a spreadsheet. The data analysis software reads the values in the table and relates them to the correct data file using the time noted in the columns, storing them as session-dependent constants. These constants then can be used in mathematical expressions ***(Figure 2.7)***.

Suppose calculation of the front dynamic ride height from the shock motion signals is desired. The equation incorporates constants for the front static ride height and motion ratio. At the start of the test, the front static ride height is 58 mm. Before the car's next outing, this is modified to 56 mm. The time at that moment is 13h00 as noted in the setup sheet. The analysis software modifies the appropriate session constant and recalculates the dynamic ride height channel for all files recorded after 13h00.

Occasionally, old data files need to be referenced. Maintaining a qualitative record of the car configuration and ambient conditions helps with quickly finding information.

Mathematical Channels

A software feature that is mentioned throughout this book is the creation of mathematical channels. Calculations are performed on the logged data so that the results can be plotted and analyzed as separate channels. The way these channels are created can vary between different software packages, but the following operations should be possible:

- add/subtract,
- multiply/divide,
- saine/cosine/tangent,
- differentiate/integrate, and
- average.

The software often features the capability to include constants that can be used in the math expressions ***(Figure 2.8)***.

Data Overlays

One of the most powerful features of data analysis software is overlaying graphs from separate laps. If a software package does not support this option, buy another system! This technique is extremely useful in analyzing setup changes, driver consistency, and performance changes due to varying ambient conditions. In multicar teams or in

***Figure 2.4** Stretching the graph's y-axis can make signal variations easier to detect.*

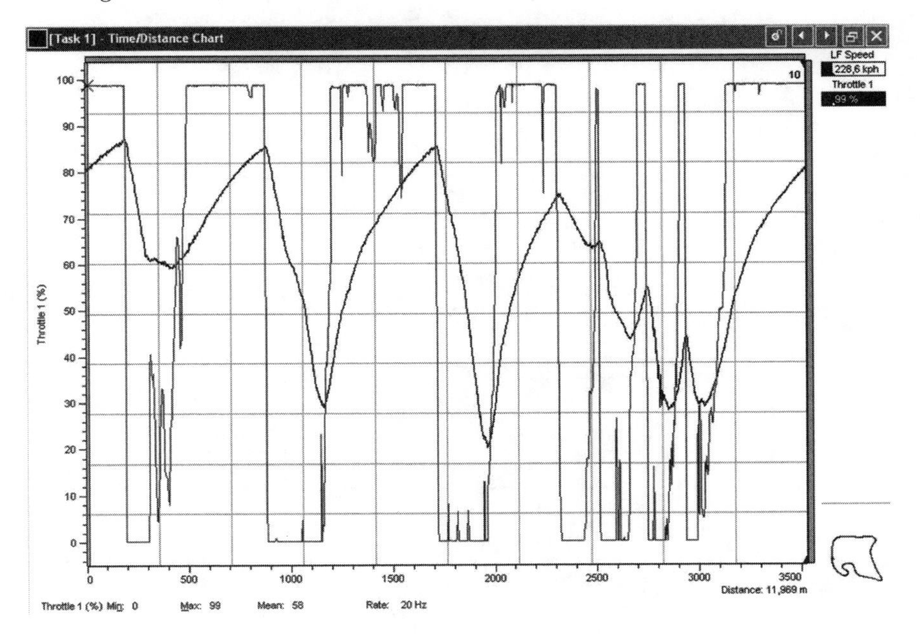

***Figure 2.5** Longitudinal against lateral g-force data from a rallycross heat around Circuit Duivelsberg at Maasmechelen in an Opel Corsa. The graph represents this vehicle's traction circle.*

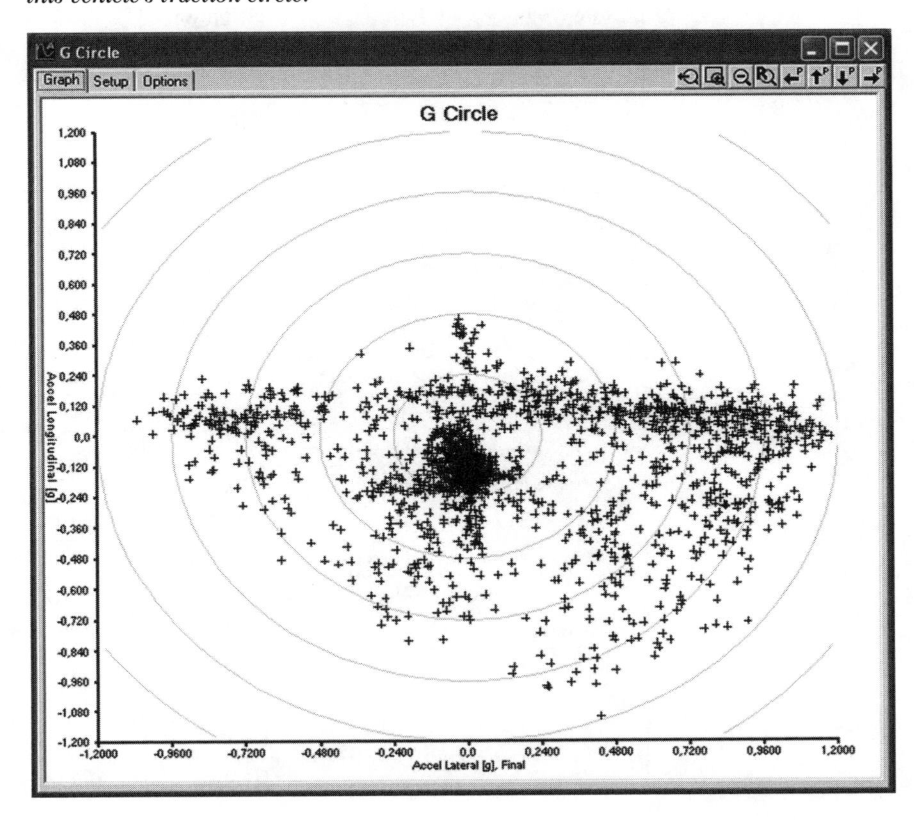

Table 2.1 *MoTeC session constant setup sheet*

MoTeC Device:	ADL
Serial No:	834
Vehicle:	Viper C11
Filename:	ADL834.mcs 12/10/2004
Last Written:	15:50:43

MoTeC Session Constant Setup Utility

			Date	Initial	12/okt/04	12/okt/04
			Time	Initial	13h00	13h17
Long Name	Short Name	Channel ID	Units		Test Spa: Outing 1	Test Spa: Outing 2
Vehicle General						
Wheelbase	Wbase	20400	mm	2418		
Wheelbase—Left	WbaseL	20442	mm	2418		
Wheelbase—Right	WbaseR	20443	mm	2418		
Track—Front	TrackF	20401	mm	1654		
Track—Rear	TrackR	20402	mm	1722		
Total Weight—Front	TWghtF	20403	kg	577		
Total Weight—Rear	TWghtR	20404	kg	697		
Static Ride Height—Front	SRHghtF	20409	mm	58	56	
Static Ride Height—Rear	SRHghtR	20410	mm	97		
Static Weight—Front Left	SWghtFL	20426	N	2894		
Static Weight—Front Right	SWghtFR	20427	N	2766		
Static Weight—Rear Left	SWghtRL	20428	N	3502		
Static Weight—Rear Right	SWghtRR	20429	N	3335		
Ackerman Factor	AckFact	20444	%			
Suspension Constants						
Roll Bar Rate—Front	RBR F	20405	kg/mm	60.21		
Roll Bar Rate—Rear	RBR R	20406	kg/mm	21.57		
Spring Rate—Front Left	SprngFL	20407	kg/mm	28		32
Spring Rate—Front Right	SprngFR	20408	kg/mm	28		32
Spring Rate—Rear Left	SprngRL	20440	kg/mm	32		
Spring Rate—Rear Right	SprngRR	20441	kg/mm	32		
Static Roll Center Height—Front	SRCHtF	20430	mm	46.19		
Static Roll Center Height—Rear	SRCHtR	20431	mm	50.49		
Motion Ratio—Front Left	MRatFL	20434		1.373		
Motion Ratio—Front Right	MRatFR	20435		1.373		
Motion Ratio—Rear Left	MRatRL	20436		1.725		
Motion Ratio—Rear Right	MRatRR	20437		1.725		
Motion Ratio Roll Bar—Front	MRatRBF	20438		1.495		
Motion Ratio Roll Bar—Rear	MRatRBR	20439		1.550		
Front Antidive	FrADive	20423	%	36.7		
Rear Antisquat	RrASqt	20424	%	73.8		
Driveline						
Diff Ratio	DiffR	20411		4.10		
Gear Ratio 1	GearR1	20412		8.57		

endurance racing where multiple drivers share the same car, data overlays can indicate differences in driving style.

In ***Figure 2.9,*** speed traces are overlaid from two laps around Silverstone Circuit. When comparing two laps, begin with the speed traces because the intention of every change in setup or driver activity is to influence the vehicle speed. First find where the gains and losses are, and then find out why they occur.

When overlaying different traces, it is preferable to plot them against covered distance. When the x-axis is time, the two traces tend to diverge over the duration of the lap. Different lap times mean different times were measured to get to a given point on the racetrack.

The software usually calculates the time difference between the two laps being compared. The Pi Toolbox software from which Figure 2.9 was taken calculates the cumulative time difference between two laps. Lap overlays are covered further in ***Chapter 3.***

Filtering

Data filtering or, more appropriately smoothing, is a process in which data points are averaged over a given time interval. This suppresses the higher frequencies in the signal and enhances the lower frequencies. Filtering removes noise from the signal or conditions the signal for the analysis of slower movements. Filtering is a useful but sometimes dangerous tool because the risk exists that relevant high-frequency events are removed from the signal. In general, use filtering as little as possible.

Figure 2.10 shows a damper signal logged at 50 Hz (dark gray-colored trace). The lighter trace represents the same damper signal filtered at four samples. This means that every data point is replaced by the average of this point and the four samples at either side of it.

***Figure 2.6** Histogram of chassis heave motion*

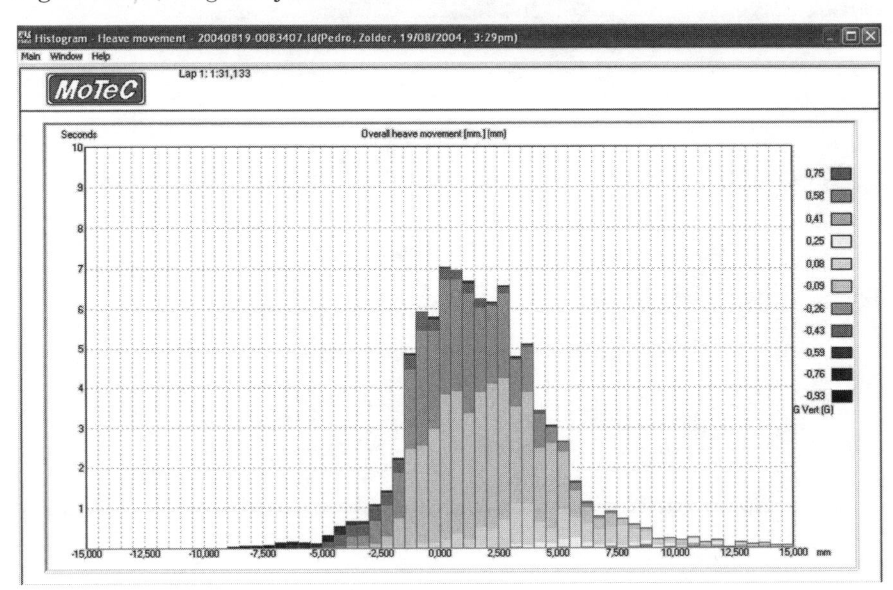

Table 2.1 (continued)

Gear Ratio 2	GearR2	20413		5.94		
Gear Ratio 3	GearR3	20414		4.64		
Gear Ratio 4	GearR4	20415		3.74		
Gear Ratio 5	GearR5	20416		3.14		
Gear Ratio 6	GearR6	20417		2.78		
Gear Ratio 7	GearR7	20418				
Aero Constants						
Rear Wing Angle	RWngAng	20425	deg	0		
Splitter Height Front	SpltHF	20432	mm	65		
Splitter Angle Front	SpltAgF	20433	deg	0.6		
Ambient Conditions						
Static Ambient Temperature	SAmbT	20419	C	15		
Static Ambient Pressure	SAmbP	20420	kPa	100		
Track Temperature	TrkTmp	20422	C	17		

The filter time can be calculated using ***Equation 2.1***. In this equation, *n* is the number of samples in the filter interval. The filter time in the example is 8 / 50 Hz = 0.16 s.

$$\text{Filter time} = \frac{2 \cdot n}{\text{sampling frequency}} \qquad (Eq\ 2.1)$$

The lighter trace has the advantage in that the slow movement of the damper (induced by chassis movement) is much clearer. On the other hand, information dealing with high-speed movement, such as road irregularities or nonsuspended mass effects, is lost. The maximum deflection of the damper also seems smaller. Preferably, display filtered graphs as shown in Figure 2.10, with the filtered signal in a light color and the original trace in a darker background color to view what is occurring at the higher frequencies. If channels are included in mathematical expressions, use the raw channel instead of the filtered one. Afterwards, the math channel can be filtered if necessary.

***Figure 2.7** The setup constants in Table 2.1 can be used in the MoTeC analysis software in mathematical expressions.*

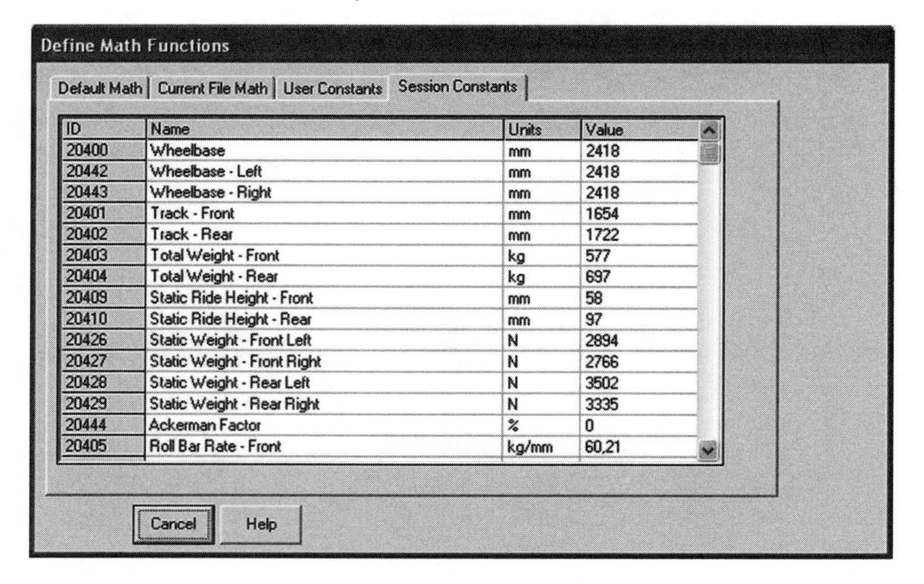

***Figure 2.8** Examples of mathematical channel definitions*

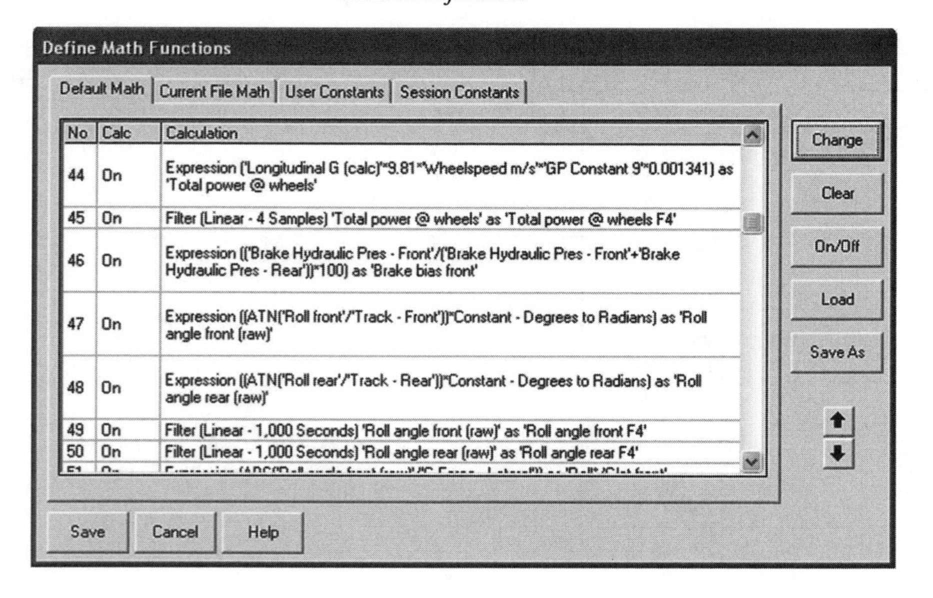

Exporting Data to Other Software Packages

Sometimes data logging software does not handle every analysis requirement. It may be required to export data into other software packages such as spreadsheets, mathematical software, and lap time simulations. Most software packages can export logged data in ASCII (American Standard Code for Information Interchange) or CSV-format (comma-separated values). These are essentially text formats that can be read by basic text editors, spreadsheets, or mathematical analysis software.

Applications for data export can deal with fuel strategy, running logs for the car, Fourier analyses on shock absorber motion, and lap time simulation reference laps. ***Table 2.2*** is an example of how the MoTeC Interpreter software exports the wheelspeed and throttle position signals of a lap segment to a CSV file.

Getting Organized

Most popular software packages provide a number of ways to configure the software to suit different needs. This configuration should be done before arriving at the racetrack. A good impression is not made when the display templates still need to be organized when a printout of speed, throttle, and RPM data is needed. Preparation is everything.

Channel Grouping

Display templates are often preprogrammable, allowing users to choose which channels they want to display together. This is a matter of preference, but all signals related to the same type of analysis should be grouped together. The following list can be used as a guide:

1. **Vital Functions**
   engine RPM, engine water and oil temperature, oil pressure, gearbox and differential temperature, battery voltage
2. **Gearing**
   vehicle speed, engine RPM, throttle position, gear ratio
3. **Fuel Consumption**
   fuel pressure, fuel level, fuel used, fuel per lap
4. **Engine Performance**
   speed, engine RPM, manifold air pressure, air temperature inlet manifold
5. **Lambda**
   engine RPM, throttle position, lambda
6. **Driver Activity**
   speed, throttle position, steering angle, brake pedal position
7. ***g*-force**
   speed, *g*-force lateral, longitudinal, vertical, combined *g*-forces
8. **Braking**
   speed, *g*-force longitudinal, brake pedal position, brake line pressures
9. **Damper Position Raw**
   speed, damper position channels
10. **Roll and Pitch Angle**
    speed, lateral and longitudinal *g*-forces, roll angle, pitch angle
11. **Wheel Load**
    speed, lateral weight transfer, longitudinal weight transfer
12. **Understeer/Oversteer**
    speed, throttle position, front lateral *g*-force, rear lateral *g*-force
13. **Open Template**
    Use when putting signals together that do not fall under the preprogrammed display templates.

Channel Colors

Assign specific colors to specific channels. For example, all sensor signals related to one wheel share the same color. A channel that comes back in different display templates should have the same color. Choose appropriate background colors. Make things easy.

Sensor Prep

Before arriving at the track, make sure that all sensors are properly calibrated, the dashboard programmed, and the correct sampling frequencies set. Suspension potentiometers, strain gages, steering angle sensors, accelerometers, and brake pressure sensors should be zeroed when the car is on the setup pad. Also, remember that the development of mathematical functions is not a trackside job!

***Figure 2.9** Overlay of two laps around Silverstone Circuit performed with the Pi Toolbox package. The time compare channel shows where time is gained or lost.*

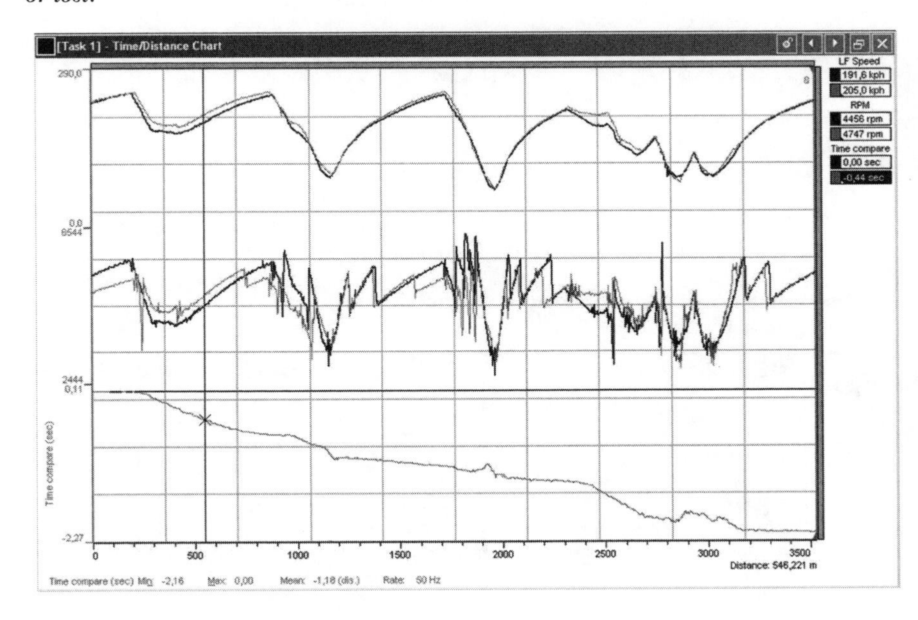

***Figure 2.10** Raw and filtered signal*

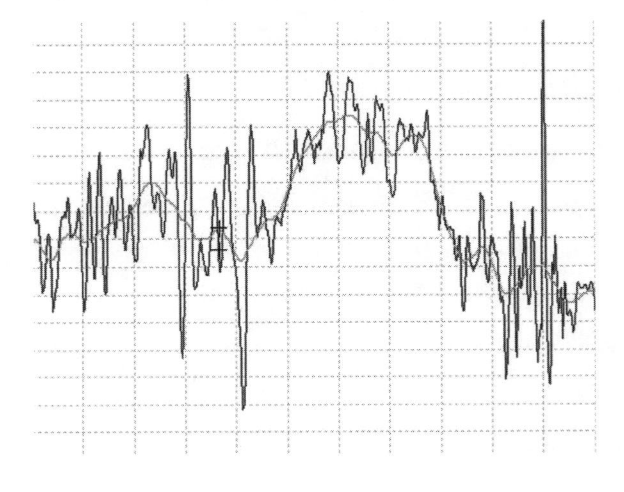

Example Checklist

Figure 2.11 is a checklist that was duplicated from the Pi System 6 software manual. With the exception of a couple of items, it can be used for any data acquisition system.

Pitbox Setup

Set up the computer equipment in the pitbox at a location where it does not disturb the mechanics ***(Figure 2.12)***. A table with a couple of chairs in a corner of the pitbox are just fine for preparing analysis work and holding discussions with the drivers. Mainly a matter of preference and budget, the following equipment is recommended:

- laptop with an external mouse,
- color printer,
- data download cable,
- installation CD of the data acquisition software,
- a USB drive (convenient to transfer data between different users), and
- a digital weather station.

Basic Tips

When transporting a car to the racetrack, it is normal to take spare parts. The same goes for the data acquisition system. Not only should critical sensors be within reach in case of failure but so should cables, connectors, and other spare equipment. Budgets have a say in this, but it is always better to have backup solutions.

Download the data logger every time the car enters because even rollout laps can provide relevant information. When analyzing the data, begin

Figure 2.11 *Trackside checklist for Pi System 6 software (Courtesy of Pi Research)*

Typical procedures: Version 6 Professional

	Procedure	when	section
Setup engineer	Set PC communication	first time	Server
	Name data logging system	first time	Setup
	Set track and driver details	first time / each track	Setup
	Set download directories	first time / each track	Setup
	Choose Junction boxes	first time	Setup
	Calibrate sensors and channels	first time / change sensors	Setup
	Set System wheelspeed and RPM	first time / change sensors	Setup
	Choose LCU channels and logging criteria	first time / each track	Setup
	Choose MRC channels and logging criteria	first time / each track	Setup
	Set driver display	first time / each track	Setup
	Set alarms and telltales	first time / each track	Setup
	Set end-of-lap and real-time telemetry	each track	Setup
	Set microwave telemetry	first time / each track	Setup
	Create spreadsheet or database setup sheet	each track	(Excel or Access)
	Set fuel strategy	each track	Setup
	Reset Engine Log Book	first time / new engine	Setup
	Send setup to logging system	each track / change sensors	Server / Setup
	Test sensors	each track / change sensors	Setup
	Copy setup and calibrations	first time / change sensors	Setup
	Log data		
Data analyst	Download data	each outing	Server
	Make and adjust Map	each track	Analysis
	Define and adjust beacons	each track	Analysis
	Create and edit Math channels	first time / each track	Analysis
	Create and edit User Math Functions	first time / each track	(C++) / Analysis
	Integrate vehicle dynamics information	each track	Analysis
	Define global constants in Excel or Access	each track	Analysis
	Produce reports and graphs	each outing	Analysis
	Export data for engine and chassis experts	each outing	Analysis
Engine technician	Analyse Engine Log Book data	dependent on engine performance	Analysis
	Analyse end-of-lap telemetry data	each outing	Telemetry
	Analyse microwave telemetry data	each outing	Analysis / Telemetry
	Analyse Engine Manufacturers' channels	each outing	Analysis

Table 2.2 *Example of how data is exported into a CSV file*

```
FormatMoTeC CSV File"
Venue,"Spa"
Vehicle,"Dodge Viper GTS/R"
User,"Anthony"
Data Source,"MoTeC ADL 2222"
Comment,"R LSB+1 HSB+1 Reb+2 RARB H/S"
Date,"17/03/2004"
Time,"14:56:21"
Sample Rate,"10.240"
Duration,"14.648"
Segment,"Lap 1 - 2:19.941"

WSpd FL,"TP",
Wheel Speed - Front Left,"Throttle Position",
km/h,"%",
170,"2",

211.0,"80.8",
211.1,"79.8",
210.9,"79.5",
212.5,"80.1",
212.9,"95.2",
212.6,"100.0",
212.0,"100.0",
212.6,"100.0",
214.0,"100.0",
214.6,"100.0",
214.9,"100.0",
215.9,"95.9",
215.9,"92.7",
217.1,"91.4", . . .
```

with the vital channels: engine and driveline temperatures, pressures, and battery voltage. Ensure that no strange things are happening before addressing the performance data, even if the driver is standing nearby, shouting that the car is not drivable!

Listen to the driver; it is much easier to know what to look for. Then, observe what the car is doing (e.g., speed, acceleration, driver activity) before trying to understand why it is doing this (e.g., shock data, strain gages). Remember that the speed trace is where it all happens. A stopwatch determines whether a lap is a slow or quick one, but the speed graph tells where time is gained or lost.

Do not focus only on the fastest lap in a session; analyze all laps and look for consistencies and inconsistencies. Remember that traffic on the racetrack can have a considerable influence on lap time.

One last tip: Look for the obvious first. Determine which channels are expected to show differences after a setup change. The effects of aerodynamic changes most likely show up in the speed and wheel-load traces.

***Figure 2.12** Try to find separate space in the pitbox to do analysis work. (Courtesy of GLPK Racing)*

CHAPTER 3

THE BASICS

Much of the workload for the data acquisition engineer consists of comparative analysis. Comparing data from different laps or runs with previously collected data reveals the effect of setup changes or driver performance. Most data analysis packages offer similar techniques for comparing different data sets. This chapter covers these techniques and provides a basic interpretation of the patterns showing up in the most often used sensor signals.

Check the Car's Vital Signs

When analyzing data, reliability and safety are the first priority. There is not a lot of performance in a car that's standing still when compared to one on the racetrack! Make sure all pressures, temperatures, and voltages are safe before analyzing performance. A deficiency in these signals may explain a lack of performance.

The most important channels to check are engine oil pressure, engine water and oil temperatures, transmission oil temperatures, battery voltage, fuel pressure, and (maximum) engine RPM. To these, reliability indicators such as tire pressures, brake pressures, clutch pressures, and engine knock signals should be added.

An easy way to check the vehicle's vital signs is by using the tabular report shown in ***Figure 3.1***. This table shows the minimum, maximum, and average values for each logged channel for the duration of the run or a single lap. Some software packages can highlight values when they exceed a user-defined alarm value.

Some care must be taken when interpreting the values in this table. In Figure 3.1, the minimum value for engine oil pressure is zero, but this value was obtained when the car halted in the pits at zero engine RPM. This is a pretty straightforward conclusion, but these values may be hiding another problem in the data. That is why one cannot rely solely on a statistics table to check vital signals. The graphs must be examined to ensure that everything is working like it should.

To check for deficiencies in the signals, begin with a graph covering the complete run (**Figure 3.2**). The data was obtained from the first stint of a 24-hour race in Zolder, Belgium, from a Dodge Viper. This example shows engine RPM, oil and fuel pressure, engine oil and water temperature, gearbox and differential oil temperature, and battery voltage. The advantage of examining a complete run is that trends in the signals can be recognized easily. In Figure 3.2, all fluid temperatures rise to a maximum value, after which they stabilize. Engine oil pressure is slightly higher at the beginning of the stint when the engine has not reached operating temperature. After that, it stabilizes to around 5.5 bar. Everything seems to be working fine here. When something abnormal is discovered, focus on the specific event and investigate the problem.

In ***Figure 3.3***, a graph shows the results of a qualifying run around the Silverstone Circuit. The alternator belt fails on this car and as a result the battery voltage gradually drops. At the beginning of the run, the battery voltage is approximately 10 V, which is already too low. Investigations of the previous run indicate the problem began there (and that it should have been dealt with already!).

Figure 3.1 *Statistics table with minimum, maximum, and average values for each channel*

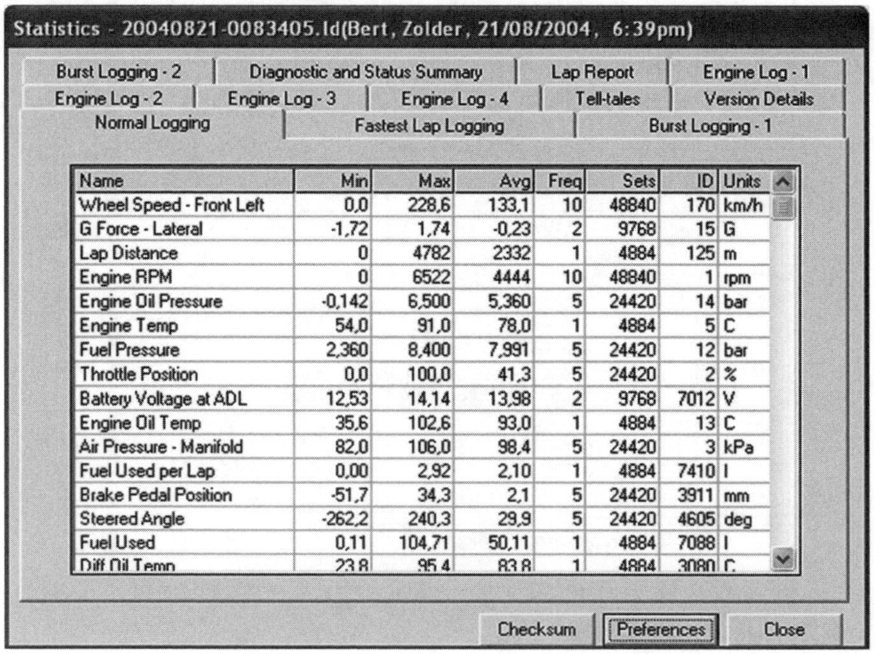

Statistics - 20040821-0083405.ld(Bert, Zolder, 21/08/2004, 6:39pm)

Burst Logging - 2 | Diagnostic and Status Summary | Lap Report | Engine Log - 1
Engine Log - 2 | Engine Log - 3 | Engine Log - 4 | Tell-tales | Version Details
Normal Logging | Fastest Lap Logging | Burst Logging - 1

Name	Min	Max	Avg	Freq	Sets	ID	Units
Wheel Speed - Front Left	0,0	228,6	133,1	10	48840	170	km/h
G Force - Lateral	-1,72	1,74	-0,23	2	9768	15	G
Lap Distance	0	4782	2332	1	4884	125	m
Engine RPM	0	6522	4444	10	48840	1	rpm
Engine Oil Pressure	-0,142	6,500	5,360	5	24420	14	bar
Engine Temp	54,0	91,0	78,0	1	4884	5	C
Fuel Pressure	2,360	8,400	7,991	5	24420	12	bar
Throttle Position	0,0	100,0	41,3	5	24420	2	%
Battery Voltage at ADL	12,53	14,14	13,98	2	9768	7012	V
Engine Oil Temp	35,6	102,6	93,0	1	4884	13	C
Air Pressure - Manifold	82,0	106,0	98,4	5	24420	3	kPa
Fuel Used per Lap	0,00	2,92	2,10	1	4884	7410	l
Brake Pedal Position	-51,7	34,3	2,1	5	24420	3911	mm
Steered Angle	-262,2	240,3	29,9	5	24420	4605	deg
Fuel Used	0,11	104,71	50,11	1	4884	7088	l
Diff Oil Temp	23,8	95,4	83,8	1	4884	3080	C

Checksum | Preferences | Close

Figure 3.2 This graph shows the car's vital signs for a complete run.

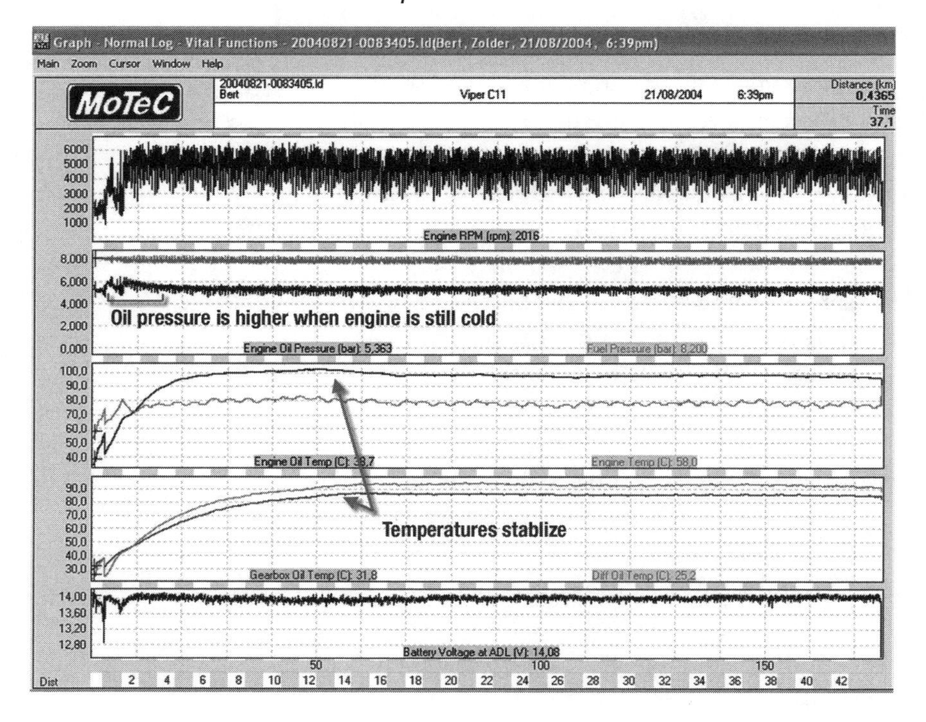

Figure 3.3 Effects of a broken alternator belt

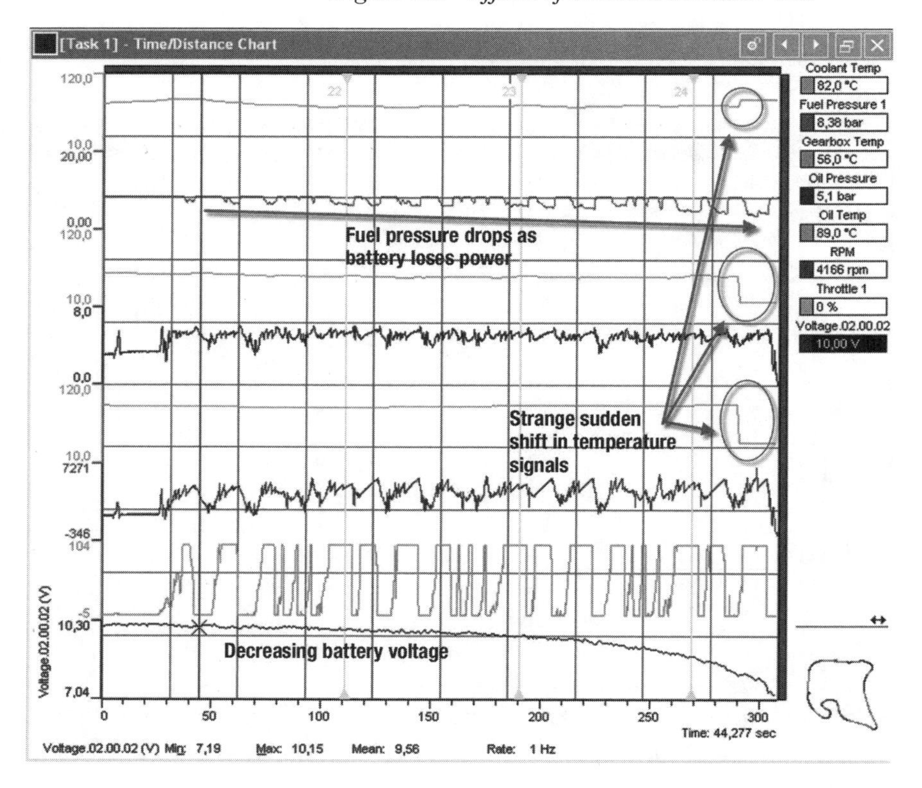

The decreasing power of the battery had some side effects that become clear from this graph as well. The fuel pumps cannot maintain the desired fuel pressure. The more the battery voltage drops, the bigger the drop in the fuel pressure signal.

At the end of the run, there is a sudden change in the three pictured temperature channels (engine oil, water, and gearbox temperature). They change to default values that the ECU uses when a sensor is not working. Obviously, the drop in battery voltage causes the sensors to malfunction.

The problem in this example was easily detectable. Sometimes it is not that obvious, and the diagnostic signs are much more subtle and hidden in the data. The data engineer should develop a feel for the normal patterns of the car's vital signs so that potential problems can be detected early. The car in this particular example halted next to the track, meaning there were two problems:

- The data acquisition engineer did not notice the problem in the data from the previous run.
- The driver did not notice the problem during driving or did not get a warning.

Mathematical channels can help visualize alarm values in the car's vitals. Conditional expressions can be created to give a value of one if a vital channel exceeds or drops below a predetermined value. For instance, a low oil pressure alarm channel assuming a value of one as soon as oil pressure drops below 3 bar while the engine RPM is greater than 3,000 can be defined as shown in ***Equation 3.1***.

$$\text{LowOP} = \left(\text{oil pressure} < 3\right) \cdot \left(\text{engine RPM} > 3000\right)$$

(Eq. 3.1)

The resulting channel is shown in ***Figure 3.4***.

Most racecar data acquisition systems can configure the driver display by adding alarm messages in case of a problem or sending an output signal to a warning light. Make sure that if something goes wrong, the driver gets a clear warning. Also think about priorities when determining the right alarm values. In the example shown in Figure 3.4, the driver gets fuel pressure alarm that overrides the low-battery warning.

Lap Markers and Segment Times

Performance analysis usually begins with figuring out where on the track time is gained or lost before actual events are investigated. A quick way to assess this is to investigate lap segment times.

Lap times are determined by the analysis software measuring the time it takes for the car to pass the lap beacon. This beacon represents the location on the track where a lap ends and the next one begins. It can be an infrared pulse logged by the data system or a manually entered beacon point in the data. Most data analysis packages offer the option of placing additional virtual beacons around the track at certain distances from the start/finish beacon. Lap segment times are determined by measuring the elapsed time between two consecutive beacons. Placement of these beacons depends on what needs to be analyzed. Engine performance can be evaluated best on a straight track segment where the car is accelerating. Corners can be defined as separate segments, but the corner itself can also be divided into *entry, apex,* and *exit* to investigate cornering performance.

In ***Figure 3.5,*** an example is given from the Istanbul Grand Prix track. The track was divided in seven segments, separated by six manually entered segment beacons and an infrared lap beacon.

The tabular report in this example gives segment times for all covered laps for each track sector. The fastest and slowest segment time is highlighted in the table. The last column gives the difference in lap time to the fastest lap in the outing.

Furthermore, the software calculated two performance indicators from this table—the theoretical fastest lap and the fastest rolling lap, which attempt to indicate the true performance potential of a given configuration. As often happens during a lap, the driver was delayed by a slower car on the track or a mistake was made by the driver. In this case, the concerned segment time was slower. The theoretical fastest lap is the sum of the fastest sector times in an outing and represents the time that could have been achieved in an ideal lap. In the example shown in Figure 3.5, this ideal lap would have been 0.235 sec quicker than the fastest lap. Very good drivers should get close to their theoretical fastest lap time.

Great care should be taken in placing importance on the theoretical fastest lap because a slower time in one segment can result in a faster time in the next. The driver may take a driving line through one corner that is faster in the particular corner but compromises speed in the next one. Missing braking points also can cause inconsistent segment

Figure 3.4 *Low oil pressure alarm channel*

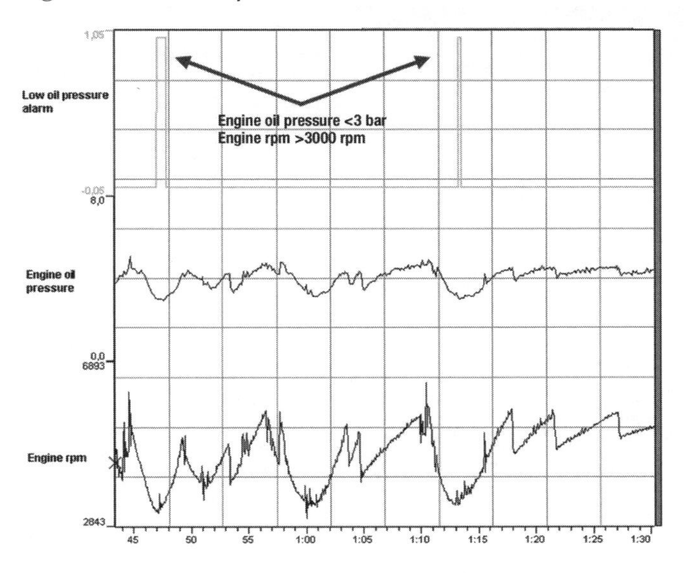

Figure 3.5 *Segment times report using "virtual" beacons*

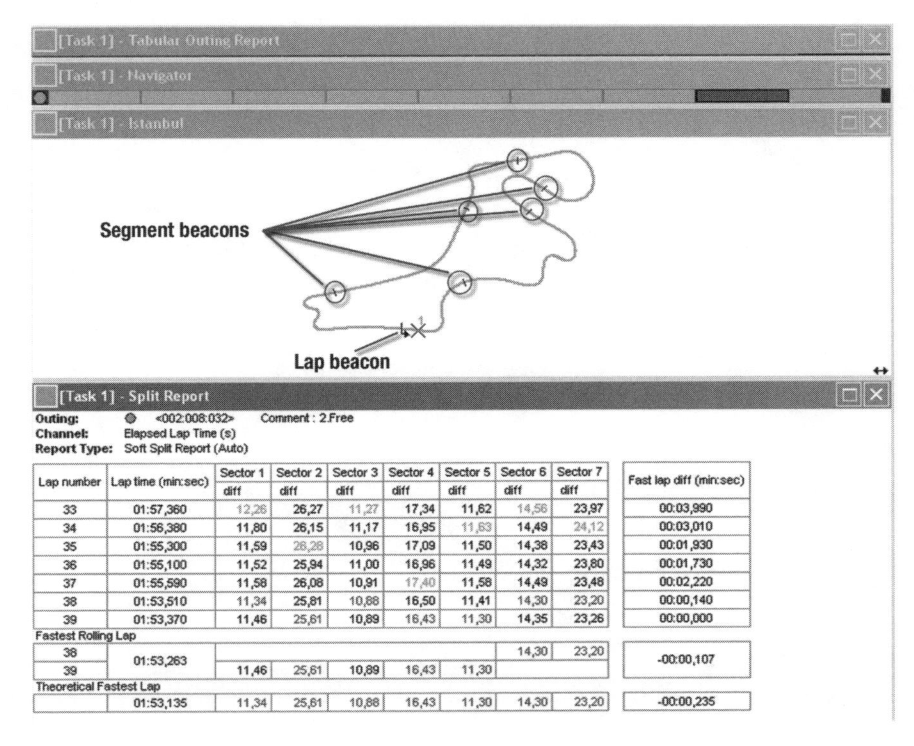

Outing: <002:008:032> Comment : 2.Free
Channel: Elapsed Lap Time (s)
Report Type: Soft Split Report (Auto)

Lap number	Lap time (min:sec)	Sector 1 diff	Sector 2 diff	Sector 3 diff	Sector 4 diff	Sector 5 diff	Sector 6 diff	Sector 7 diff	Fast lap diff (min:sec)
33	01:57,360	12,26	26,27	11,27	17,34	11,62	14,56	23,97	00:03,990
34	01:56,380	11,80	26,15	11,17	16,95	11,83	14,49	24,12	00:03,010
35	01:55,300	11,59	26,26	10,96	17,09	11,50	14,38	23,43	00:01,930
36	01:55,100	11,52	25,94	11,00	16,96	11,49	14,32	23,80	00:01,730
37	01:55,590	11,58	26,08	10,91	17,40	11,58	14,49	23,48	00:02,220
38	01:53,510	11,34	25,81	10,88	16,50	11,41	14,30	23,20	00:00,140
39	01:53,370	11,46	25,61	10,89	16,43	11,30	14,35	23,26	00:00,000
Fastest Rolling Lap									
38	01:53,263						14,30	23,20	-00:00,107
39		11,46	25,61	10,89	16,43	11,30			
Theoretical Fastest Lap									
	01:53,135	11,34	25,61	10,88	16,43	11,30	14,30	23,20	-00:00,235

times. The confidence level of this performance indicator also depends on the location of the segment beacons. Segment times in areas bordered by beacons placed at the apex of a corner are more sensitive to inconsistencies than those bordered by beacons placed at the middle of a straight.

Figure 3.6 *The Circuit de Spa Francorchamps track divided in seven track segments*

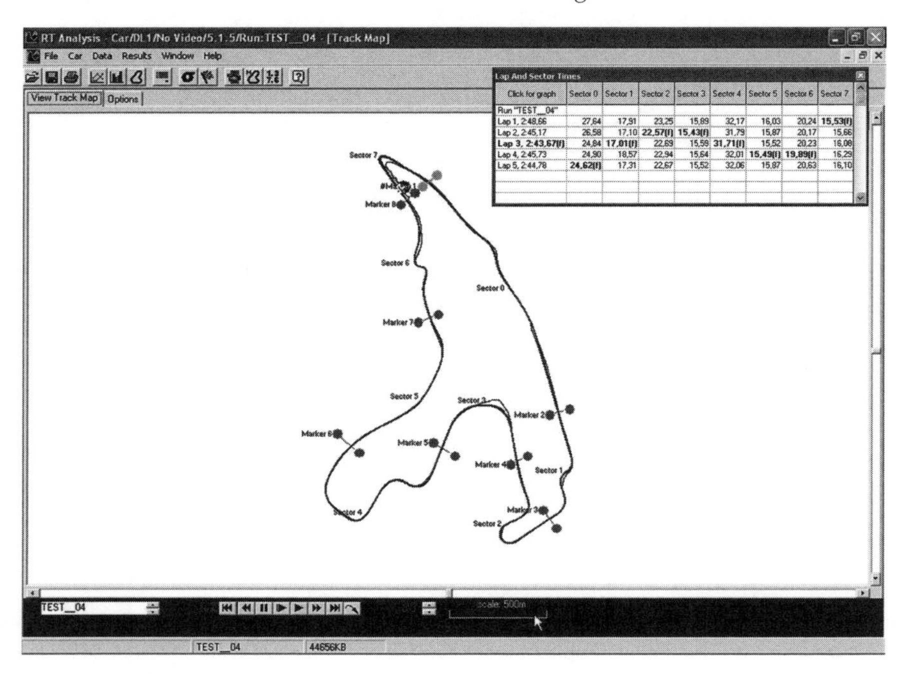

Click for graph	Sector 0	Sector 1	Sector 2	Sector 3	Sector 4	Sector 5	Sector 6	Sector 7
Run "TEST__04"								
Lap 1, 2:48,66	27,64	17,91	23,25	15,89	32,17	16,03	20,24	**15,53(f)**
Lap 2, 2:45,17	26,58	17,10	**22,57(f)**	**15,43(f)**	31,79	15,87	20,17	15,66
Lap 3, 2:43,67(f)	24,84	**17,01(f)**	22,69	15,59	**31,71(f)**	15,52	20,23	16,08
Lap 4, 2:45,73	24,90	18,57	22,94	15,64	32,01	**15,49(f)**	**19,89(f)**	16,29
Lap 5, 2:44,78	**24,62(f)**	17,31	22,67	15,52	32,06	15,87	20,63	16,10

Figure 3.7 *The Spa track divided into three segments*

Click for graph	Sector 0	Sector 1	Sector 2
Run "TEST__04"			
Lap 1, 2:48,66	1:07,14	1:06,43	35,09
Lap 2, 2:45,17	1:04,35	1:05,81	**35,01(f)**
Lap 3, 2:43,67(f)	**1:02,80(f)**	**1:05,44(f)**	35,43
Lap 4, 2:45,73	1:04,69	1:05,79	35,25
Lap 5, 2:44,78	1:03,06	1:05,63	36,09

The fastest rolling lap is the lap time achieved between a beacon that is not necessarily at the end of a lap—an indication of the performance potential when there is heavy traffic on the track. It is a lap time the driver actually achieves. In the example, if the end-of-lap beacon was the split between segment five and six the fastest lap time would be 0.107 sec faster than the fastest lap in this outing.

Another issue that requires attention when calculating the theoretical fastest lap is the number of track segments. The more track segments incorporated, the faster the ideal lap is. There must be an optimum number of segments that provide a realistic theoretical fastest lap. Sound judgment is necessary here, both in choosing the right number of segments and having confidence in this performance indicator. The following example illustrates the effect of the number of track segments on the theoretical fastest lap.

This example discusses a run in a Porsche 996 around the Circuit de Spa Francorchamps recorded with a Race Technology DL1 GPS data logger. The fastest lap by this vehicle in this run was 2'43"67. First, the track was divided into seven segments and the segment times for the full run were calculated by the analysis software. The results were given in the table in ***Figure 3.6;*** the fastest sector time per segment is highlighted. The sum of these fastest segment times resulted in a theoretical fastest lap of 2'42"25 or a difference of 1.42 sec compared to the fastest performed lap.

In ***Figure 3.7,*** the same run was taken, but the track was only divided into three segments. The difference is obvious, with a theoretical fastest lap time of 2'43"25. The difference here compared to the fastest real-life lap is only 0.42 sec.

In most cases, begin with fewer segments. Variation in segment time may be greater with fewer sections because there is much more potential for mistakes or other problems in longer sectors. If that happens, increase the number of segments when it is necessary to pinpoint problem areas.

Comparing Laps

The most powerful tool in any data acquisition software package is overlaying and comparing different laps. Most analysis performed on racecar data is comparative. When something is changed on the car, comparing a run to previous ones indicates the difference of that change.

By overlaying two traces as a function of distance, the performance of the vehicle and the driver can be compared at the same point on the track. Overlaying against time does not bring any meaningful conclusions because events at the same time probably happen at other locations, and the traces tend to diverge over the length of the lap.

Figure 3.8 shows an example of two overlaid laps around the Nürburgring. It illustrates how the vehicle speed can be compared directly for every location on the track. Comparing speed is often the first step in the analysis because this channel is the results graph. An increase in speed inevitably decreases lap time.

The respective lap times in this example are

Lap 6 (dark trace) 1'57"334
Lap 9 (light trace) 1'56"065

This is a difference of 1.269 sec between the two laps. Try to find in Figure 3.8 where this difference in lap time was created. Once the locations where time was gained or lost are pinpointed, they should be further investigated to find out what exactly happened and where the speed difference originated.

To pinpoint more efficiently areas on the track where time is gained or lost, mathematical functions can be created. These are created automatically by the analysis software. Some subtle differences between software packages are possible, but the idea is always the same.

Continuing with this example, ***Figure 3.9*** shows the same speed trace overlay with two extra graphs that determine the difference in lap time between the two laps. The MoTeC analysis software calls these *variance* that can be graphed either as *instantaneous* or *cumulative*. Instantaneous variance is defined in ***Equation 3.2***.

$$\text{Variance}_{\text{Inst.}}(d) = t_{\text{LAP1}}(d) - t_{\text{LAP2}}(d) \qquad (Eq.\ 3.2)$$

with d = distance, going from zero to the length of the track measured between two beacons (lap distance)

$t_{\text{LAP1}}(d)$ = running lap time of Lap 1 as a function of distance

$t_{\text{LAP2}}(d)$ = running lap time of Lap 2 as a function of distance

This is the difference in running lap time plotted against the distance of both laps. In Figure 3.9, instantaneous variance is shown as the upper trace. The trace measuring less than zero means an advantage for the lap in which the black speed trace was recorded. This gives the user the opportunity to locate areas where variations between the two laps occur.

Cumulative variance can be expressed mathematically as ***Equation 3.3***.

$$\text{Variance}_{\text{Cumm}} = \int_{d=0}^{d=\text{lapdistance}} \text{Variance}_{\text{Inst.}} \qquad (Eq.\ 3.3)$$

This is the sum of the time differences between the two laps. The last sample in this graph is the total difference in lap time. The lower trace in Figure 3.9 shows the cumulative variance for the example. A positive value indicates an advantage for the lap in which the gray trace was recorded. Cumulative variance shows how a difference in lap

Figure 3.8 Overlay of the speed trace of two laps around the Nürburgring

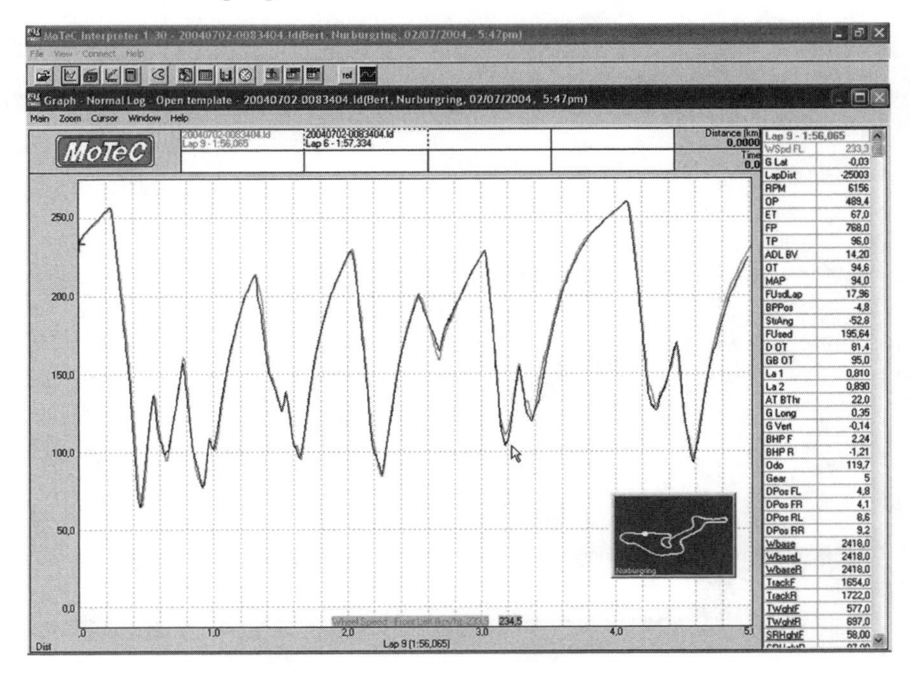

time developed over the duration of the lap. Effects of corner exit speed, changes in gearing, and braking points on lap time can be determined easily from the graph.

Figure 3.9 ***Instantaneous and cumulative variance between two laps around the Nürburgring***

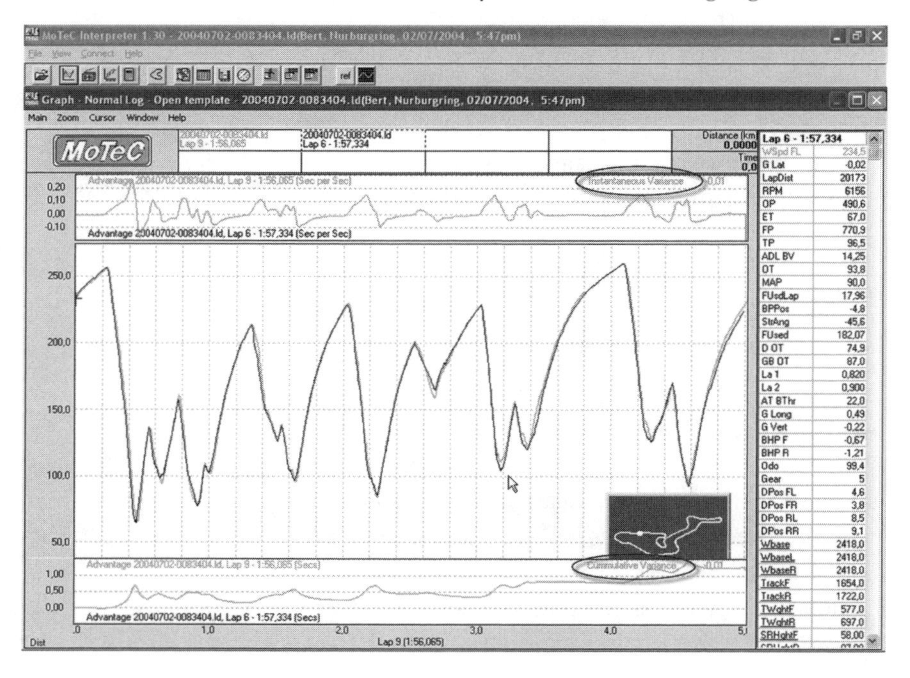

Figure 3.10 ***Time slip rate and time slip between two laps around Zolder***

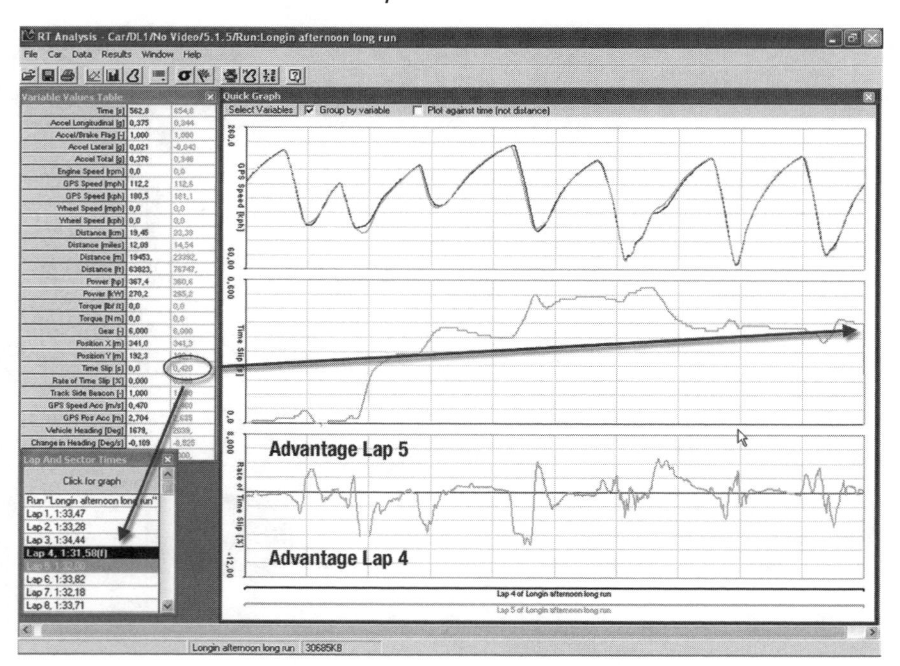

Because speed equals distance per unit of time, variance calculated by the subtraction of the two speed signals gives the same result. As an example, Race Technology's analysis software calculates two similar variables—time slip rate and time slip.

Time slip rate is the difference in vehicle speed, expressed as a percentage. If in one corner the car is doing 100 km/h during the reference lap and in the next lap it is doing only 80 km/h, it is 20% slower. Time slip rate at that point on the track is 20%. Therefore, the higher the time slip rate, the slower the car is at that point. A negative time slip rate means the speed at that point is higher than that of the reference lap.

Time slip is the sum of all time slip rates. Multiplied by the elapsed lap time, it provides exactly the same result as the cumulative variance. ***Figure 3.10*** gives an example of two laps around Circuit Zolder. At the end of the lap, the time slip value is exactly the difference between the two lap times.

Distance usually is calculated by integrating the speed signal. Therefore, by using this speed signal directly to calculate the difference between two lap times as a function of distance, the accuracy is a bit better.

The quality of an overlay and the variance function depends greatly on the accuracy of the distance calculation. Most of the time, distance is determined by integrating the wheel-speed signal and, therefore, is subject to a number of potential errors. Locking brakes and wheel spin may alter the relationship between the RPM of the wheel and the vehicle speed temporarily. The rolling radius of the tire changes as a function of the load put on it. The accuracy of the timing beacon also can be a source of error. Finally, the path the driver follows differs somewhat from lap to lap, so the lap distance also varies as a result.

All of this means care with overlays is required because it might mean events that occurred at the same location are not being compared. Most modern qualitative software packages incorporate algorithms to reduce the effect of wheel lockup by comparing wheel speed to the longitudinal *g*-force channel. When the values of the speed channel change too much with reference to the integrated longitudinal *g*-force channel, these values are corrected by interpolation.

Measuring suspension travel helps evaluate the error in the distance channel and even can serve to align the data. As long as the car follows the same line around the track, the road profile creates an enormous amount of small peaks in the suspension potentiometer data, and this serves as a fingerprint of the track. The bigger the offset between these peaks, the larger the error is in the distance function. Suspension travel often is measured at a high sampling frequency, so it is an ideal channel to align the distance function of two overlaid laps. An example is illustrated in ***Figures 3.11–3.13***.

Figure 3.11 shows an overlay of two laps around the Istanbul track. Speed trace and left-front suspension potentiometer data are pictured. On first glance, this looks like a clean overlay. Two areas are indicated in the graph. Area 1 is situated at the beginning of the lap, while Area 2 is more to the end of the lap, with the preceding corner showing a wheel lockup event in one of the laps under comparison.

In Figure 3.12, the potentiometer data zooms in on Area 1. The spikes in the data created by road surface irregularities a synchronized for the two laps. At this point during the two laps, there is no significant difference in the distance function and the overlay can be considered aligned.

Figure 3.13 takes a closer look at Area 2. As shown in Figure 3.11, this area was preceded by a corner where in the gray trace a wheel lockup event occurred. This means that at this point, measured wheel speed is much lower than the true vehicle speed. As a result, the potentiometer data is not synchronized after this event. The gray trace lags the black one by 6.2 m. At this point, the accuracy of the overlay was compromised, which could lead to wrong conclusions.

Most software packages can offset the timing beacon with a certain distance or time interval, allowing a manual correction to be performed on the data. This does, however, change only the starting point of the lap and has no influence on the distance errors created during the lap. If one is zooming in on a specific area, the data must be checked to ensure it is aligned at that location and the beacon is offset as necessary.

Some software writers use algorithms in their software to align data using the track pattern. This method was developed by William. C. Mitchell[1] and produced effective results.

***Figure 3.11** Overlay of two laps around the Istanbul Park Racing Circuit. The upper trace shows the left-front suspension potentiometer channel. The lower trace is the left-front wheel speed.*

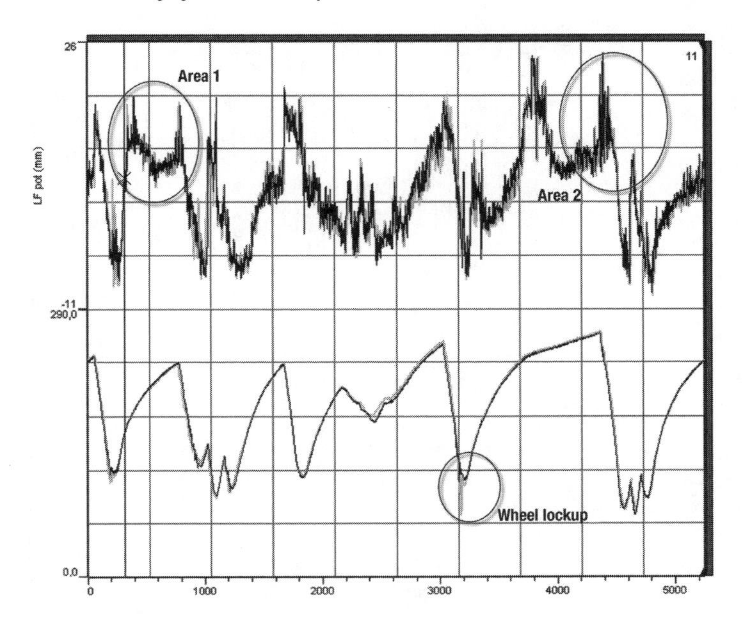

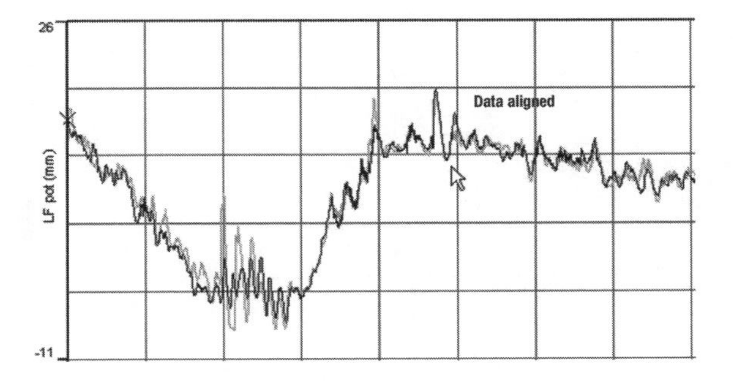

***Figure 3.12** Zoom of left-front suspension potentiometer signal on Area 1. At this point in the lap, there was no significant offset in road profile data.*

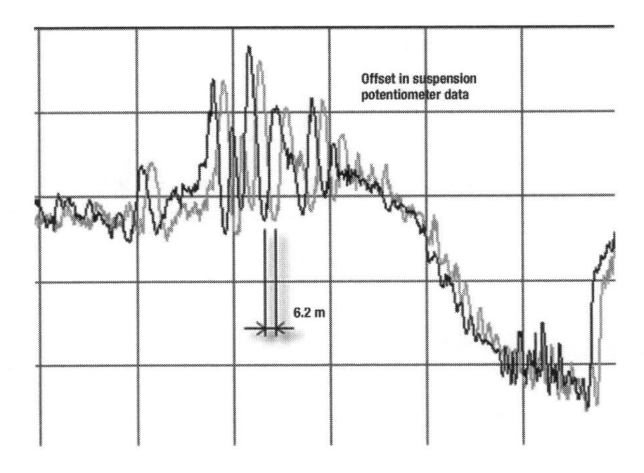

***Figure 3.13** Zoom of left-front suspension potentiometer signal on Area 2. The signal shifted 6.2 m.*

Measurement of the road surface also can help when a beacon is missed or when the signal is not present at all. If the software allows the user to insert beacons manually, the peaks in the potentiometer signal can serve as reference points to place beacons into the data.

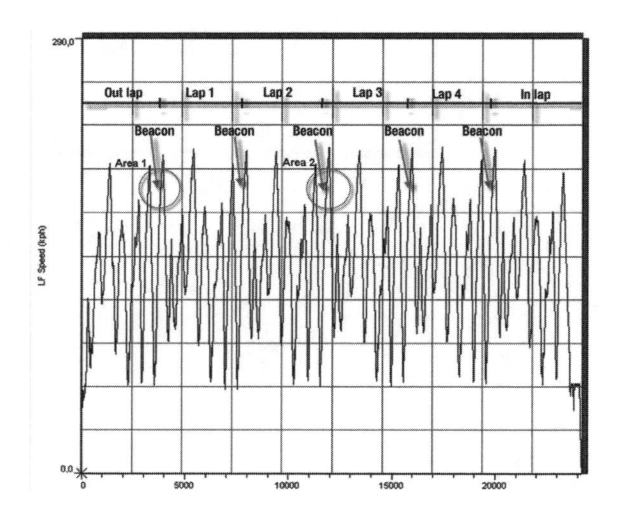

Figure 3.14 *Speed trace around Zolder without a beacon signal. The arrows indicate the locations where beacons are desired.*

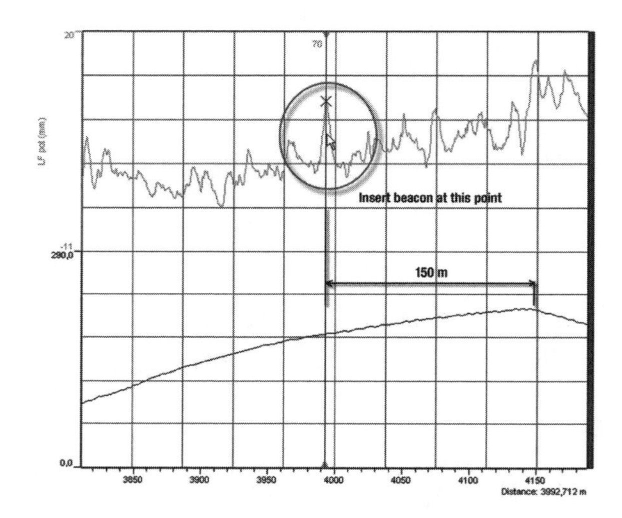

Figure 3.15 *Area 1 zoomed in. A beacon was manually inserted at a distinctive suspension potentiometer peak.*

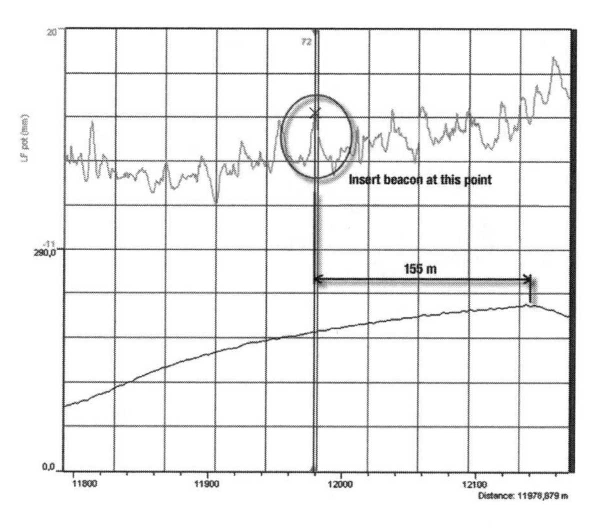

Figure 3.16 *Area 2 zoomed in. The same distinctive potentiometer peak returns and, again at this point, a beacon was manually inserted.*

In ***Figure 3.14,*** a speed trace recorded around Zolder is shown. Because a beacon channel was not recorded, lap time data is not present. The arrows in the illustration indicate where the beacon normally would be recorded if it had been present, which was somewhere at the start/finish straight in front of the pits. A quick look at the graph shows that during this run the car did 6 laps.

When a suitable suspension position peak can be found near the locations indicated by the arrows, this peak can serve as a virtual beacon. ***Figures 3.15*** and ***3.16*** provide two examples for the Zolder speed trace. The first graph shows speed and left-front potentiometer traces for the location indicated in Figure 3.14 as Area 1. The road profile is evident from this signal, and a distinctive peak in the trace was selected as the location for the beacon. As a reference, the distance to the next braking point is 150 m.

The same was done in Figure 3.16 for Area 2. The same distinctive peak returned in this trace, now at 155 m from the next braking point.

This procedure is followed for every required beacon, and at the conclusion the following lap times were achieved during this run:

Out Lap	2'20"065	Lap distance = 3,763 m
Lap 1	1'52"835	Lap distance = 4,000 m
Lap 2	1'52"725	Lap distance = 4,004 m
Lap 3	1'51"730	Lap distance = 4,000 m
Lap 4	1'50"450	Lap distance = 3,991 m
In Lap	2'28"655	Lap distance = 4,363 m

To validate the quality of the manually inserted beacons, two laps from this run can be overlaid and the road profile compared. ***Figure 3.17*** illustrates an almost perfect fit in the area just after the start/finish beacon for Laps 3 and 4.

The previous example used a run that did not have a beacon signal. When a run only misses part of the required beacons, the distance from a recorded beacon signal to a distinctive peak in the road profile can determine where a missing beacon should be inserted, as illustrated in ***Figure 3.18***.

Inertial Track Mapping

Track maps are graphical representations of the location at which logged data was recorded. It is a helpful software feature for lap navigation and a visual aid for drivers and engineers that can be used while analyzing the data. To draw a track map, three signals should be present: wheel speed, lateral acceleration, and a lap beacon. Wheel speed integration gives the covered distance, and combined with lateral acceleration the heading of the vehicle can be calculated. The lap beacon indicates the start and finish point of the lap. This technique is called inertial mapping.

It is, however, just a graphical representation of the racetrack and has its limitations. It can offer a clear and quick illustration of events occurring on the track, as shown in ***Figure 3.19***. In this example, braking and acceleration zones are indicated by different colors. Top speeds on the straights and minimum cornering speeds are given with their corresponding engine RPM. Accuracy limitations, however, mean that it is no more than that—a visual aid.

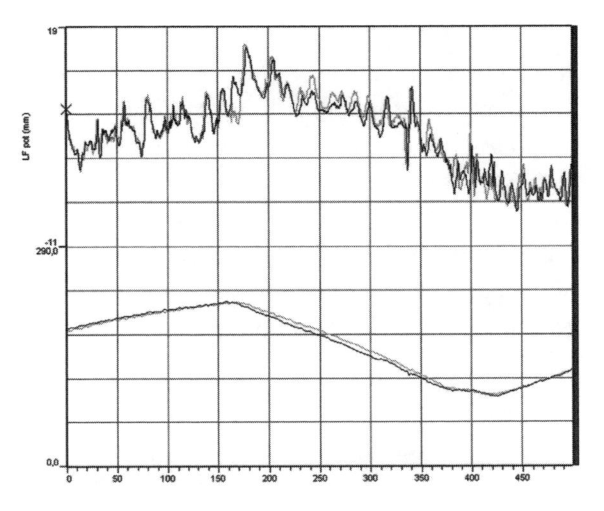

Figure 3.17 *Overlay of Lap 3 and 4 at the area just after the manually inserted beacon. The two traces are perfectly aligned.*

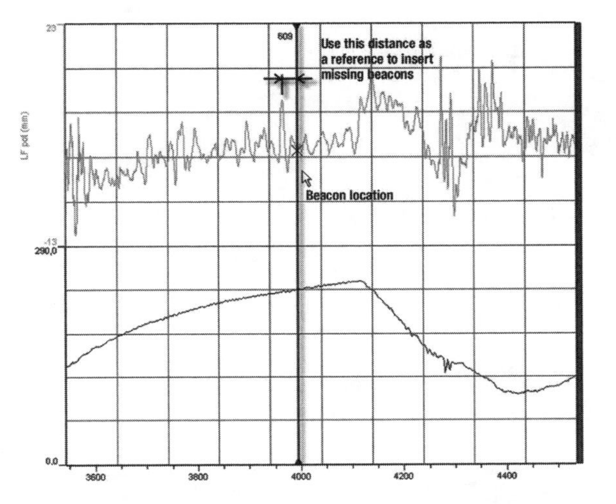

Figure 3.18 *Use the distance from a recorded beacon to a distinctive road surface irregularity as a reference to insert missing beacons in the data.*

GPS and Track Mapping

With the use of GPS becoming more popular in racecar data acquisition, the accuracy of track maps generated by analysis software packages has increased significantly. Maps now offer some interesting features.

By combining the GPS position and velocity signal with lateral and longitudinal acceleration, track maps that are accurate to within 1 m can be generated. This makes it possible to overlay track maps of different laps and compare the driving lines (see ***Chapter 12***).

Inertial mapping does not work on tracks that are not closed (e.g., a special stage in a rally). For motorcycles, special algorithms are necessary to calculate a track map often detrimental to accuracy. For boats, inertial mapping also is not an option. GPS mapping gives excellent results here. GPS in itself does not produce highly detailed track maps, but in combination with inertial mapping the results increase in quality.

Another advantage of GPS mapping is that a trackside beacon is not necessary. A coordinate beacon can be defined in the software, which is more accurate than using an infrared beacon. How many times are beacons not placed before a session or left out on the track after the day is finished?

Figure 3.19 *Inertially mapped track map of Zolder indicating top speeds, minimum cornering speeds, and corresponding engine RPM*

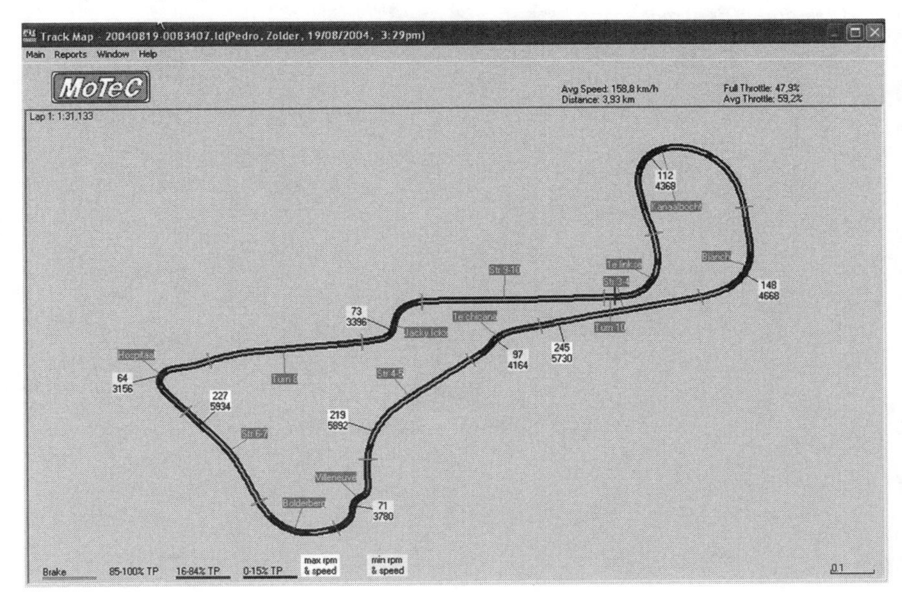

Figure 3.20 shows a track map of Spa generated by Race Technology's DL1 GPS data logger. The graph covers a complete outing. Zooming in on separate corners reveals the driving lines taken during this outing, as the boxed section (the Raidillon corner) clearly illustrates. Even an off-road deviation due to a missed braking point is recorded, making it easy to determine where the grass in the radiator came from.

Figure 3.20 *Track map of Spa created using GPS*

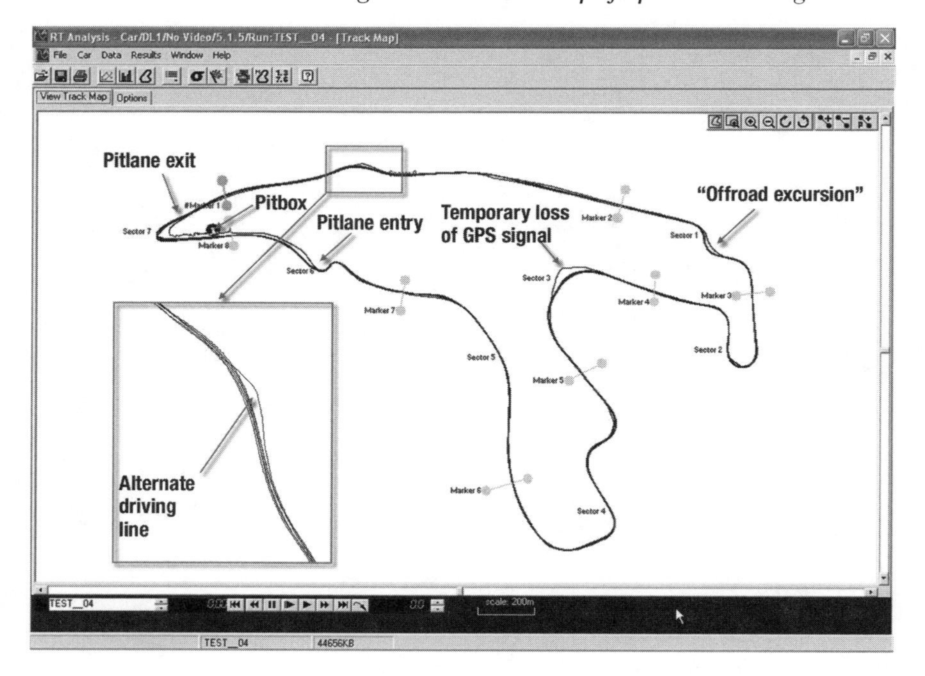

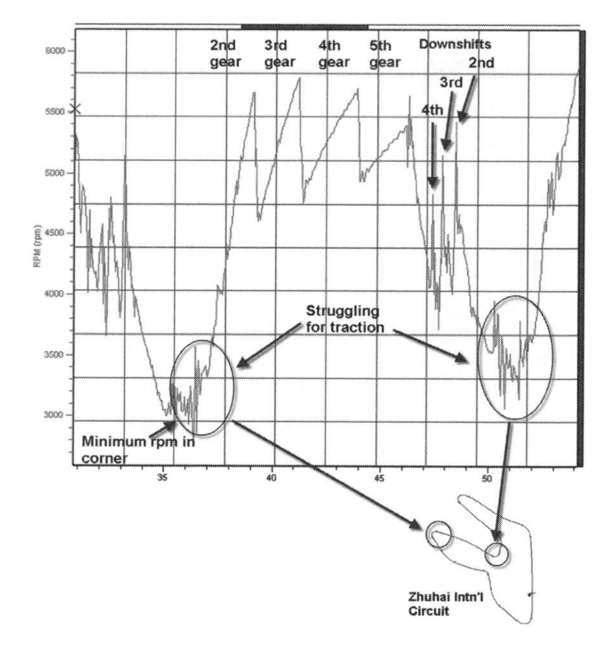

Figure 3.21
Engine RPM trace

GPS does not increase only the accuracy of track maps. The speed measurement of a GPS is not influenced by wheel spin, lockup, or changing a tire radius due to load or wear. This means that the distance function becomes more accurate, thereby increasing the quality of lap overlays. More confidence can be put in the variance function.

Segment beacons are determined coordinates created in the analysis software. They are not dependent on any distance calculation, which increases the segment time calculation accuracy, and, therefore, offer a more accurate theoretical fastest lap and fastest rolling lap.

The Beginner's Data Logging Kit

As mentioned in Chapter 1, any data logging system intended for the analysis of racecar and driver performance should log at least the six basic signals: engine RPM, vehicle speed, throttle position, steering angle, and lateral and longitudinal acceleration. These signals already contain a vast amount of information to analyze. Even in a state-of-the-art data acquisition package with numerous sensors, these six signals remain the most important and most used data resource for the engineer. The next logical step is to add suspension potentiometers to the system. In this section, the traces created by these sensor signals are explored and a feel for reading the graphs developed.

Logging Engine RPM

Engine RPM often is recorded from a magnetic sensor placed near a teethed trigger wheel on the engine's crankshaft. This sensor counts the pulses generated by this trigger wheel and converts them into the number of crankshaft revolutions per unit of time. The produced graph typically resembles the one in ***Figure 3.21***. Next to an engine-related analysis, this graph reveals the shifting activities of the driver as well as information on chassis balance.

This trace was logged on the Zhuhai International Circuit. The illustration shows the area from and between the two indicated corners. The first corner is a right-hand hairpin turn taken in second gear followed by a short straight. The second corner is a tight lefthander, again taken in second gear. The typical sawtooth pattern indicates upshifts as the driver goes through the gearbox from second to fifth gear. Shift RPM varies between 5,700 and 5,800 RPM and the engine reaches 5,400 RPM in fifth gear before the driver applies the brakes. Downshifting is indicated by the steep upgoing spikes in the graph as the driver blips the throttle to synchronize the engine and gearbox.

The minimum corner RPM is somewhat difficult to read from the graph, as the jaggedness of the trace indicates a car struggling to find traction out of the corner. The hairpin turn was exited at an engine speed of approximately 3,100 RPM. A quick glance at the power curve of this car's engine indicates if a lower gear should selected or another ratio be used. Gearing is investigated more thoroughly in ***Chapter 6.***

Logging Vehicle Speed

The speed trace is the results graph. It is the best way to conclude if a change on the car or in driving style produced any result. This is the reason why most analysis work is done with the speed graph as a reference. It is also the easiest trace to use for track navigation. Because this graph represents a typical layout of each track (indicating corners between acceleration and deceleration zones) and the signal is used so much in analysis work, an experienced data engineer looks more closely at the speed trace rather than at a track map to find a location on the track.

Figure 3.22 tracks the vehicle speed of a GT car during a lap around Zhuhai. A rising line represents acceleration, while a downward slope means the car is losing speed. Minimum cornering speed and top speed on the straights are important performance measures of this graph, and comparing these to previous outings shows where time was gained or lost. Use this graph to evaluate braking points, aerodynamic configurations, and engine tweaks.

A general glance at the speed trace helps establish some track characteristics. Is it a high-speed track with many high g-force corners or is it primarily a Mickey Mouse track with numerous tight corners and heavy braking zones? What kind of base setup should be used for the track in question? Another way to visualize this is to picture the speed data in a histogram. Notice the differences in achieved speeds for the same car between Zhuhai ***(Figure 3.23)*** and Monza ***(Figure 3.24)***.

Figure 3.22 *Vehicle speed trace*

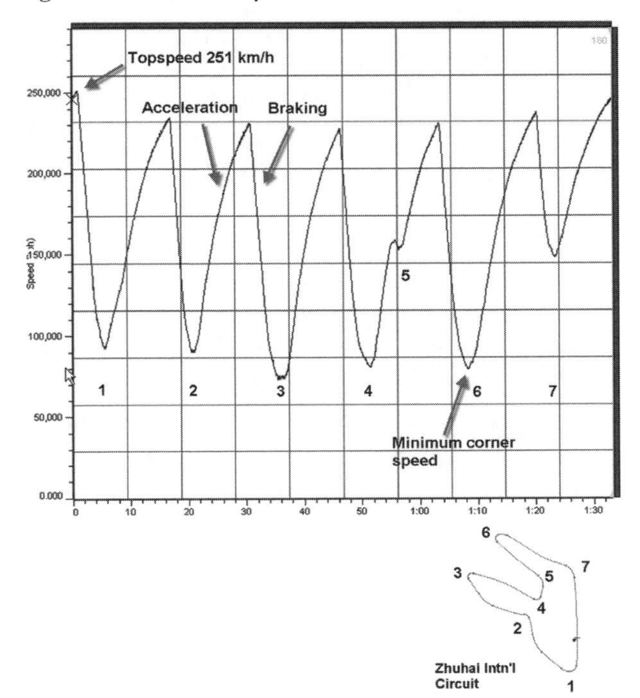

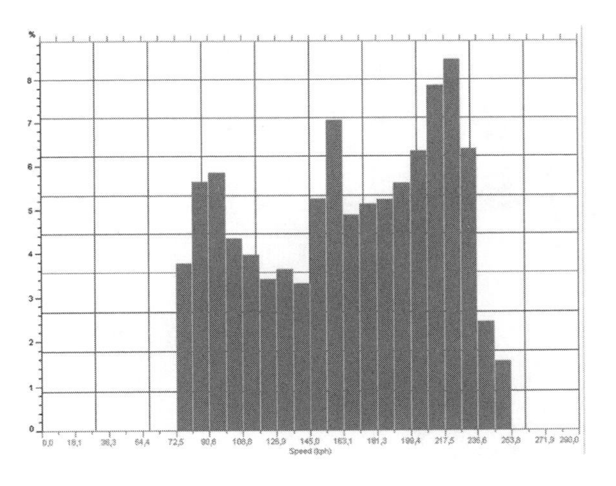

Figure 3.23
Speed histogram of Zhuhai

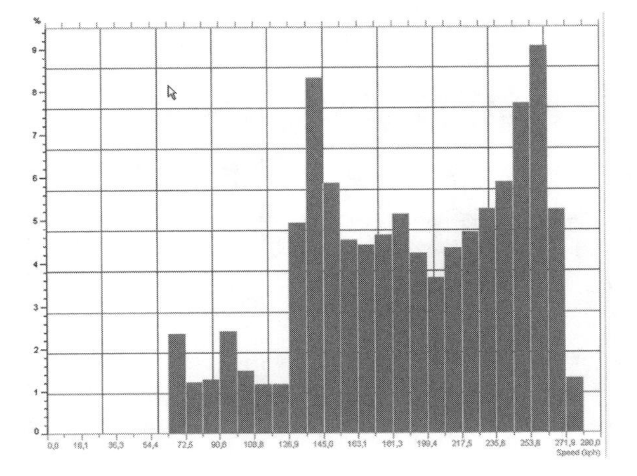

Figure 3.24
Speed histogram of Monza Autodrom

Logging Throttle Position

The throttle position signal measures what the driver is doing with his right foot on the accelerator pedal. Throttle position usually is expressed as a percentage, with 0% meaning the driver is completely off the accelerator pedal and 100% meaning full throttle.

Looking at the throttle trace in isolation is not very illustrative. It is, however, one of the most important channels for diagnosing chassis or driver issues. When the throttle is opened partially, it usually indicates a problem with the car or the driver.

Figure 3.25 shows an example of a throttle position trace. This channel becomes an important analysis tool when viewed with other channels, as numerous examples in the remainder of this book show.

Figure 3.25 *Throttle position trace*

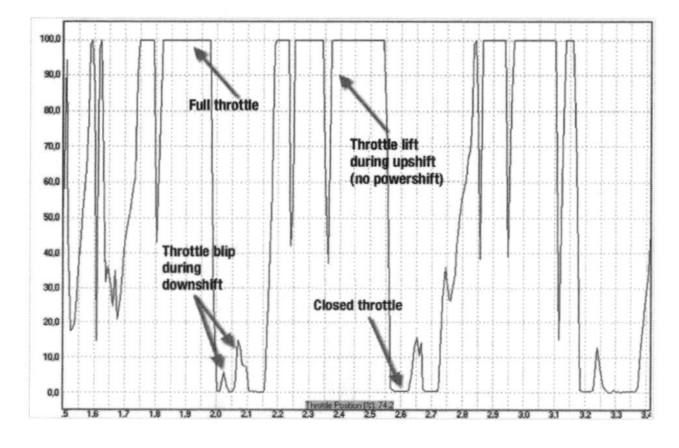

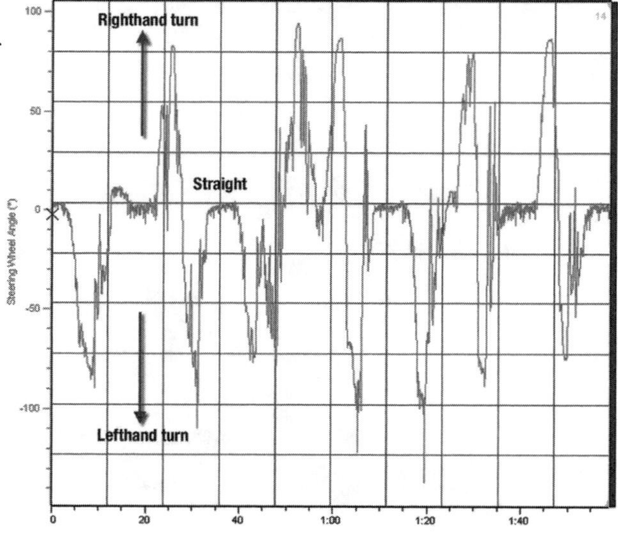

Figure 3.26 *Steering angle trace (expressed as angle of the steering wheel in degrees)*

Logging Steering Angle

As with throttle position, steering angle is a driver activity channel. It records the angle at which the steering wheel is turned and, just like throttle position, is an invaluable diagnostic tool. Steering angle can be expressed as degrees turned by the steering wheel, spindle, or rim, as well as steering rack travel in millimeters. The shape of the graph is the same in all cases ***(Figure 3.26)***.

If the sensor is properly calibrated, a 0-deg steering angle means that the car is traveling straight. The SAE Vehicle Axis System in Reference 2 defines a positive steering wheel angle for right-hand turns. The way the steering angle trace is pictured in graphs depends on the way the sensor is mounted, but it usually is also user-definable in the software. Note that there are examples in this book in which this sign convention was not followed.

Logging Lateral Acceleration

Lateral *g*-force is the channel logged as the acceleration perpendicular to the car's centerline, and strictly speaking it measures cornering force. This channel usually is displayed in units of *g*-forces (1 *g*-force = 9.81 m/s^2).

Sign convention, according to the SAE Vehicle Axis System for this trace, is the same as with the steering angle channel (i.e., positive for a right-hand turn). In ***Figure 3.27,*** a positive value indicates a right-hand turn, while a left-hand turn produces a negative lateral *g*-force value. Note that there are examples in this book in which this sign convention is not followed.

The maximum values for cornering acceleration depends on the available grip (known as the friction coefficient between road and tire) and the normal load working on the tires. Physically, ***Equation 3.4,*** where V = vehicle speed and R = corner radius, applies.

$$G_{lat} = \frac{V^2}{R} \quad \text{(Eq. 3.4)}$$

Therefore, the higher the speed at which a corner with a given radius is negotiated, the higher the lateral acceleration is. This means that a car generating a certain amount of grip and downforce at a certain speed has a theoretical maximum speed through a corner with a given radius.

The lateral *g*-force trace helps in analyzing handling behavior and absolute cornering power and is also a parameter in numerous mathematical channels used in this book.

Logging Longitudinal Acceleration

Longitudinal *g*-force is the acceleration logged along an axis parallel to the car's centerline, (i.e., perpendicular to the lateral *g*-force). It is basically the acceleration created by the engine's power or the deceleration due to application of the brakes. A positive value is used for acceleration. For deceleration, the sign for the longitudinal *g*-force is negative.[2]

An example of longitudinal acceleration trace is given in ***Figure 3.28,*** from which some standard features can be read. Maximum braking effort is displayed as the minimum value of the downward dips as the car decelerates. This value is higher if braking commences at a higher speed because the effect of aerodynamic drag adds to the braking effort. Maximum forward acceleration decreases as speed increases, also an effect of aerodynamic drag. On long straights, forward acceleration is close to zero when engine power output matches the aerodynamic resistance. The short downward spikes occurring during forward acceleration represent upshifts into a higher gear (see Chapter 6).

If a longitudinal *g*-force sensor is not present, another solution is to differentiate the speed channel. Speed and longitudinal acceleration are related through ***Equation 3.5***.

$$G_{long}(t) = \frac{\partial V(t)}{\partial t} \qquad (Eq.\ 3.5)$$

Most analysis software packages allow differentiation of a channel. Express the speed channel used as input for the differentiation in meters per second (1 km/h = 0.278 m/s), so that the output is in m/s^2. Then convert to g-forces if necessary.

A calculated longitudinal acceleration trace is less accurate because differentiating is basically filtering, and it depends mainly on the differentiation time used by the software and the sampling frequency used to log the speed channel. This way, events such as gearshifts may not be visible in the data.

Longitudinal acceleration displayed with lateral *g*-forces in an *X-Y* graph form the popular traction circle, a useful visualization technique illustrating how the potential of the tires is used. This graphical representation is covered in ***Chapter 7.***

Logging Suspension Travel

The six basic signals covered in the previous sections already give the engineer a significant amount of information about chassis and driver performance. Equipping the car with four suspension travel potentiometers helps in further diagnosing vehicle dynamics. Because suspension travel is

Figure 3.27 *Lateral acceleration trace*

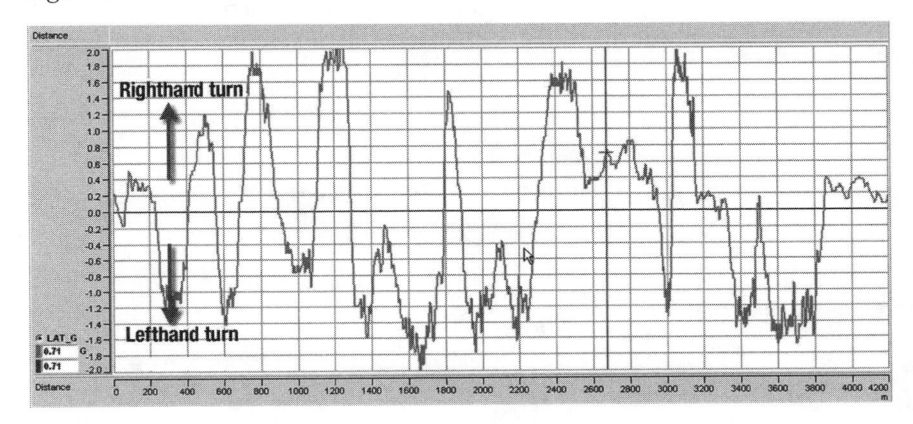

Figure 3.28 *Longitudinal acceleration trace*

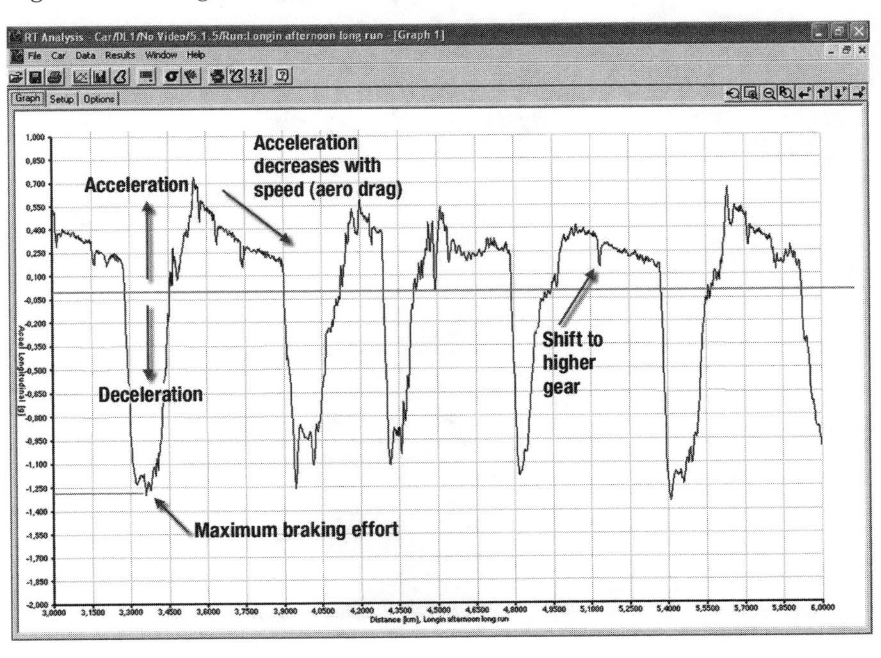

used extensively in the following chapters in mathematical channels, a review of basic properties of the signal is needed.

Suspension movement typically is measured as shock absorber displacement. With this signal, a sign convention and a short explanation of nomenclature is necessary. All ingoing shock travel from static ride height is considered positive. When the wheel goes up relative to the chassis, the sensor signal has a positive sign. This motion is called *bump*. The opposite is true for all outgoing shock travel from static ride height. This is called *droop*. Shock absorber specialists use the names *bump* and *rebound*, but this actually refers to the gradient of the sensor signal. A positive gradient (ingoing shock travel, regardless of static position) is called *bump travel*, while a negative gradient (outgoing shock travel, regardless of static position) is referred to as *rebound travel* ***(Figure 3.29)***.

Pictured is the signal from a damper potentiometer measuring suspension travel at the front left wheel. The upper trace is the raw, unfiltered signal.

***Figure 3.29** Suspension movement trace*

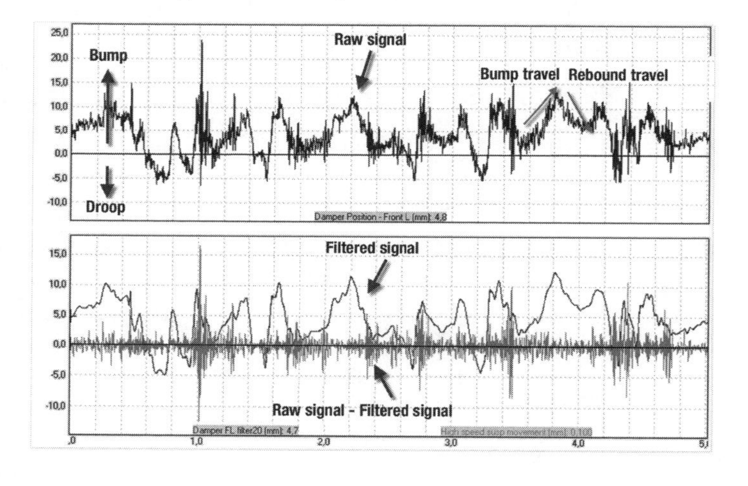

This measurement combines two different categories of suspension movement in one signal.

1. Low-speed Movement

This includes the suspension movement in response to chassis attitude changes due to weight transfer (pitch and roll) and to the varying aerodynamic loads at different speeds. The lower portion of Figure 3.29 shows the suspension movement signal after a filter (20-sample moving average filter) is applied. This trace represents the low-speed movement of this suspension corner.

2. High-speed Movement

The suspension movement induced by road irregularities and curbs take place at a higher frequency than the low-speed movement. In Figure 3.29, this movement is separated from the raw signal by subtracting the filtered signal from it. It is this portion of the data that was used to insert missing timing beacons in the data earlier in this chapter.

Remember that suspension travel is being measured here, not wheel travel ***(Figure 3.30)***. When the motion ratio of the suspension is known, one can calculate the wheel travel from the suspension travel signal using ***Equation 3.6***.

$$MR = \frac{x_{wheel}}{x_{suspension}} \qquad (Eq.\ 3.6)$$

with MR = motion ratio

x_{wheel} = wheel movement

$x_{suspension}$ = suspension movement

The motion ratio can be measured statically by jacking up the wheel to a certain distance and measuring the stroke of the shock absorber (or recording it electronically). If a good suspension geometry software package is available, this ratio can be calculated as well. To get the wheel movement, create a mathematical channel that multiplies the suspension movement channel with the motion ratio.

Ensure that the potentiometers actually measure shock absorber travel. Sometimes the sensors are mounted in such a way that there is a motion ratio between sensor travel and shock absorber movement. This should be corrected in the software.

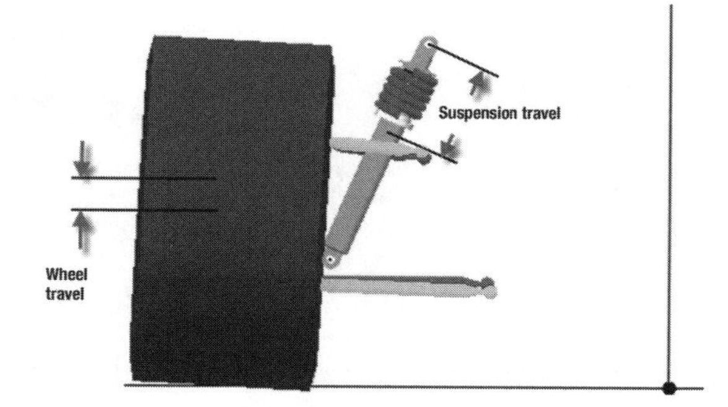

***Figure 3.30** Suspension travel versus wheel travel*

CHAPTER 4

STRAIGHT-LINE ACCELERATION

Cornering is one thing, but as soon as the racecar exits a turn the next challenge the driver faces is covering the following straight in the least possible time. In this chapter, analysis tools are provided to evaluate the performance of a racecar in straight-line acceleration.

Torque and Horsepower

The torque and horsepower delivered to the driven wheels has been always of interest to the race engineer. Evaluation of engine changes often is performed on an engine dynamometer. However, this requires removal of the engine from the racecar, and it only shows the torque and power values measured at the flywheel.

Measuring the car's longitudinal acceleration, vehicle velocity, and engine RPM makes it possible to calculate the torque (and, therefore, the horsepower) delivered to the wheels with relatively reliable accuracy.[3] As mentioned in Chapter 3, longitudinal acceleration can be derived from the velocity signal, so only RPM and speed are necessary to perform the calculation.

Torque delivered to the driven wheels must conquer primarily the external forces acting on the vehicle. These are (a) rolling resistance and (b) aerodynamic drag.

(a) Rolling resistance is created when a tire in contact with the road surface faces a distortion in its footprint. This is called *tire drag* and is characterized by a nondimensional rolling resistance coefficient (R_x).

A modern radial-ply tire on a passenger car typically has an R_x value of approximately 0.03. For racecar tires, this value can be as little as 0.005. The rolling resistance is given in ***Equation 4.1,*** where *m* is the total mass of the car, and g the gravitational acceleration (9.81 m/s²).

$$F_{rolling} = R_x \cdot m \cdot g \qquad (Eq.\ 4.1)$$

(b) The aerodynamic drag of a vehicle depends on its frontal area, drag coefficient, and local air density and is a function of the vehicle velocity squared ***(Equation 4.2),*** where ρ is the density of air (1.187 kg/m³ at 101325 Pa/20°C), C_D the drag coefficient, A the frontal vehicle surface, and V the vehicle velocity.

$$D = 0.5 \cdot \rho \cdot C_D \cdot A \cdot V^2 \qquad (Eq.\ 4.2)$$

Equations 4.1 and 4.2 possibly imply that some coefficients need to be estimated. However, it is also possible to determine the total external resistances through a coastdown test, which is covered in ***Chapter 11***.

The torque required to overcome the total external force on the vehicle is given in ***Equation 4.3,*** with $r_{rolling}$ the driven tires' rolling radius.

$$T_{ext} = \left(F_{rolling} + D\right) \cdot r_{rolling} \qquad (Eq.\ 4.3)$$

In addition to the torque required to overcome the external forces, the amount of torque available to accelerate the vehicle is given by ***Equation 4.4,*** where G_{long} is the longitudinal acceleration of the vehicle measured by the data logging system.

$$T_{mass} = m \cdot M_f \cdot G_{long} \cdot r_{rolling} \qquad (Eq.\ 4.4)$$

M_f is a factor that takes into account driveline rotational inertias and the mass factor ***(Equations 4.5*** and ***4.6)***.

$$M_f = \frac{M + M_r}{M} \qquad (Eq.\ 4.5)$$

$$M_f = 1 + 0.04 + 0.0025 \cdot i_{total}^2 \qquad (Eq.\ 4.6)$$

with M = translational mass = m/g
M_r = equivalent rotational mass

Following this, the driven-wheel torque can be calculated with ***Equation 4.7***.

$$T_{wheel} = \frac{T_{mass} + T_{ext}}{i_{total}} \qquad (Eq.\ 4.7)$$

And finally, the driven-wheel power is given by ***Equation 4.8*** and defined by ***Equation 4.9***.

$$P_{engine} = \left(F_{rolling} + D + F_{mass}\right) \cdot V \qquad (Eq.\ 4.8)$$

$$F_{mass} = \frac{T_{mass}}{r_{rolling}} \qquad (Eq.\ 4.9)$$

Aerodynamic Drag

At higher vehicle speeds, aerodynamic drag becomes the dominating factor and the equation for air drag force incorporates the vehicle speed squared. To calculate the power at the wheels, drag force is again multiplied by speed. Simply, the power required to overcome aerodynamic drag is closely related to the vehicle speed cubed. This means that to double the speed of a vehicle eight times the engine power is needed. This is why engine modifications have only a small impact on top speed. An old engine that is down on power might accelerate slowly but still be able to reach close to its original top speed. For example, the top speed of a car needs to be increased by 10%. To do this, 33% more power ($1.10^3 = 1.33$) is required. Do not get stuck too much on top speed.

Figure 4.1 *Dodge Viper GTS-R (Courtesy of GLPK Racing)*

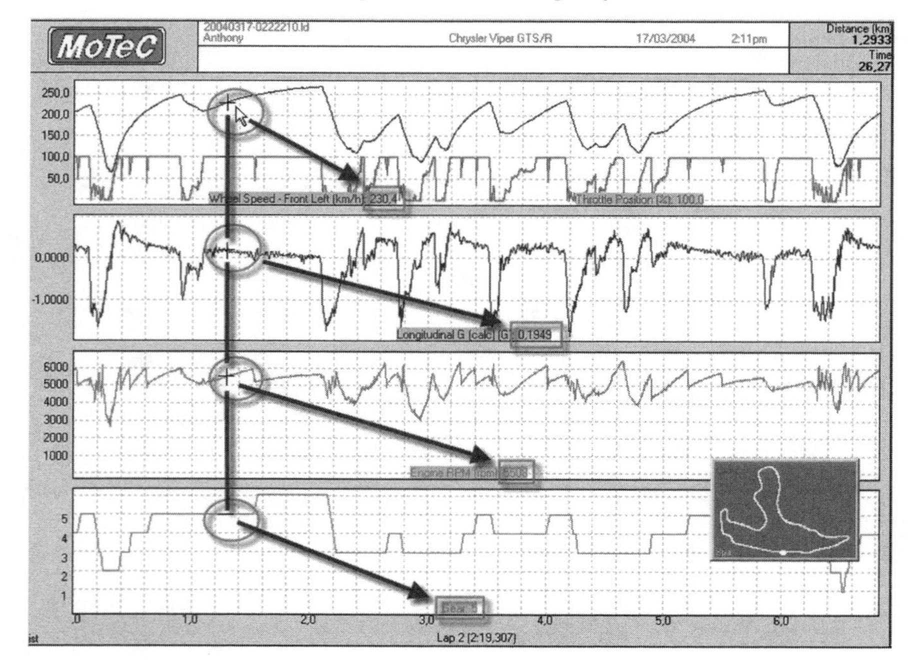

Figure 4.2 ***Distance graph illustrating a lap around Spa. Channels provided are front wheel speed, throttle position, longitudinal acceleration (in this case calculated by deriving the vehicle speed channel), engine RPM, and the gear position.***

The following example is a case study on the Dodge Viper GTS-R pictured in ***Figure 4.1***.

From the distance graph pictured in ***Figure 4.2,*** the following calculation data is taken at the point indicated by the cursor:

engine RPM = 5,508 RPM
vehicle speed (V) = 230.4 km/h
longitudinal acc. (G_{long}) = 0.195 G

The necessary vehicle properties to perform the calculation are the following:

vehicle total weight (m) = 1,323 kg
gear = 5th (i_{total} = 3.14)
$r_{rolling}$ = 0.365 m
R_x = 0.025
C_D = 0.601
A = 2.3 m²

The data was taken from a test run performed on the Circuit de Spa-Francorchamps. The top speed during that particular lap was 267 km/h. The calculation example concentrates on a random data point along the straight leading to the Les Combes chicane. At this point, the car is accelerating on a straight line. Note that at this section on the track the road runs uphill, which influences the calculation (the longitudinal acceleration here is lower than expected). If the slope of the track is known, the longitudinal force component can be calculated as an extra external resistance working on the car (see Chapter 10). For simplicity, track slope is not taken into account in this calculation ***(Equations 4.10–4.16)***.

$$F_{rolling} = 0.025 \cdot 1323 \cdot 9.81 = 325 \text{ N} \quad \text{(Eq. 4.10)}$$

$$D = 0.5 \cdot 1.187 \cdot 0.601 \cdot 2.3 \cdot \left(\frac{230.4}{3.6}\right)^2 = 3352 \text{ N} \quad \text{(Eq. 4.11)}$$

$$T_{ext} = (325 + 3352) \cdot 0.365 = 1342 \text{ Nm} \quad \text{(Eq. 4.12)}$$

$$M_f = 1 + 0.04 + 0.0025 \cdot 3.14^2 = 1.06 \quad \text{(Eq. 4.13)}$$

$$T_{mass} = 1323 \cdot 1.06 \cdot 1.912 \cdot 0.365 = 979 \text{ Nm} \quad \text{(Eq. 4.14)}$$

$$T_{wheel} = \frac{1343 + 979}{3.14} = 739 \text{ Nm} \qquad \textit{(Eq. 4.15)}$$

$$P_{engine} = (325 + 3352 + 2682) \cdot 64 = 406976 \text{ Nm/s} \qquad \textit{(Eq. 4.16)}$$

with $1 \text{ Nm/s} = 0.001341 \text{ hp}$
$P_{engine} = 546 \text{ hp}$

Figure 4.3 represents a dynamometer run with this engine prior to the test from which the data was taken to perform the calculation above. Note that below 1,500 RPM the power and torque values were not measured because this engine speed is out of the measurement interval of the dynamometer. At an engine speed of 5,500 RPM, the graph shows a torque measured at the flywheel of 784 Nm and a power of 614 hp. Comparing this to the result of the preceding calculation, this means that the loss of torque due to internal friction and inertia is 45 Nm, or approximately 6%.

If the data acquisition system can export data to a mathematical analysis package or a spreadsheet, it is possible to create a torque and power curve directly from the logged data with the equations used to perform the previous calculation.

Calculating Torque and Power at the Wheels

The procedure for calculating the torque and power at the wheels is time consuming (unless performed using a spreadsheet or other mathematical software package). For quick analysis purposes and to have an idea of the engine's power output, one should define a mathematical channel that calculates the power the engine is using to accelerate the car. Equation 4.4 gave the torque used to accelerate the vehicle. Converting this to horsepower gives ***Equation 4.17***.

$$P_{mass} = G_{long} \cdot m \cdot V \cdot 0.001341 \qquad \textit{(Eq. 4.17)}$$

For the lap in Figure 4.2, this channel looks something like that illustrated in ***Figure 4.4***. Interestingly, this graph also shows the power the brakes are utilizing to decelerate the vehicle, so this channel also can be used to analyze the braking effort of the car-driver combination.

Traction and Longitudinal Slip

To brake or accelerate a vehicle, Newton's second law indicates that a longitudinal force needs to be developed between the ground and the tire footprint.

This longitudinal force is created because of tire mechanics, where the front of the footprint is compressed under the driving torque. The compressed part adheres to the road surface, resulting in forward stress. This stress reverses in the back part of the footprint as the tire radius recovers. In this part of the footprint, sliding occurs between the tire and the road, which is defined as *slip*. Slip means that the angular velocity of a driven wheel is always greater than that of a free-rolling wheel.

The longitudinal slip velocity is calculated by ***Equation 4.18,*** where V is the linear velocity measured at a driven wheel, and V_0 the linear velocity of one of the free-rolling wheels.

$$V_{slip} = V - V_0 \qquad \textit{(Eq. 4.18)}$$

Note that linear speeds are compared where angular velocities should be compared. However,

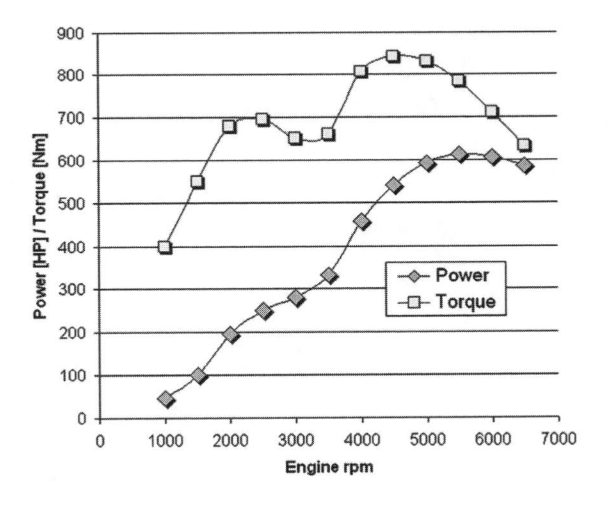

Figure 4.3 *Torque and power curve of the Viper V10 engine, measured on a dynamometer*

Figure 4.4 *The power the engine is delivering to the wheels to accelerate the car*

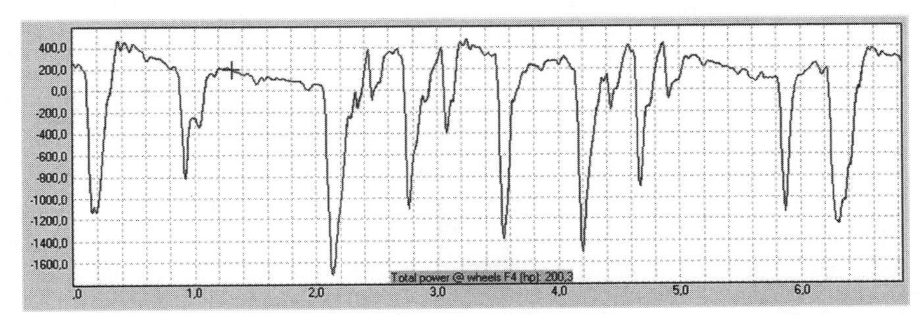

for simplicity, the effective radius of the tire is assumed to be constant.

Further, the slip ratio (SR) can be defined by ***Equation 4.19***.

$$SR = \left(\frac{V}{V_0}\right) - 1 \qquad (Eq.\ 4.19)$$

For a free-rolling wheel SR = 0 and for locked braking SR = –1. Wheel spin generally is defined as SR > 1.[4]

Any given tire develops its maximum coefficient of friction and, therefore, its greatest longitudinal force at a given percentage of slip ***(Figure 4.5)***. When the slip ratio exceeds this value (i.e., where more torque is being applied than the tire can transmit to the road surface), the traction capacity decreases. This is true for acceleration and braking.

Multiple wheel-speed signals are required to measure slip and calculate the slip ratio for a given racecar. Various configurations are possible:

- Measure one free-rolling and one driven wheel speed, preferably on the side that is loaded most of the time, on the racetrack (left-hand side on a right-hand side racetrack and vice versa).
- Measure one free-rolling and two driven wheel speeds. This makes it possible to calculate separate slip values for left- and right-driven wheels and to evaluate differential work.
- Measure four wheel speeds (the best setup). Reference speed V_0 can be averaged between the two free-rolling wheels.
- Measure at least one free-rolling wheel speed and calculate the speed of the driven wheels from the engine RPM, gear ratios, and tire radius. This is the least accurate option but useful in championships where restrictions exist for the number of used wheel-speed sensors to ban traction control systems (TCSs).

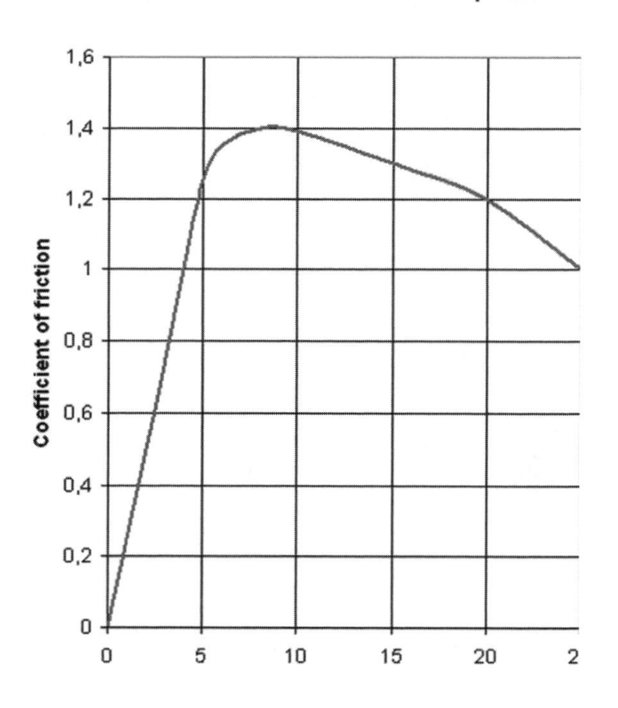

Figure 4.5 *Tire coefficient of friction versus slip ratio, expressed as a percentage*

Figure 4.6 illustrates data taken from a GT car around the Spa track, zoomed in on the La Source hairpin where the car decelerates from 225 km/h to approximately 55 km/h (the slowest point on the track) and follows with a hard acceleration phase.

The car was equipped with two Hall-effect sensors to measure the driven-wheel velocities and one to measure the undriven front-left wheel speed.

Longitudinal slip ratio is calculated as a percentage using ***Equation 4.20***.

$$SR = 100\% \cdot \frac{0.5 \cdot \left(WSPD_{RL} + WSPD_{RR}\right)}{WSPD_{FL}} \qquad (Eq.\ 4.20)$$

where $WSPD_{RL}$ = measured wheel speed rear left
$WSPD_{RR}$ = measured wheel speed rear right
$WSPD_{FL}$ = measured wheel speed front left

During the braking phase, the aerodynamic drag acting on the car was decreasing rapidly due to the decrease in vehicle speed, so it became more difficult for the driver to modulate the brake pedal to not lock the front wheels. During the first part of the braking phase, the slip ratio went down to approximately –3%, indicating that the tires could handle the brake torque. At a speed just below 80 km/h the slip ratio suddenly—but momentarily—drops to –15%. When one investigates using ***Equation 4.22,*** this means that at least one of the rear wheels was about to lock. At this point the driver, getting a warning signal from the tires that lockup was about to occur, eases off a little bit on the brake pedal, restoring the slip ratio to about – 5%. From the point where the driver pushes the throttle pedal, the slip ratio becomes positive, peaking at 5%. Note that maximum slip ratio is reached when the throttle pedal is half-open. After this, the slip ratio drops off to 3%, a similar value seen during the braking phase. At higher speeds, the slip ratio becomes zero.

In general, the graph can be read as follows:

SR positive	**During braking phase**	**Front wheel (about to) lock**
	During acceleration	**Rear wheel (about to) spin**
SR negative	**During braking phase**	**Rear wheel (about to) lock**
	During acceleration	**Front wheel (about to) spin**

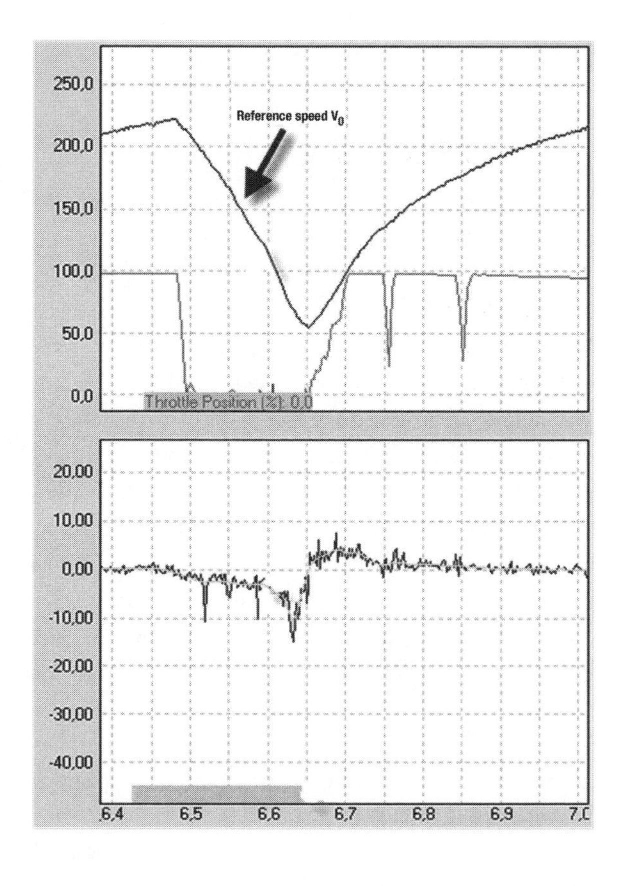

Figure 4.6 *Driver Magnus Wallinder negotiating the Spa La Source hairpin corner in a GT car. Both raw and filtered slip percentage channels are illustrated.*

Traction problems most often occur when the driver is exiting the turn, basically in a transient situation. Therefore, loss of traction or wheel spin most likely takes place first on the driven wheel that has the least load on it (i.e., the wheel at the inside of the corner). For this reason, it may be useful to calculate separate slip ratio channels for both driven wheels. These channels are pictured in ***Figure 4.7***.

This data is taken from the same car as in Figure 4.6. This distance graph shows the separate slip ratio values for left and right rear wheels. The lower graph shows the speed difference between the two rear wheels in km/h.

The slip ratios for the two driven wheels are particularly useful for analysis. The conclusion drawn from Figure 4.7 is that the inside rear right wheel (La Source being a right-hand corner) was against the blocking limit under braking. This wheel has less load on it than the outside wheel due to the weight transfer resulting from cornering. Traction occurring out of the corner seems to be pretty good as the driver progressively puts on more throttle.

The channels covered in this paragraph require the use of multiple wheel-speed sensors, using up to four digital inputs from the data logging unit. Even logging at moderate sampling rates, the channels use up a considerable portion of the unit's

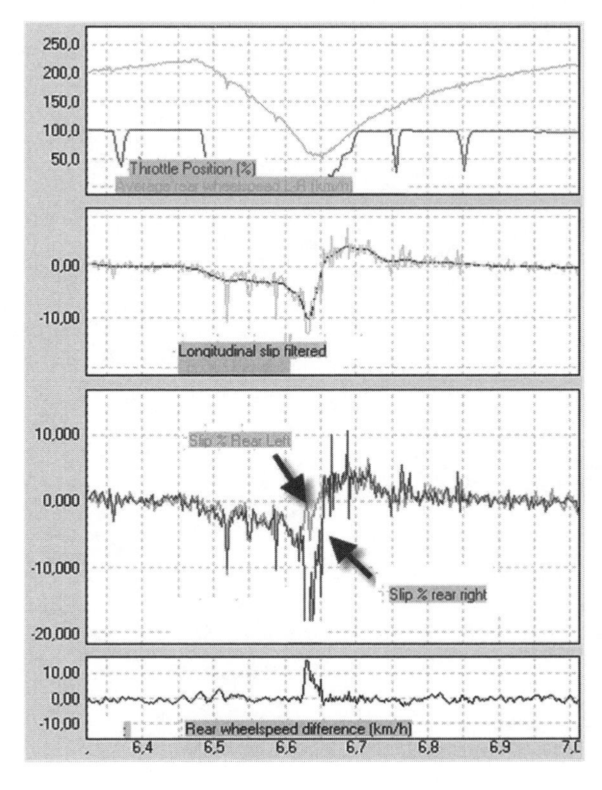

Figure 4.7 *Again Wallinder around the Spa La Source corner, with separate channels for slip ratios of the rear-driven wheels illustrated and the difference in speed between left- and right-driven wheels*

memory. However, analyzing slip ratios with driver activity and lateral/longitudinal G channels makes pinpointing handling problems much easier. It also provides the engineer good insight on the performance of the differential.

ABS/TCS and Slip Ratios

For any tire, there is a given slip ratio where the tire develops its maximum longitudinal force. The driver uses the accelerator and brake pedals to control slip ratio. ABSs and TCSs are designed to keep the tire at its maximum longitudinal force with the slip ratio as a controlled variable.

The ABS monitors the speed of the wheels and regulates the hydraulic pressure to the calipers accordingly to maximize the braking effort. By preventing the wheels from locking and keeping them at an optimal slip ratio, the system enables the driver to maintain steering control and brake the car in the shortest possible distance under most conditions. Traction control evaluates the amount of engine power transferred to the driven wheels. This can be done either by directly limiting the engine output (including cutting the fuel supply or retarding or suppressing the spark to one or more cylinders) or by applying the brake to the wheel that exceeds the optimal slip ratio.

***Figure 4.8** MoTeC Engine Management software with traction control function*

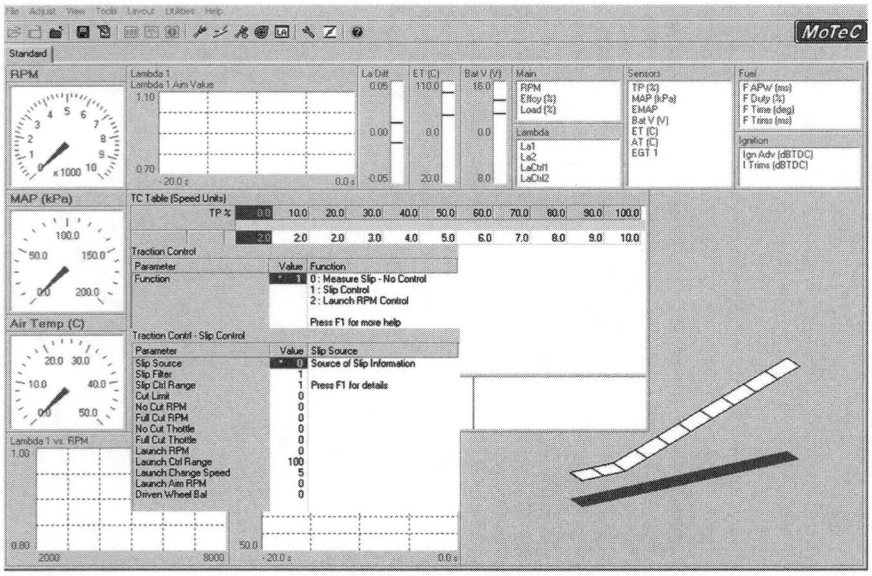

While the ABS and brake-controlled TCS require separate actuation units, most current motorsport engine management systems limit engine output power in case of excessive slip. In most cases, the only requirement is to wire the necessary wheel speed signals to the engine ECU and do the programming.

Figure 4.8 shows an example of a simple engine-controlled TCS. The user can enter the desired amount of slip, in this case expressed as the difference between the linear velocity of the driven wheels and a reference speed, for different engine loads. The engine management measures the different velocities and calculates the slip. As soon as this value exceeds the aimed slip value at the given load site, the software starts progressively cutting the ignition.

In some series, technical regulations only allow the use of one sensor to specifically measure the vehicle's velocity. A backdoor to this rule is (when the software allows) to use a calculated channel as the driven-wheel speed for the slip ratio calculation and put a wheel-speed sensor on one of the free-rolling wheels. The speed of the driven wheels can be calculated using ***Equation 4.21***.

$$V = \frac{2 \cdot \pi \cdot r_{rolling} \cdot n_{engine}}{i_{total}} \qquad \text{(Eq. 4.21)}$$

This method may not be as accurate as using separate sensors on each wheel but can be useful when regulations limit speed measurements.

This description of the TCS and ABS is very limited. Keep in mind that electronically maintaining the optimum slip value of four tires is more difficult than it looks because the slip value where maximum traction occurs is not a constant and is likely to vary even during one race lap. Measuring and evaluating slip values using the method mentioned in this paragraph may provide a useful tool in TCS and ABS development.

The following example represents data taken from an endurance racecar equipped with a Bosch engine-controlled TCS on the French Magny-Cours racetrack. The displayed channels are the following:

- Ignition angle during ASR (acceleration slip regulation, another commonly used name for traction control). This channel shows how much the TCS interferes with the engine ignition to decrease the slip of the driven wheels. The values for this channel are taken from a user-definable table in the engine ECU and are expressed as percentages of the original ignition angle.

- Nominal slip value represents the maximum allowable slip value for each particular instant. Channel values for this also are taken from an engine ECU table with speed, throttle position, and lateral G as variables.
- Reference speed
- Slip of drive train is the actual slip experienced by the driven wheels calculated according to Equation 4.20.
- Throttle position

Figure 4.9 shows the car during a corner exit phase. The ignition angle during the ASR trace indicates when the TCS is active. The graph shows a couple of short downward spikes indicating brief TCS activation. One spike is indicated with an arrow and occurs during a moment when the driver applies the throttle pedal. At this moment, the actual slip value shortly peaks at 6%, whereas the ECU concludes that no more than 4% is tolerable. As soon as the actual slip ratio drops below the nominal value, the TCS deactivates.

As the driver applies more throttle, the system needs to correct more to stay within nominal slip values. The application of traction control is also very noticeable from the short upward spikes in the actual slip ratio. An example is given in ***Figures 4.10*** and ***4.11,*** the first with traction control and the second without.

Time Versus Distance

"I catch him in the corners, but he runs away from me on the straights . . ." is an often-heard comment from racecar drivers that could be just the illusion of a time gap versus distance. Suppose the driver is closing to within 0.4 sec to the car directly in front of him ***(Figure 4.12)***. At this point, both cars have a cornering speed of 65 km/h or 18.1 m/s. Here, 0.4 sec translates to a distance between both cars of 7 m. The cars are exactly matched in horsepower and aerodynamics and, while exiting

Figure 4.9 *Traction control example*

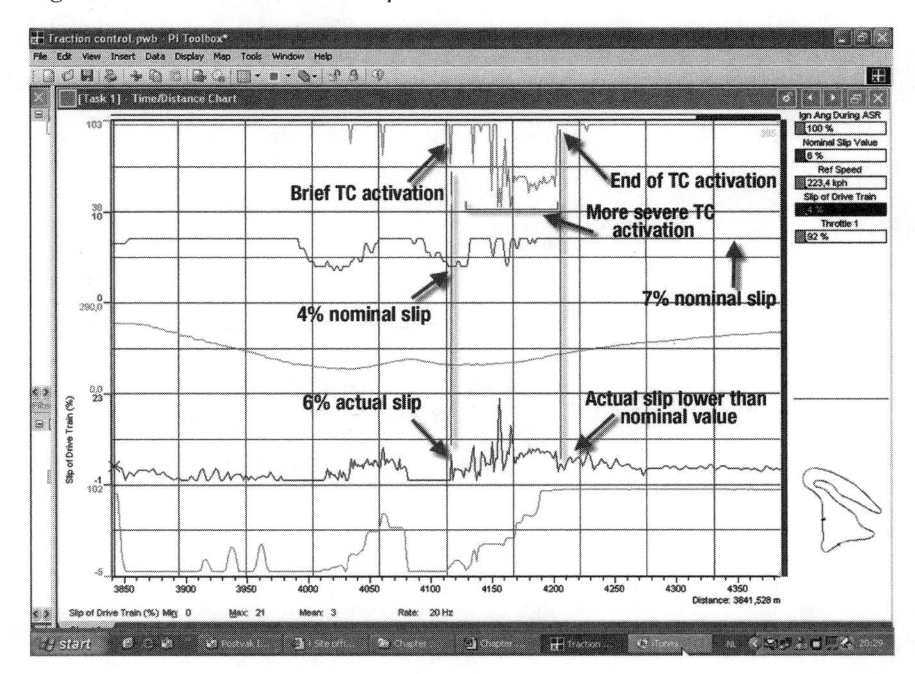

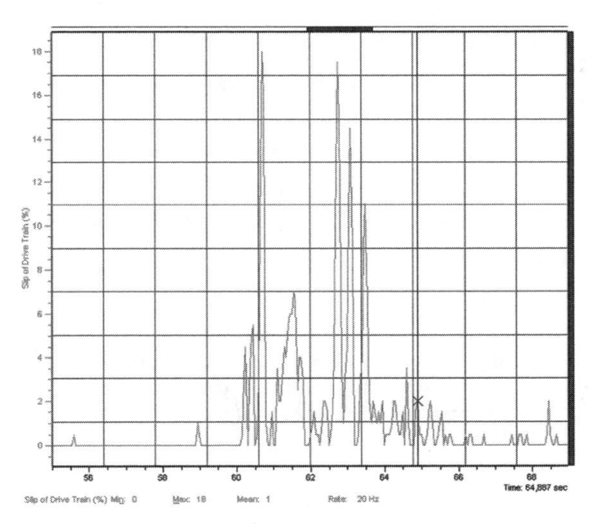

Figure 4.10
Slip ratio trace with traction control

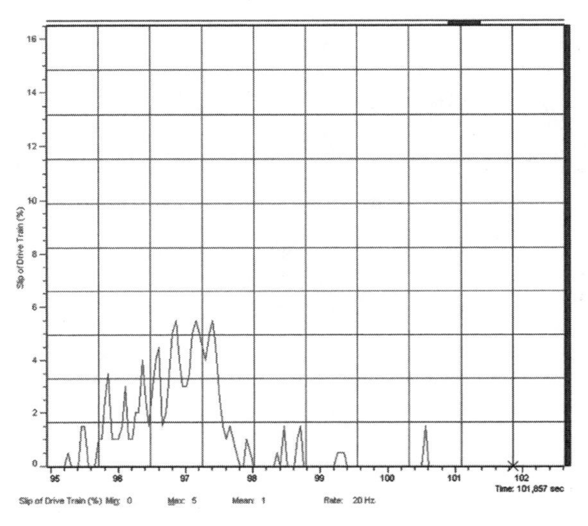

Figure 4.11
Slip ratio trace without traction control

the corner, both accelerate at exactly the same rate (as the speed trace in Figure 4.12 shows), keeping 0.4 sec between them. Halfway up the straight the cars were doing 180 km/h or 50 m/s. Now 0.4 sec translates to 20 m, a difference that reaches its maximum at the end of the straight (27 m at a speed of 265 km/h). To the driver, it seems that his opponent is running away from him, but in reality he is not. The time separation never changes; it is just an illusion.

The Importance of Corner Exiting Speed

Straights on racetracks can account for 70–80% of the total track length. A racecar often spends much more time on a straight at maximum acceleration than anywhere else. However, every straightaway starts with exiting the preceding corner, and maximizing the speed at which the car comes out of this corner can minimize the time until the next braking zone.

In racing literature and driving schools, corner exit speed is often a bit overrated, but the previous principle still remains true. ***Figure 4.13*** shows two overlaid laps of the Spa track, focusing on the world-famous Raidillon curve and the following straight.

Figure 4.12 *Time versus distance: both cars had a constant gap of 0.4 sec between them, but in distance the gap increases with speed. The driver in the gray car probably entered the pits and asked for a new engine.*

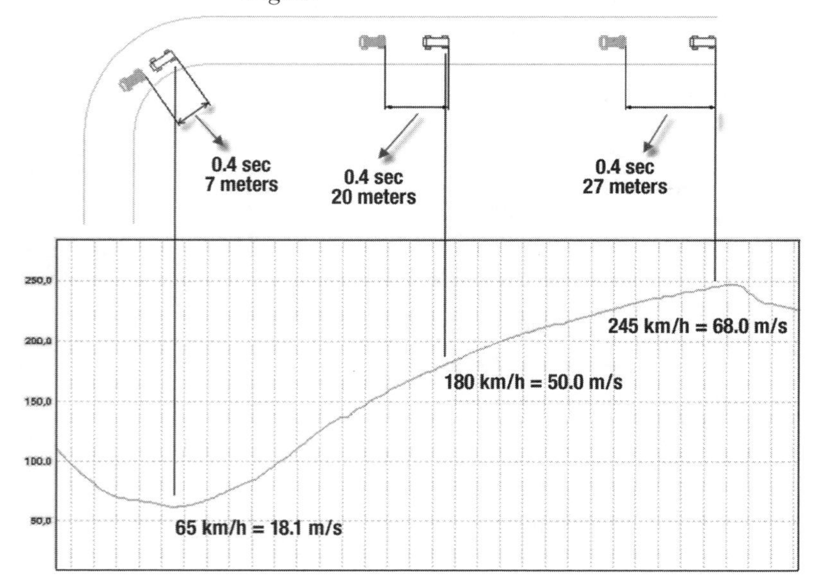

Both laps were run in the same car with the same driver on the same day. An aerodynamic setup change was performed on the car to cure a high-speed understeer problem, and the gray speed trace in Figure 4.13 represents the result. The driver is able to go on the throttle more fluently with the result that cornering speed increases by 2.9 km/h and the driver successfully negotiates the exit of the corner.

The straight following this corner has a length of 990 m. The lap illustrated by the black speed trace gives an average speed along this straight (calculated by the data acquisition software) of 246.41 km/h (or 68.446 m/s) and a sector time of 14.464 sec. Assuming that the car has the potential to keep the 2.9 km/h difference until the next braking point means an average speed of 249.31 km/h (or 69.252 m/s) and a sector time of 14.296 sec. By exiting this corner 2.9 km/h faster, this car gains a sector advantage of 0.17 sec. In reality, however, this was not the case.

The second lap (gray trace) has an average speed on the straightaway sector of 248.32 km/h (or 14.353 m/s) giving a sector time of 14.353 sec, a difference with the black lap of 0.11 sec. Figure 4.13 clearly shows that most of this time is gained on the first half of the straight. Aerodynamic drag increases with the square of speed. During the second lap, speed is higher; thus, there is more drag and less acceleration. In the slower lap, acceleration is higher and the difference in speed reduced. At the end of the straight, the car approaches the point where engine output is matched by drag and the car can no longer accelerate. In this case, a high-speed corner is followed by a long straight. Optimizing corner exit speed for this corner gives a net gain of just over a tenth of a second, which is a considerable advantage.

Drag Racing Specifics

A drag race is an automotive acceleration contest from a standing start between two vehicles over a predetermined distance.[5] The nature and rules of the game make drag racing a mathematical exercise, and data logging can enhance effectively the success of this exercise.

The accepted standard for the distance of a drag race is a quarter mile or an eighth mile. The race begins using an electronic device commonly referred to as the *Christmas tree*, which displays a visual countdown to the driver through a series of lights. When competitors leave the starting line, it activates a timer that is stopped when they cross the finish line. The time between these two events is the elapsed time (ET).

A drag race is a tournament-style elimination race between two vehicles. The losing vehicle in each round is eliminated, while the winner progresses into the following round of competition.

ET Bracket Racing

Some classes use a handicapped form of competition called *ET bracket racing*[5]. This makes it possible for vehicles of varying performance abilities to compete with each other on an even basis. Each competitor has to predict the ET at which the vehicle will run. This is called *dial-in*. In a run between two cars, the slower vehicle receives a head start equal to the difference in ET.

Here's an example. Car 1 has previously covered the quarter mile in 16.68, 16.71, and 16.73 sec. The driver feels that a dial-in of 16.70 sec is appropriate. Car 2 has been timed at 13.98, 13.99, and 14.02 sec, and a dial-in of 14.00 sec is selected. This means that Car 1 receives a headstart of 2.70 sec when the Christmas tree counts down to each car's start. The result of the race is determined as follows:

- If both cars run a higher ET than their dial-in, the win goes to the vehicle that crosses the finish line first.
- If both cars cover the quarter mile in their exact dial-in times, the win goes to the driver that reacted quickest to the starting signal (reaction time).
- If a car goes quicker than its dial-in, it is disqualified. This is called *breakout*.
- If a car reacts to the Christmas tree too quickly, it is disqualified. This is called *red light*.
- If both cars break out, the one that runs closest to its dial-in is the winner.
- If one car breaks out and the other car jumps the Christmas tree, the one that breaks out wins.

In sanctioned events, interval times are available to the competitors at 60, 330, 660, and 1,000 ft. The 60-ft interval time is a measure of the launch from the starting line, and it often determines how quickly the rest of the run is.

Determining the appropriate dial-in consistency is of vital importance. Starting strategy, gearing, and tire pressures are some issues that need careful attention and are where a data logging system can be of assistance. Weather circumstances also play a major role in determining the vehicle's

***Figure 4.13** Corner exit speed and its consequences on the following straight*

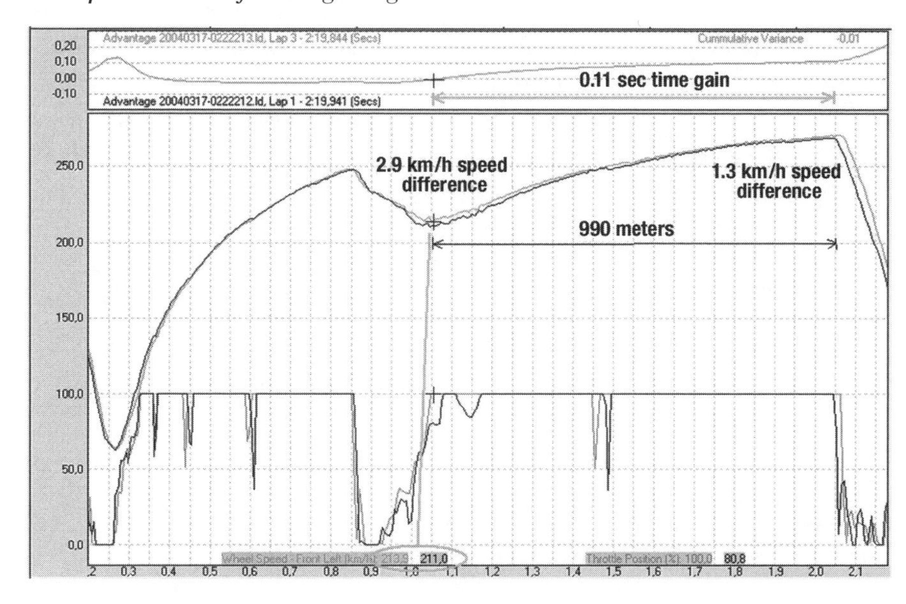

***Figure 4.14** Drag racing is a specific form of motor-sport in which data acquisition also can be an advantage.*

potential ET. Straightline acceleration was covered earlier in this chapter, and the influence of the weather on engine and aerodynamics is investigated in ***Chapter 11.***

Figure 4.15 shows a run of a Top Fuel racer on the Willowbank quarter-mile drag strip. Top Fuel represents the highest class in drag racing, and the cars accelerate from 0 to 160 km/h in less than 0.8 sec. Top speeds can exceed 530 km/h. The following signals are shown: engine RPM, drive shaft RPM, throttle position, clutch fluid pressure, and speed (front wheel speed and corrected speed). These traces were created using MoTeC's i2 analysis software in which a specific drag racing template can be created.

The vertical lines indicate the different sequences of the run. The official result sheet showed the following intermediate times for this particular car:

60 ft	1.020 sec
330 ft	2.420 sec
660 ft	3.446 sec
1,000 ft	4.315 sec
1,320 ft	5.073 sec

Figure 4.15 Engine and drive shaft RPM, throttle position, clutch fluid pressure, and speed of a Top Fuel drag racer going down the quarter mile

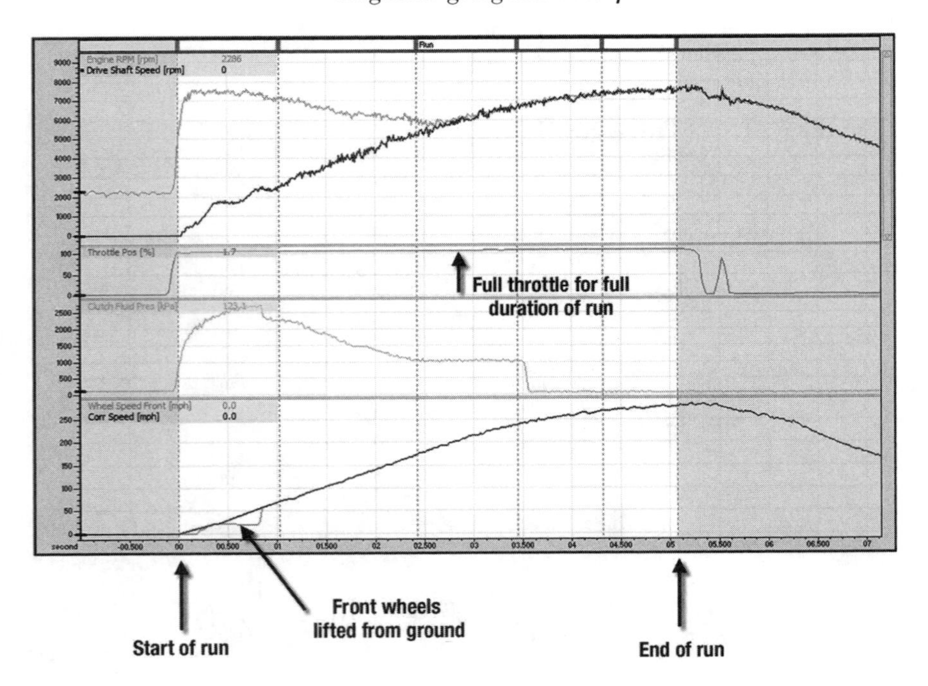

These results can be entered manually in the analysis software after which they are indicated in each graph.

The software calculates a corrected speed signal to determine the covered distance. In this case, the front wheel speed is corrected using the longitudinal acceleration. During the first 60 ft of the run, there is quite a difference between the measured wheel speed and the calculated corrected speed. This occurs because the front wheels of the vehicle are not in contact with the ground at this stage.

Measuring the wheel speed at the rear axle generates huge errors in the logged data because of the inherent soft spring rates of the rear tires of a Top Fuel racecar. In fact, this kind of car uses the variation in tire radius due to the wheel's centrifugal forces to increase the gear ratio as drive shaft-RPM increases. GPS-based speed measurement is a suitable solution for this problem.

In this example, the car reaches a maximum velocity of 283 mph, but it achieves this speed after the finish. The speed as the car crosses the finish line is 277 mph. Up to the 1,000-ft point, the speed trace is nearly linear, indicating a constant longitudinal acceleration.

Another obvious fact is that the throttle pedal is on the floor for the complete duration of the run. The amount of engine power transferred to the driven wheels is determined by the clutch. The clutch and its release mechanism of a Top Fuel dragster is a complex system, and its tuning is vital for a fast ET. The clutch fluid pressure trace does not look as though it is influenced by only the driver's foot. Although electronic closed-loop systems are banned in the Top Fuel class, a mechanical system—partly hydraulic, partly pneumatic—takes care that the clutch is engaged in a controlled way over a predetermined period.

Another run of an undefined dragster is shown in ***Figure 4.16***. This example concerns a bracket race. The driver in this case needs to lift the throttle in the last 300 ft to prevent a breakout. The driver covers the drag strip in 6.138 sec. This trace also shows the braking effect of the parachute that opens after the car crosses the finish line.

The concerned car was equipped with a 3-speed gearbox with the following ratios:

1st gear	**1.60**
2nd gear	**1.28**
3rd gear	**1.00**

However, gear changes cannot be detected here from the engine RPM graph as is the case in a typical road racing RPM trace. Two small, rough driveshaft RPM areas hint at a gear change; in this case, use a gear position sensor to determine the used gear.

The difference in engine RPM (which remains relatively constant) and driveshaft RPM is influenced completely by the clutch. To determine the amount of slip in the clutch, use Equation 4.22.

$$\text{Clutch slip} = \frac{n_{engine} - \left(n_{driveshaft} \cdot i_{total}\right)}{n_{engine}} \qquad (Eq.\ 4.22)$$

with n_{engine} = engine RPM
$n_{driveshaft}$ = drive shaft RPM
i_{total} = total gear ratio

The resulting mathematical channel is pictured in Figure 4.16. As in the previous example, this race is driven with the clutch instead of the throttle pedal.

Figure 4.16 ***This dragster covers the quarter mile in 6.138 sec, but decelerates in the last section to prevent a breakout.***

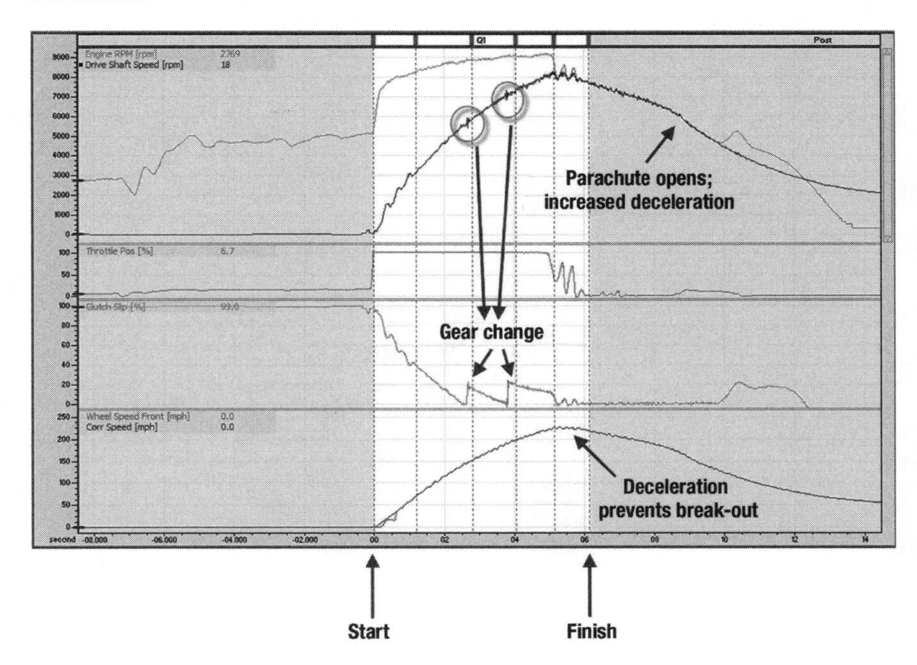

CHAPTER 5

BRAKING

Acceleration is required to minimize the lap time of a given racecar. Improving the acceleration capabilities of the vehicle results in faster laps. Optimizing the braking action of car and driver does not directly have the same effect. However, when the time spent under deceleration can be decreased, the difference can be applied to acceleration, which results in faster lap times.

A simplified example is offered here. ***Table 5.1*** shows the percentage of time spent under braking and acceleration during two different laps performed by the same car and driver around Circuit Zolder.

Table 5.1 *Acceleration and deceleration percentages for two different laps around Circuit Zolder*

	Lap A	Lap B
Acceleration	68.24%	70.78%
Deceleration	31.76%	29.13%
Lap time	1'33"212	1'32"178

Between the two runs, a fresh set of tires is put on the car, which creates a gain in other areas as well. However, a decrease in the total duration under braking over completion of a lap brings an advantage almost every time. This chapter demonstrates how the racecar's braking system as well as the capabilities of the driver to slow the car down can be analyzed and optimized.

Maximizing Braking Speed

The act of braking begins when the driver hits the pedal. To waste as little time as possible, brake pressure must be built up as quickly as possible. An analysis of braking speed is achieved using the longitudinal acceleration channel to measure the time it takes to get to maximum braking force ***(Figure 5.1)***.

Anything under 0.5 sec is considered fast braking. This does not mean that the driver has to slam the pedal, as this probably locks up the front wheels. There needs to be a balance between a controlled buildup of braking effort and aggressiveness.

The longitudinal G channel can be differentiated (by calculating the slope of the signal trace) to give a measure of braking speed. To brake the vehicle in Figure 5.1 to a maximum braking effort of 1.49 G within 0.48 sec, this differentiation shows a peak of 3.10 G/s.

Braking Effort

How does one know what the vehicle's maximum braking potential is? The simplest way to answer this question is to test it. A straight-line test indicates the highest negative longitudinal acceleration achieved under braking. The speed signals indicate when the wheels tend to lock up.

Author Buddy Fey presents a target value for the car's peak braking deceleration by comparing it with the car's cornering potential.[6] For cars producing up to 2 G of cornering power, maximum longitudinal Gs should be approximately 95% of the maximum lateral Gs.

Figure 5.1 *The time it takes for the driver to achieve maximum braking force*

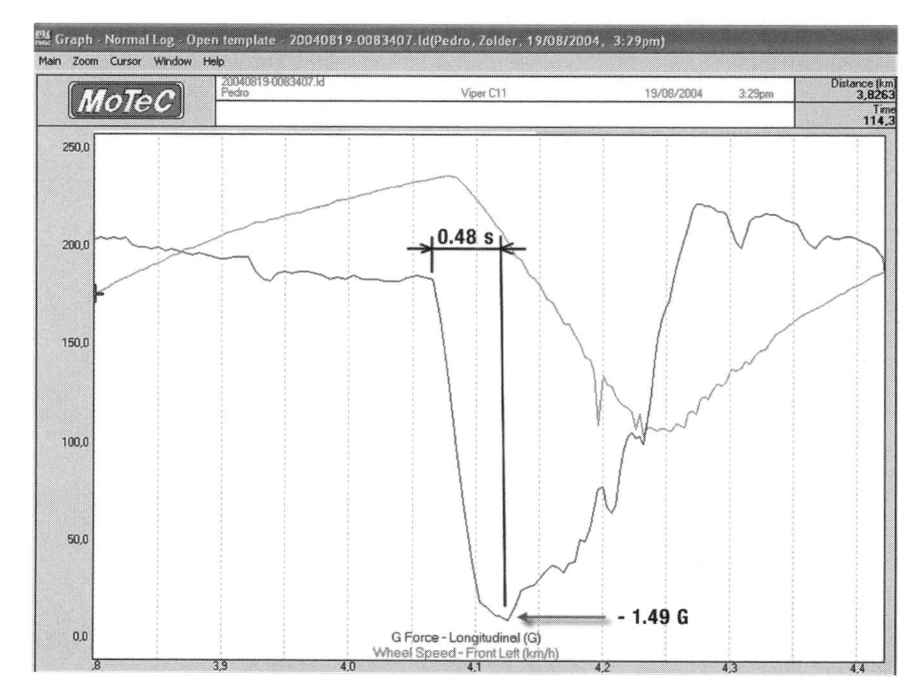

This value varies slightly with the configuration of the vehicle. The following corrections should be applied where necessary:

front engine	−2%
rear engine	+2%
square tire contact patch	+2%
average tire contact patch	+0%
wide tire contact patch	−2%

As an example, consider the Dodge Viper GTS-R, front-engined and equipped with wide racing tires. This car is capable of cornering at 2.0 G. Therefore, the maximum braking effort for this car is the following:

95%	lateral G
−2%	front engine
−2%	wide tires
+91% · 2.0 G = 1.8 G	

The maximum longitudinal deceleration that the car actually reaches can vary somewhat from corner to corner. Uphill or downhill braking increases or decreases peak longitudinal Gs, respectively. Track surface, tire temperature, wear, and compound have an influence as well.

Once the braking activity coincides with entering a corner, longitudinal Gs decrease while the tires need grip to build up cornering force. ***Equation 5.1*** calculates combined acceleration and can be used to determine the maximum deceleration under braking while turning.

$$G_{combined} = \sqrt{{G_{lat}}^2 + {G_{long}}^2} \quad \text{(Eq. 5.1)}$$

To find the longitudinal G, rearrange the equation as shown in ***Equation 5.2***.

$$G_{long} = \sqrt{{G_{combined}}^2 - {G_{lat}}^2} \quad \text{(Eq. 5.2)}$$

The Viper has a braking threshold of 1.8 G. Consider that, while braking, this car is cornering at 0.5 G. Substituting these values in Equation 5.2 gives the maximum achievable braking deceleration as given in ***Equation 5.3***.

$$G_{long} = \sqrt{1.8^2 - 0.5^2} = 1.73 \text{ G} \quad \text{(Eq. 5.3)}$$

Figure 5.2 Peak longitudinal acceleration as a function of cornering acceleration

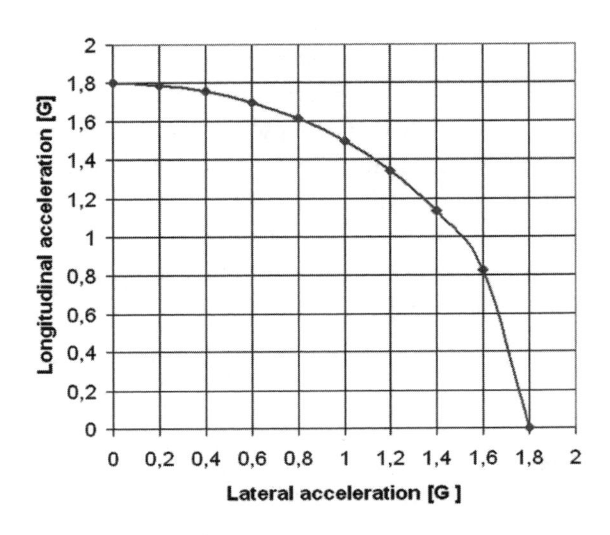

Figure 5.2 shows this relationship graphically. Note that this is a quarter of the traction circle (see ***Chapter 7***). From the graph, it is obvious that between 0 and 0.5 lateral G braking capabilities are not affected strongly.

To evaluate if the driver is braking adequately, compare how closely he approaches the target set by this calculation. Not braking hard enough may not be a driver-related issue. It may be possible that the brake balance is set up improperly or that there is another problem in the vehicle configuration.

Figure 5.3 Effect of the speed at which a corner is approached on the braking point location

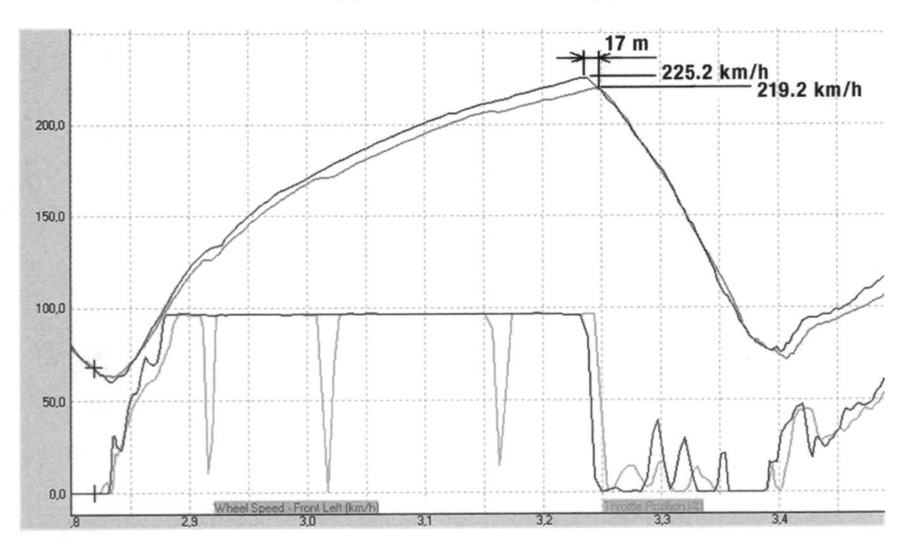

Braking Points

The discussion about braking points begins with specifying that the location of this point on the track depends on the speed at which the corner is approached. A difference in approach speed can affect significantly the braking distance ***(Figure 5.3)***. This affects the reference points the driver uses to select the braking point.

The graph shows two speed traces of the same car as it approaches a corner. The black line shows the highest top speed, 6 km/h more than the other trace. The braking point during the lap with the lower top speed has moved 17 m further down the track. At this point, the two traces converge, indicating the braking effort was the same for the two laps.

Effective braking is notable in the combined G graph as a smooth transition between the braking peak and the maximum lateral acceleration during cornering. When the driver brakes too early, this trace shows a *valley* ***(Figures 5.4** and **5.5)***, indicating that during this period the tires are not used to their fullest capacity.

Effective braking is revealed in the traction circle graph as a near-circular transition between maximum cornering G and maximum longitudinal deceleration. It is all about using the car's tires effectively.

Another indicator for early braking can be inadequate braking effort (also known as *easing* on the brakes). Determining the braking effort the driver should aim for was discussed in the previous paragraph.

Evaluating late braking is not so easy. Begin by investigating where in the corner the lateral G peaks. A peak at mid-corner or later may indicate late braking.

Lockup

Locking the brakes often is accompanied by big smoke tufts from the wheels. This results as often in one or more tires that are no longer round. When brake lockup occurs, all the wheels usually do not lock. The wheel most likely to lock is the one with the least load on it (the inside of the corner), on the axle with the highest brake bias. Other factors may apply as well. This means that wheel-speed sensors on each wheel are necessary to detect lockup during braking.

In Chapter 4, the longitudinal slip ratio was defined. During the braking phase, a positive slip ratio indicates that at least one of the front wheels is about to lock, while a negative slip ratio indicates the same problem on the rear axle.

However, the easiest way to detect lockup is the speed graphs. ***Figure 5.6*** is an example of a car with clearly too much rear brake bias. When a wheel locks, the speed trace drops nearly vertically. When the tire regains grip (because the driver senses the

***Figure 5.4** Early braking shows up in the Combined G trace as a valley*

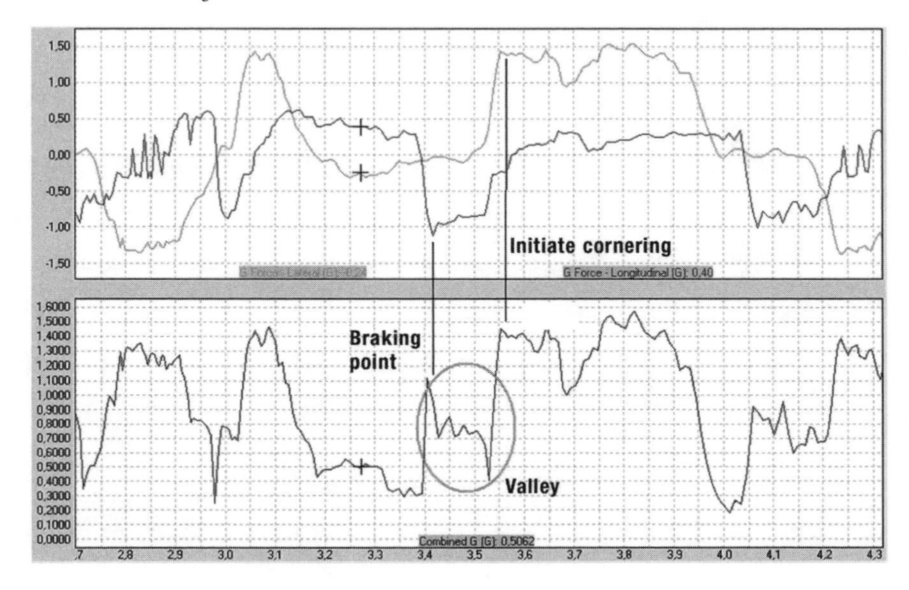

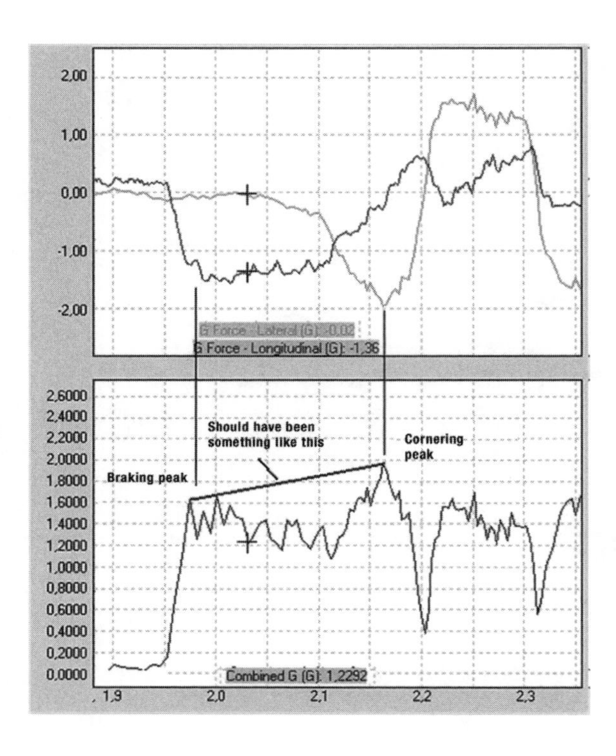

***Figure 5.5** Early braking. The driver does not use the properties of the tires to their fullest capacity.*

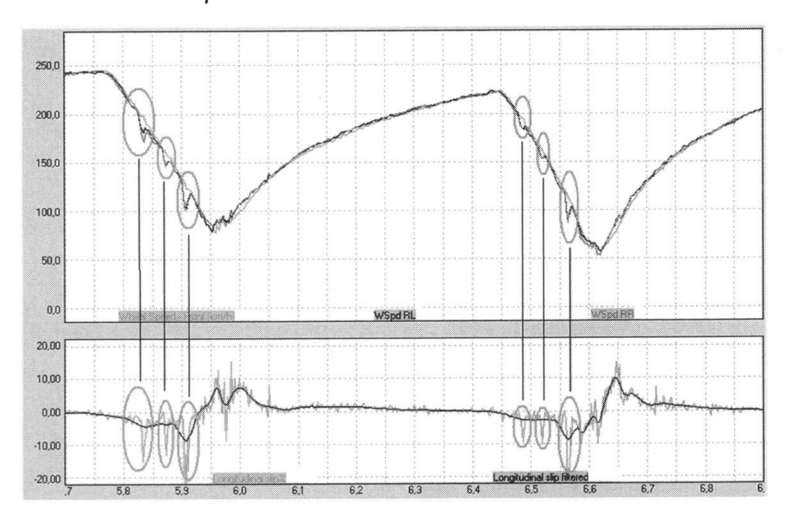

Figure 5.6 Brake lockup shows up as a downward spike in the speed graph and in the longitudinal slip percentage, depending on which axle the lockup takes place.

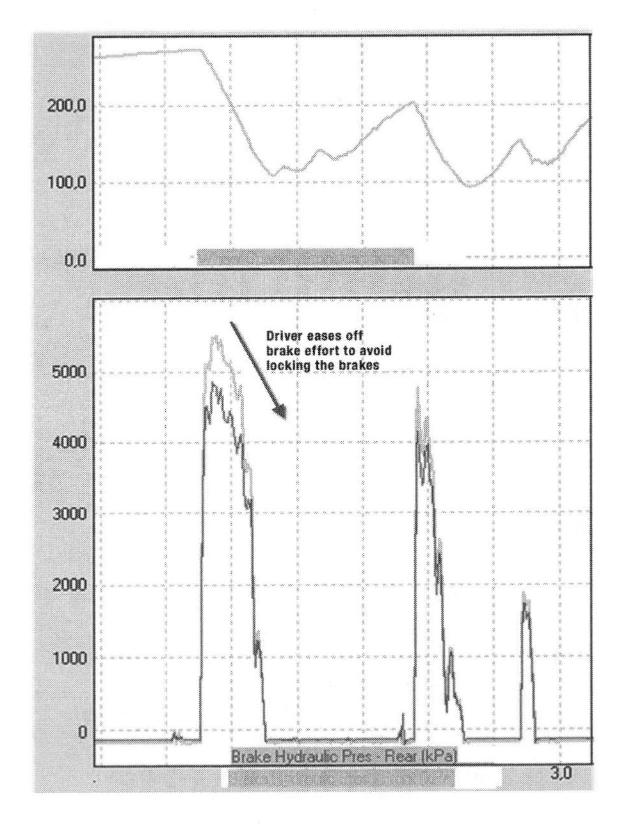

Figure 5.7 Front and rear brake line pressure

$$\text{Brake bias}_{\text{front}} = \frac{\text{Brake line pressure}_{\text{front}}}{\text{Brake line pressure}_{\text{front}} + \text{Brake line pressure}_{\text{rear}}} \cdot 100\% \qquad (Eq\ 5.4)$$

$$\text{Brake bias}_{\text{front}} = \frac{5500}{5500 + 4860} \cdot 100\% = 53\% \qquad (Eq\ 5.5)$$

wheel locking up and eases off the brake), the speed trace jumps again. Simultaneously with the downward spikes in the speed trace, the longitudinal slip percentage dives. If the front wheels are locking up, these spikes aim upward.

Brake Balance

Maximum braking deceleration only occurs when all tires simultaneously operate at their maximum coefficient of friction. Therefore, proper brake balance is vital. Corner-entry understeer followed by mid-corner understeer and the combination of low braking Gs can be diagnosed as too much front brake bias. Too much rear bias leads to corner-entry oversteer if not anticipated by the driver.

In extreme cases, too much brake bias on either axle leads to one or more wheels locking up. The car from which the data in Figure 5.6 was obtained clearly suffers from excessive rear brake bias.

Measuring brake line pressure makes it very easy to track the vehicle's brake balance. It is a great way to restore the proper balance when something in the brake system is changed. It reveals the brake-related activities of the driver, and the line pressures are variables in calculations for brake forces.

An example is given in ***Figure 5.7***. At the end of the straight, the driver slams on the brakes, resulting in peak brake line pressures of 5,500 kPa and 4,860 kPa for the front and rear axles, respectively. After this braking peak, the driver starts to ease off the pedal pressure to avoid locking the brakes.

Brake balance is defined with ***Equation 5.4***. In the previous example, the front brake bias can be calculated as ***Equation 5.5***.

Figure 5.8 is an *X-Y* graph of front versus rear brake pressure. This graph shows a small amount of scatter, indicating that for a given front brake pressure there is a variation in rear brake pressure. This undesirable effect can have various causes. Caliper deformation or freeplay in the pedalbox assembly can lead to inconsistent brake pressure distribution.

Another way to visualize compliance in the braking system is pictured in ***Figure 5.9***. This graph plots brake balance against front brake pressure. It illustrates the variation in pressure distribution for any given front brake pressure. For low pressures, this variation is high (freeplay in the pedalbox assembly), and it diminishes as brake pressure rises. Brake balance variation at higher pressures is caused by caliper or brake pad deformation or expansion of the brake hoses. A braking system with a consistent brake pressure distribution shows less than a 5% variation in brake balance from 15 bar front pressure onwards.[7] This is clearly not the case in Figure 5.9. The graph shows a brake balance variation of 10% at 20 bar and the goal of 5% is only reached at 40 bar. Between 40 and 60 bar, the graph shows a strange hill. This variation in brake balance probably is caused by brake caliper deformation. This also can be seen in Figure 5.8. This data was taken from the same session. At 40 bar, there is a shift in brake balance. Getting rid of this type of braking system compliance is vital to achieving optimum braking performance.

Pedal Travel

Brake pressure sensors are valuable for checking the braking system. Additionally, measuring the travel of the brake pedal provides useful information. If brake pressure channels are not available, logging pedal travel at least provides some information that would be revealed by the pressure channels. It tells when and how the driver applies the brakes; however, it does not reveal anything about the brake balance.

Checking pedal travel and brake line pressures provides a good indication of brake pad thickness. This can be particularly useful in endurance races. As the brake pads wear, more pedal travel is required to achieve the same brake line pressures. Pedal travel also indicates braking consistency. The previous paragraph showed that pedalbox freeplay leads to inconsistent brake balance. Figures 5.8 and 5.9 illustrated a vehicle with poor pedalbox rigidity. A plot of front brake pressure against pedal travel from the same session in this car is pictured in ***Figure 5.10***. It shows that significant brake pressure only is being built up after 10 mm of pedal travel. After that there is a significant degree of variation in brake pressure for any given pedal travel value.

Figure 5.8 *Front versus rear brake line pressure*

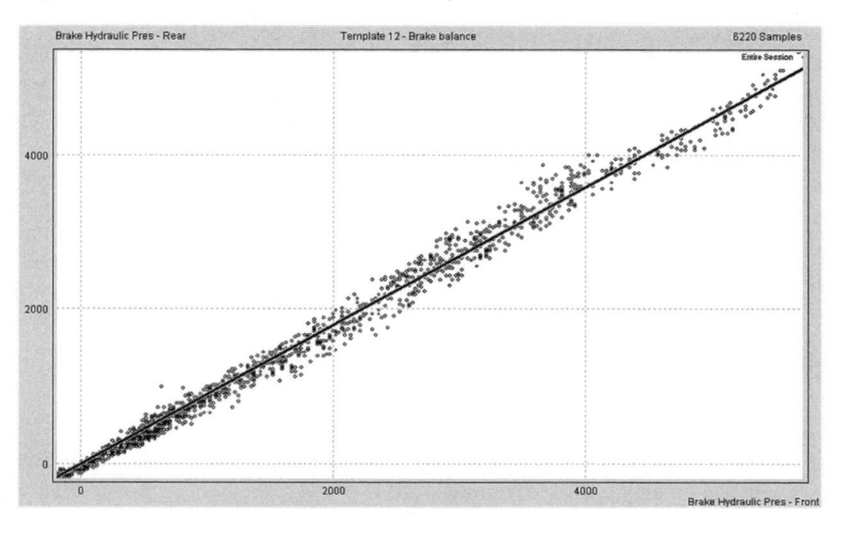

Figure 5.9 *This X-Y plot of brake balance against front brake line pressure provides a good method to check the brake balance consistency.*

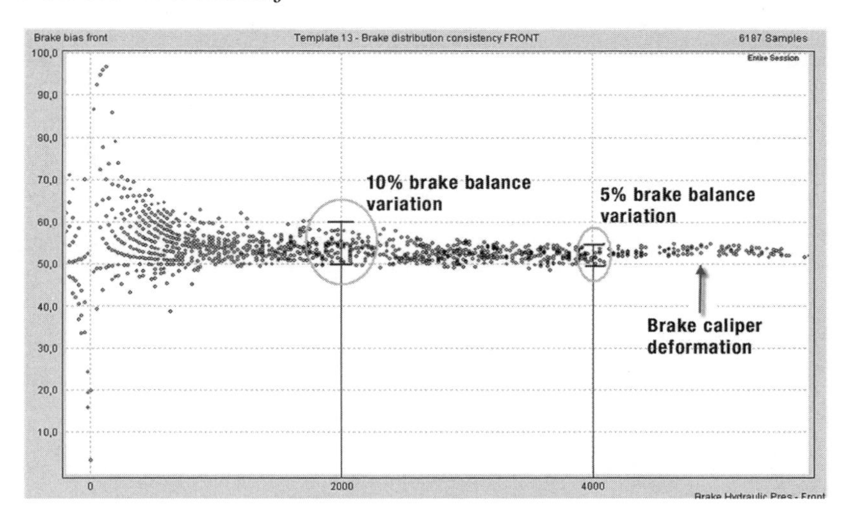

Figure 5.10 *Front brake pressure against pedal travel*

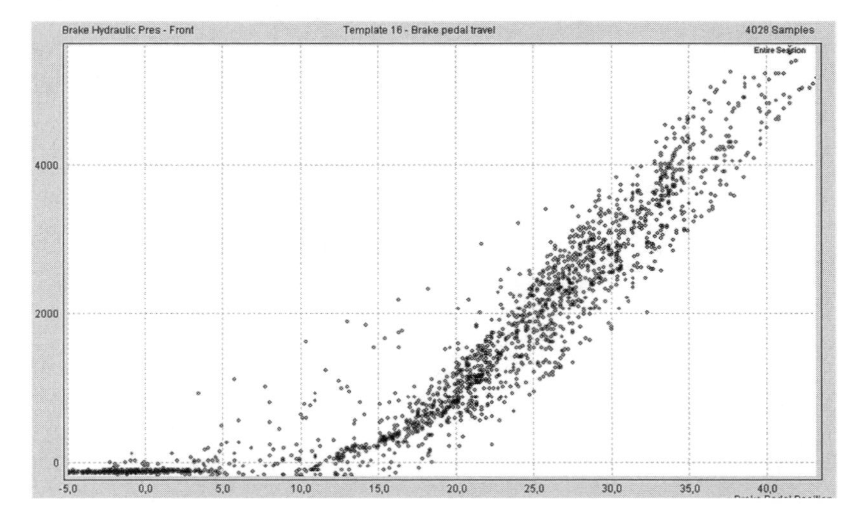

CHAPTER 6

GEARING

The gearbox is a device that allows the engine to operate within its powerband (i.e., the RPM range where the engine delivers the most of its power) for a wide speed range. This is an additional responsibility for the driver, who is already trying to balance the car based on the limits of grip produced by the tires. Data logging is very useful in evaluating the shifting techniques of the driver and can help with selecting the most efficient gear ratios for any given racetrack.

Analyzing the driver's shifting technique can be broken into two categories: upshifting, (i.e., changing to a higher gear) and downshifting (i.e., changing to a lower gear).

Upshifting

When analyzing logged data, there are two important items with regard to the upshifting procedure—the shifting point and the duration of the upshift.

Shifting Point

The shifting point depends on the shape of the engine power curve, more specifically the total area beneath the power curve situated within the engine's operating range.

As an example, take a Formula 3 single-seater car. ***Table 6.1*** gives the gearbox configuration for this car. The speeds are calculated at a shift point of 7,000 RPM. In ***Figure 6.1,*** the speed versus engine RPM is graphically represented. ***Figure 6.2*** shows the power curve of the engine.

Assume one is trying to find the optimum gearshift RPM for third gear. From third to fourth gear, the engine RPM drops approximately 1,200. So, with the earlier mentioned shift point of 7,000 RPM, the engine picks up in fourth gear at 5,800 RPM. Now the area below the power curve between 5,800 and 7,000 RPM can be calculated. To make this easy, dyno measurement points in intervals of 200 RPM are entered in a spreadsheet. These are shown in ***Table 6.2***. By taking the sum of these measurement points in the interval determined previously, a good approximation of the area below the power curve is obtained. Taking the sum of these figures gives us the following:

	RPM	hp
	5,800 RPM	207 hp
	6,000 RPM	209 hp
	6,200 RPM	208 hp
	6,400 RPM	205 hp
	6,600 RPM	201 hp
	6,800 RPM	197 hp
+	7,000 RPM	172 hp
	Total	1,399 hp

To find the right shift RPM for third gear, the interval must be determined that provides as much area as possible under the power curve. In the previous example, the power figure at 7,000 RPM is rather low compared to the lower revs. Therefore, it is necessary to change the shift point to 6,800 RPM and obtain the following results:

	RPM	hp
	5,600 RPM	205 hp
	5,800 RPM	207 hp
	6,000 RPM	209 hp
	6,200 RPM	208 hp
	6,400 RPM	205 hp
	6,600 RPM	201 hp
+	6,800 RPM	197 hp
	Total	1,432 hp

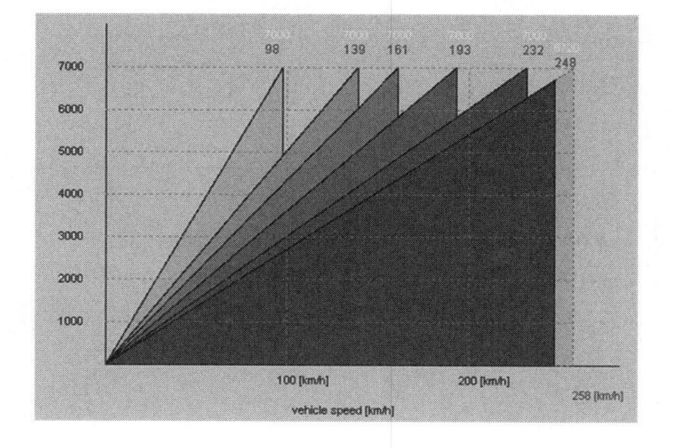

Figure 6.1
Engine RPM versus vehicle speed for an F3 car

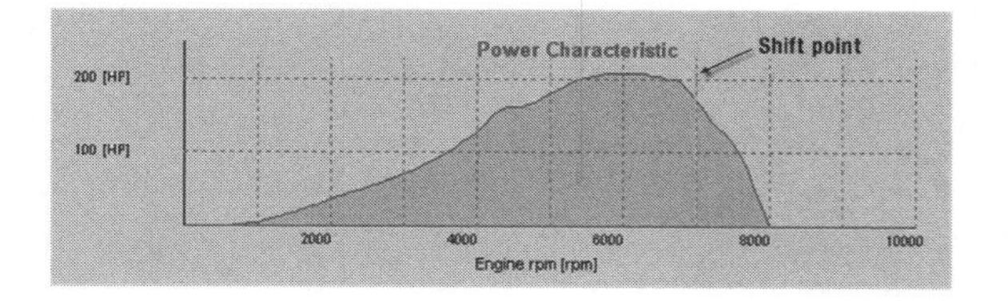

Figure 6.2
Power curve for the example F3 car

Repeating this exercise finally gives an optimum shift point at 6,500 RPM. In the same way, the required shift point in each gear can be calculated. To illustrate the importance of shifting at the correct engine speed, an acceleration test on a straight line of 1,000 m is simulated using Bosch's Lapsim software package. (See **Chapter 13** for more information on this package.) During the first run, the driver shifts at 7,000 RPM. During the second run, the shift point is decreased by 500 RPM. The results are given in ***Figures 6.3*** and ***6.4***.

Table 6.1 Gearbox properties of an F3 car, shifting point at 7,000 RPM

	Z1	Z2	I	I/step	I_{final}	V (km/h)	RPM drop
1st	33	12	2.75	2.75	7.79	98.5	
2nd	29	15	1.93	1.93	5.48	140.1	2,079
3rd	25	15	1.67	1.67	4.72	162.5	966
4th	25	18	1.39	1.39	3.94	195.1	1,167
5th	22	19	1.16	1.16	3.28	234.0	1,164
6th	26	25	1.04	1.04	2.95	260.5	713

	Z1	Z2	Ratio
Final drive	34	12	2.83
Step up gear			1

Dynamic tire radius	291	mm
Shifting point	7,000	RPM

Table 6.2 Power figures for an F3 engine, taken from a dyno test

Engine RPM	Engine Power (hp)	Engine RPM	Engine Power (hp)	Engine RPM	Engine Power (hp)
2,000	36	3,800	114	5,600	205
2,200	42	4,000	128	5,800	207
2,400	49	4,200	150	6,000	209
2,600	55	4,400	162	6,200	208
2,800	63	4,600	162	6,400	205
3,000	70	4,800	169	6,600	201
3,200	78	5,000	181	7,000	197
3,400	88	5,200	191	7,200	172
3,600	100	5,400	200	7,400	142

By decreasing the shift point for all gears by 500 RPM, the elapsed time after 1,000 m of straight-line acceleration is 0.12 sec less than before and the top speed is increased by 6 km/h. In this situation, it is necessary to decrease the shift RPM to increase the area under the engine power curve, but for different engines the opposite also can be true. In this case, ensure that increasing the shift RPM does not affect the reliability of the engine. Get the advice of the engine builder.

The longitudinal acceleration channel also provides a good indication if the shift RPM is correct. Upshifts show up in this trace as short downward spikes. A correct shift RPM results in a similar longitudinal acceleration at the beginning and end of this downward spike. ***Figure 6.5*** illustrates a properly performed upshift, while ***Figure 6.6*** shows an example of an upshift performed too early.

Shift Duration

The other important point to note in upshifting is the shift duration. Because time spent between gears is time when the car is not accelerating, minimizing this shifting time also reduces lap time. The upshift begins with disengaging the clutch, followed by moving the gear lever to the next gear and engaging the clutch. The input shaft needs to slow down to synchronize with the next gear ratio, which is achieved by backing off the throttle (or cutting the ignition) and depressing the clutch. There is a trade-off between upshift duration and reliability. Quick shifts increase the wear on dog or synchro rings but save lap time. Sound judgment is in order here.

Shift time can be determined from the logged data when the longitudinal acceleration of the vehicle is recorded. As discussed previously, shift time is the time when the car is not accelerating. This appears in the longitudinal G graph as a downward dip as shown in ***Figure 6.7***. The shift action begins where the longitudinal force drops off and ends when the car picks up acceleration again.

To get an idea about the driver's shifting technique, calculate the shift times during a lap in different gears and calculate the average.

Determining shift times can help the engineer evaluate the ability of the driver to shift gears quickly, or it indicates if the gearbox is being abused. The latter is important in endurance races, particularly when there is more than one driver in the car. ***Table 6.3*** provides some experience numbers for the upshift duration. These depend on the driver and on the state and construction of the gearbox and clutch mechanism. Sequential gearboxes and powershift systems (cutting the engine's ignition triggered by a signal from the gear lever) help minimize shifting time without sacrificing reliability.

To accurately measure shift time, the correct sampling frequency should be selected to log longitudinal acceleration. In ***Figure 6.8***, two different laps by the same car are examined. From the trace logged at 10 Hz, the time it takes the driver to change gear cannot be determined accurately. The reading of 10 Hz means there is a sensor reading every one-tenth of a second. Given the values in Table 6.3, this is clearly not accurate enough. The 50-Hz line gives a reading every 0.02 sec. In this line, the downward spikes produced by upshifts are seen clearly.

Figure 6.3 *Simulation of 1,000-m straight-line acceleration, shifting gears at 7,000 RPM*

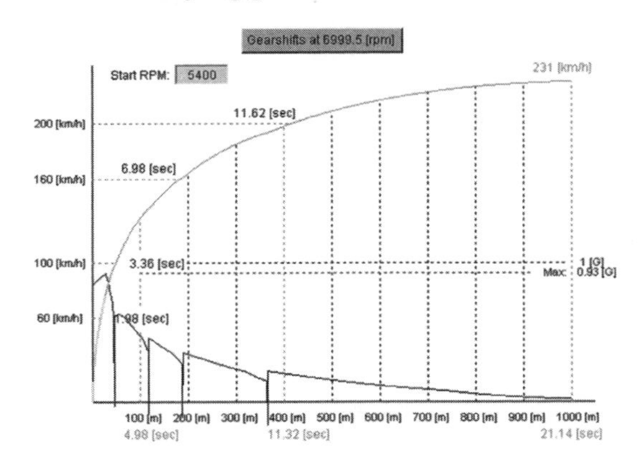

Figure 6.4 *Simulation of 1,000-m straight-line acceleration, shifting gears at 6,500 RPM*

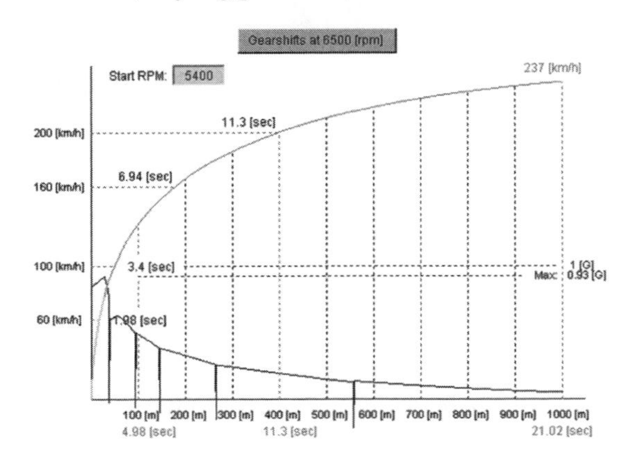

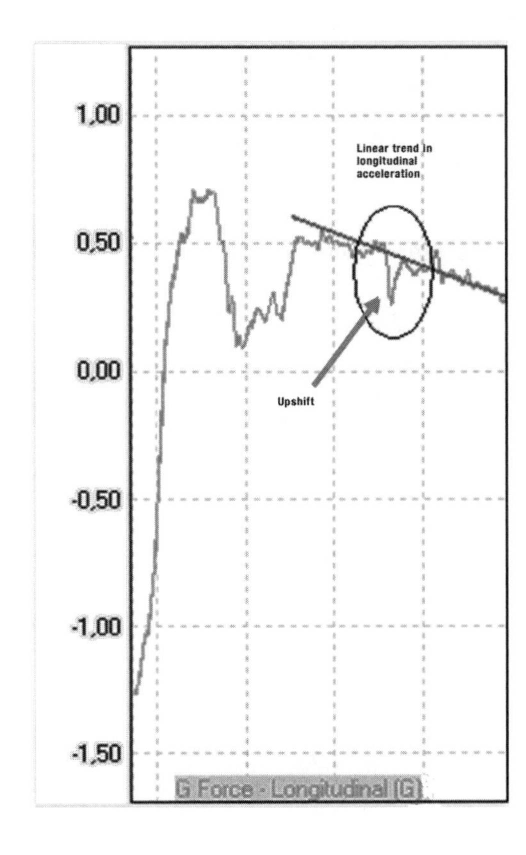

Figure 6.5
An upshift performed at the right shift RPM

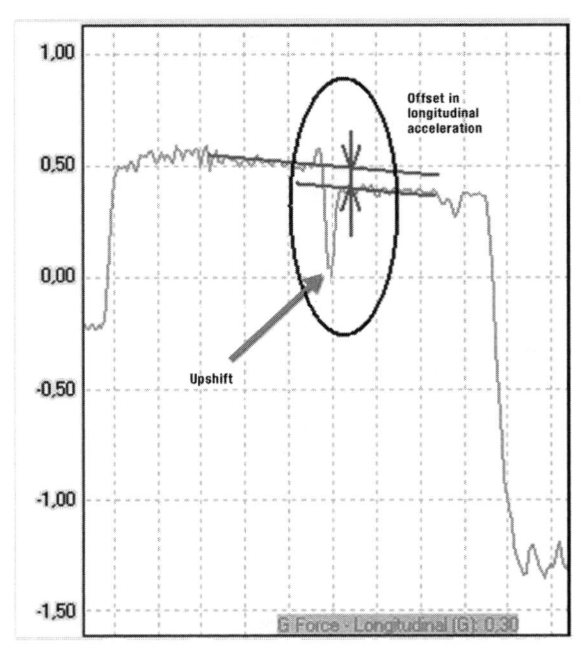

Figure 6.6
Shifting too early results in less longitudinal Gs at the right side of the upshift spike.

Table 6.3 Typical shift times for various racecars

Vehicle	Upshift duration
F3 car	0.15 sec
Porsche 911 GT2 Turbo, synchronized H-pattern gearbox	0.35 sec
Dodge Viper GTS-R, synchronized H-pattern gearbox	0.32 sec
Dodge Viper GTS-R, sequential gearbox, no powershift	0.23 sec
Dodge Viper GTS-R, sequential gearbox, with powershift	0.18 sec
LMP1 prototype, sequential gearbox, electronic paddleshift	0.10 sec

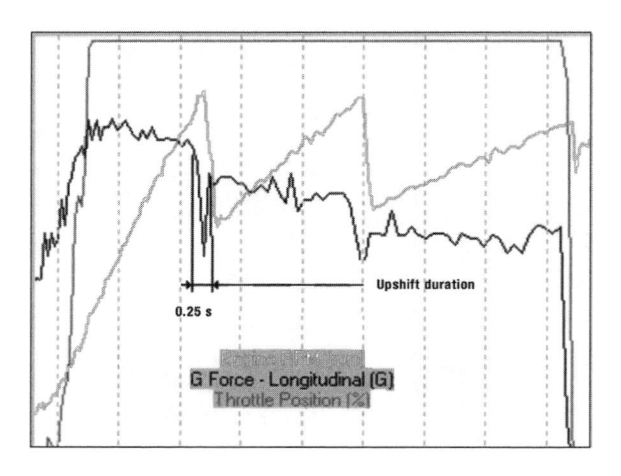

Figure 6.7 *Determining upshift duration from the longitudinal G channel*

Figure 6.8 *Longitudinal G sampled at 10 Hz and 50 Hz*

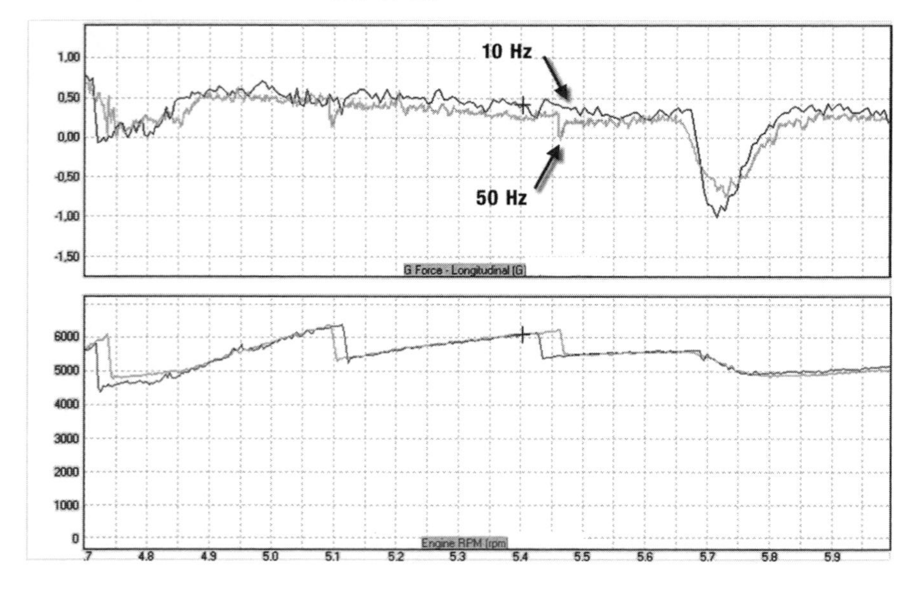

Downshifting

Upshifting and downshifting require synchronizing the engine speed with that of the transmission input speed. With upshifting, the engine passively synchronizes RPM because it slows down as the clutch disengages. This is not the case when downshifting because the engine needs to spin faster when it engages the lower gear. *Blipping* the throttle ***(Figure 6.9)*** as the transmission passes through neutral achieves this. When the engine is not sped up by the driver, it is by the driveline when the clutch is re-engaged, which upsets tire adhesion of a car cornering near the limit.

Downshifting inevitably is connected to braking, and the driver should ensure that engine and transmission RPM are synchronized properly. Failing to do so could result in snap oversteer, which is not desirable, especially when the driver is trail braking. How much blipping is necessary? Too little blipping upsets the grip at the driven wheels, while too much blipping over-revs the engine.

An example of too little throttle blipping is given in ***Figure 6.10***. The driver actuates the throttle on downshift to a maximum 10% throttle position. The engine RPM is raised less than necessary, which is illustrated by the upward spikes in the total gear ratio channel. How to create this channel is covered fully in the next section, but for now it is sufficient to know it defines the ratio between wheel speed and engine RPM. Ideally, when engine and transmission input shaft revs are equal, this channel shows the actual gear ratio. If during the downshift the engine RPM is too low when the lower gear is engaged, the transmission speeds up the engine and the gear channel momentarily increases.

Most engine over-revs occur during downshifting and usually are due to two reasons—shifting down too quickly from a too-high engine RPM and excessive throttle blipping. The trace in ***Figure 6.11*** illustrates a guaranteed way to destroy an engine!

The Gear Chart

The relationship between engine speed and wheel speed is linear under normal circumstances. This linearity is indicated by the total gear ratio. A common display format for gear ratios is the gear chart pictured in Figure 6.1. This chart displays the relationship between speed and engine RPM for each gear.

With the analysis software, it is possible to create an *X-Y* plot similar to the gear chart. ***Figure 6.12*** provides an example of this. It concerns a car with a 6-speed sequential gearbox. Only data points recorded during acceleration (longitudinal acceleration > 0) are plotted in the graph to determine from which minimum RPM the driver accelerated the engine.

The first conclusion apparent from this graph is that only five of six gears are used. No sections on the track require first gear to be used. Third gear is used the most, as indicated by the density of data points for this gear.

The maximum shift RPM for each gear is easily recognizable. In all gears, RPM ranges are detected that are well within the speed range for a lower gear. Here the driver probably is trying to use the best of the engine's torque band. Note that the engine in this case has a maximum torque output of approximately 800 Nm.

Also worth noting is the scatter of data points. Theoretically, every RPM corresponds to a predetermined vehicle speed in each gear. Therefore, all the data points should fall exactly on the straight lines plotted in Figure 6.12. The deviation from the line is greatest in the lower gear, especially in the lower rev ranges. The probable cause is wheel spin upon acceleration, but the mounting location of the wheel-speed sensor also could be an influence. A sensor mounted on the left-front wheel records a speed greater than the right-front wheel in a right-hand corner, and vice versa. When both nondriven wheels have a wheel-speed signal, the average of the two can be calculated and used as the *x*-axis in the gear chart to remove this effect, although this probably would not provide much information.

Total Gear Ratio Channel

The total gear ratio channel was used earlier in this chapter to evaluate the driver's downshifting technique. It is a mathematical channel that expresses the relationship between engine RPM and vehicle speed. In any gear, this relationship is a constant. The graph looks like a stepped line, conveniently displaying which gear was used at any part of the track.

Vehicle speed relates to engine RPM, according to ***Equation 6.1***.

$$V = \frac{2 \cdot \pi \cdot r_{rolling} \cdot n_{engine}}{i_{tot}} \qquad (Eq.\ 6.1)$$

where V = vehicle speed
$r_{rolling}$ = tire radius
n_{engine} = engine RPM
i_{tot} = total gear ratio

***Figure 6.9** Throttle blips during downshifting to synchronize engine and transmission input shaft RPM*

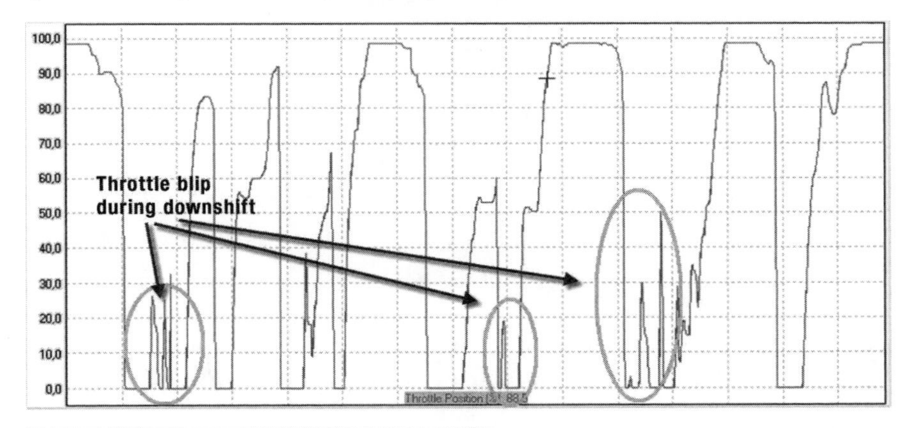

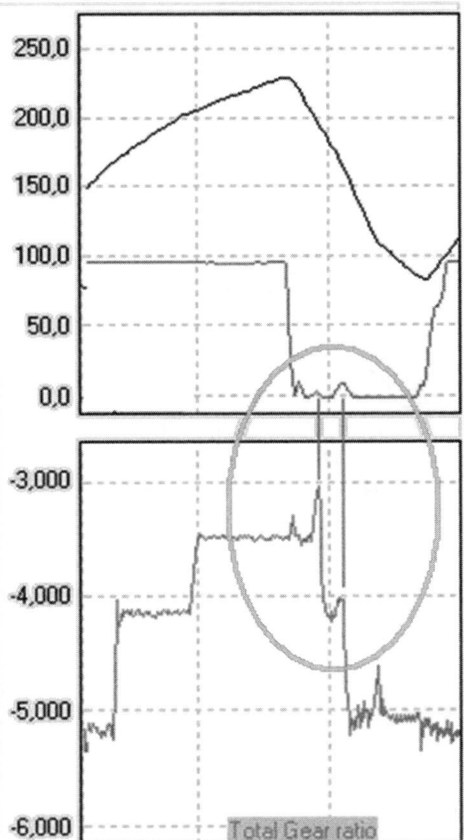

***Figure 6.10** Blipping the throttle insufficiently during downshifts shows up in the gear ratio graph as upward spikes.*

Vehicle speed and engine RPM are logged channels, so ***Equation 6.2*** is for i_{tot}.

$$i_{tot} = \frac{2 \cdot \pi \cdot r_{rolling} \cdot n_{engine}}{V} \qquad (Eq.\ 6.2)$$

Finally, if km/h are converted to m/s and RPM to revs/second, the result becomes ***Equation 6.3***.

$$i_{tot} = 0.377 \cdot \frac{r_{rolling} \cdot n_{engine}}{V} \qquad (Eq.\ 6.3)$$

***Figure 6.11** Too much throttle blipping results in over-revving the engine, as demonstrated here by an unnamed driver in a Porsche 996 around Circuit Zolder.*

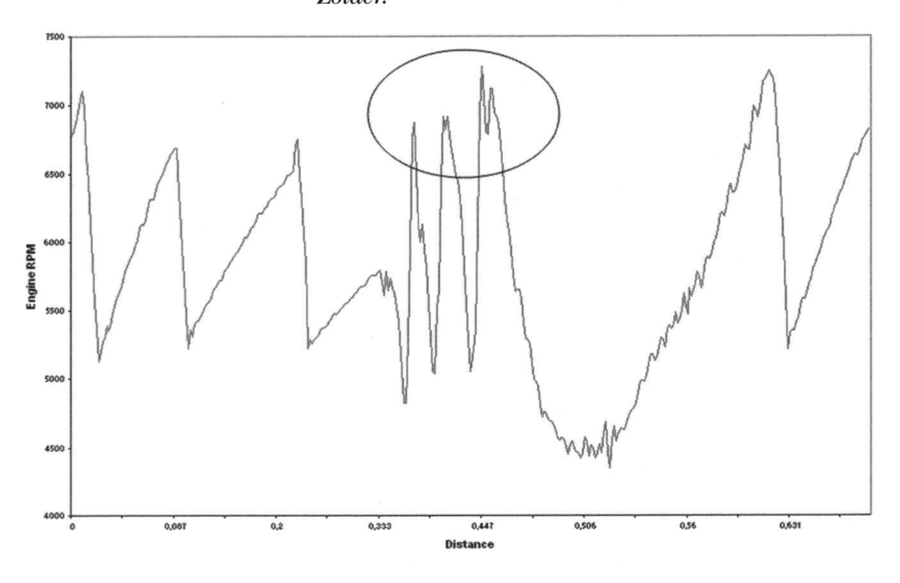

***Figure 6.12** Wheel speed versus engine RPM plot*

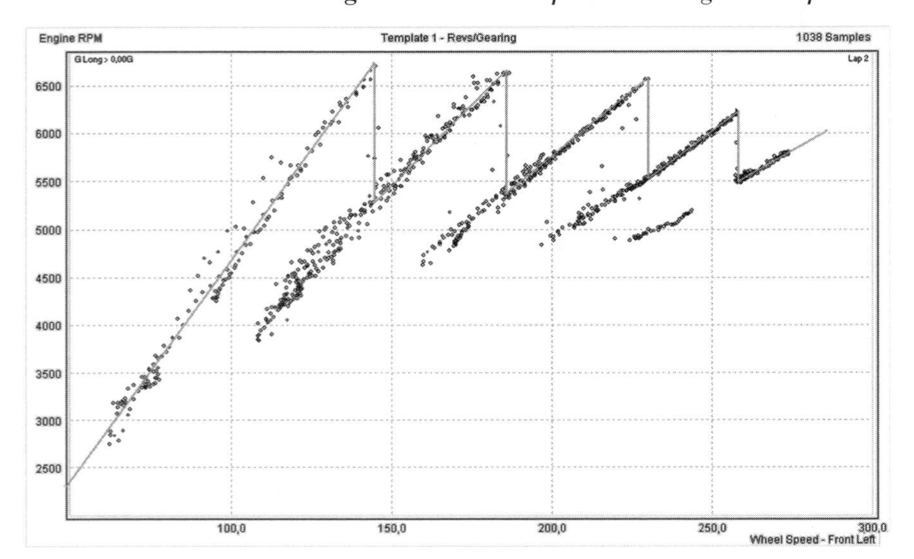

Graphically, the mathematical channel looks like the lower trace in ***Figure 6.13***. Note that Equation 6.3 was multiplied by –1 so that when a higher gear is used the line steps up, and vice versa. The upper trace was obtained from a gear position sensor in the sequential gearbox. To detect which gear the driver is using, this trace is obviously clearer. However, the mathematical channel also indicates anomalies such as wheel spin, wheel lockup under braking, and insufficient throttle blipping on downshifts.

The total gear ratio channel also may be used in other mathematical channels requiring this input. To decrease the deviation from the actual gear ratio, it is also possible to use the average wheel speed of two or more wheels when multiple sensors are present.

Determining Correct Gear Ratios

Given the large amount of literature available, the subject of selecting the proper gear ratios is not covered in this book. Author Paul Van Valkenburg describes this subject excellently.[8] Begin by finding the optimum engine RPM in top gear and optimum shift points in the intermediate gears, as discussed in the first paragraph of this chapter. Select the top gear for the longest straight on the track and the intermediate gears for the corners and most efficient acceleration.

To select the proper gear ratio for a corner exit, first find the engine RPM in each gear where maximum longitudinal acceleration occurs, as pictured in ***Figure 6.14***. Following that, find the RPM in the corner exit phase where the car accepts full throttle. This should be slightly below the revs for maximum acceleration. If not, correct this by selecting a different gear ratio.

Handling problems move the point where the car accepts full throttle further down the corner exit. Do not try to select gear ratios for a corner exit when the chassis is not balanced. Also be aware that the optimal gear for one corner may not be optimal for another. Investigate all corners leading onto a significant straight because this is where significant time can be gained.

Figure 6.13 *Gear position channel from sequential gearbox and calculated gear ratios from vehicle speed and engine RPM*

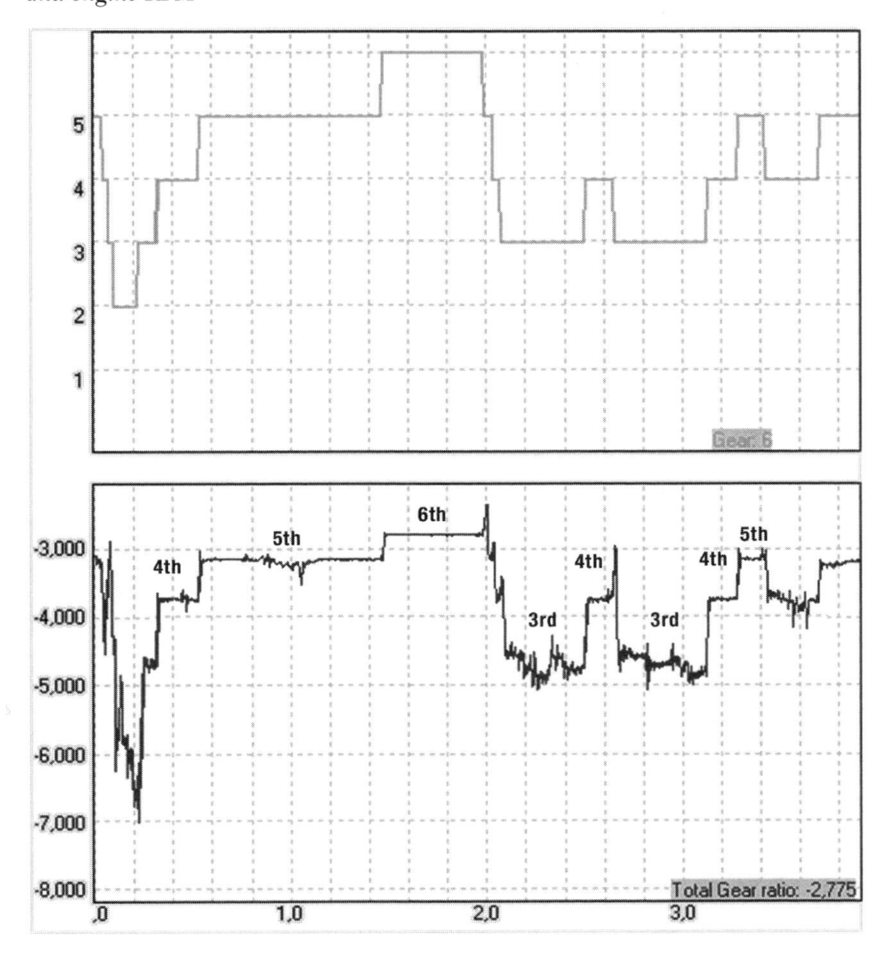

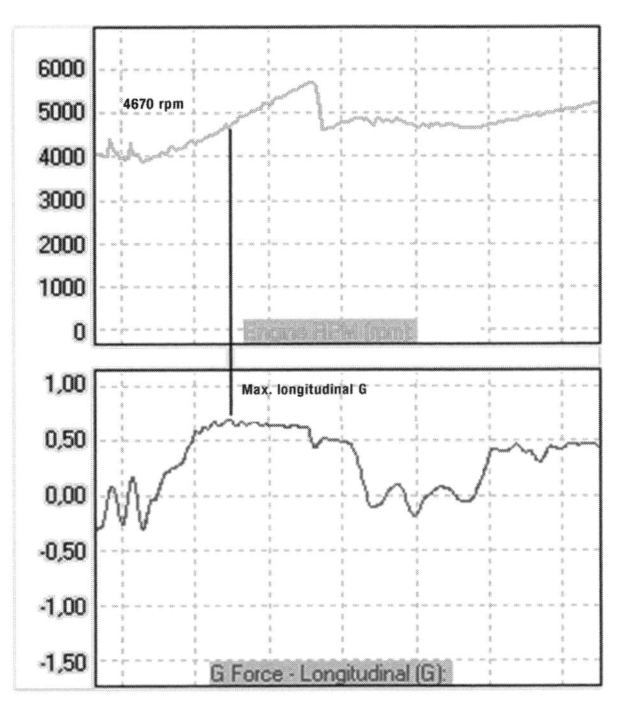

Figure 6.14 *Engine RPM at maximum longitudinal acceleration*

Except for drag racers, all racecars negotiate corners and the ability to do so as fast as possible minimizes lap time. This chapter covers the physics involved in cornering and how the cornering capability of a racecar can be investigated using data from the logging system.

The Cornering Sequence

The car-driver combination goes through various phases when taking a corner. The cornering process basically consists of the following phases:

1. Braking Point to Initial Turn-in Point

The straight-line braking phase forms an integral part of the cornering sequence because the point where the driver hits the brakes determines the location of the turn-in point and entry speed.

2. Turn-in Point to Corner's Apex

The driver usually is braking still after the turn-in point (i.e., trail braking). This period is followed by a short neutral throttle period where the driver tries to carry the speed through the corner. This is not a coasting period, and it can be very short to no time at all in duration.

3. Corner Exit

This phase begins when the driver goes hard on the throttle and exits the corner onto the following straight.

Figure 7.1 illustrates the different events taking place during cornering with the driver activity channels (throttle, brake pedal position, and steering wheel). These events are discussed in more detail in ***Chapter 12*** to analyze the racing line. For now, it is sufficient to show what the driver experiences when negotiating a corner.

For a car to get around a corner, a lateral force must exist to keep the car on the track. This force (or grip) comes from the car's tires and is twofold. The thread surface of the tire grips with the surface irregularities of the track, but there is also a molecular adhesion between both surfaces. For this reason, Newton's laws of friction do not apply for racing tires. Tires can develop more lateral force than the vertical load acting on them.

For a vehicle to change direction, all the tires assume a slip angle. This angle exists because of the resisting moment due to the elastic friction between the tire and road surface that develops when the tire is turned. Put simply, the slip angle is the difference between the direction the wheel is pointing and where it is heading.

From a physical point of view, this is what happens when a car develops lateral grip:

- On the straight preceding the corner that the driver wants to tackle, the steering wheel angle is fluctuating around zero and all tires have a very small slip angle approximately equal to their toe setting. There is no lateral acceleration.
- At the turn-in point, the driver turns the steering wheel, effectively inducing a slip angle to the front tires. The front tires develop a lateral force.
- This front lateral force causes a yaw moment around the vehicle's center of gravity, which in turn induces a rear slip angle and, therefore, rear grip.

Figure 7.1 Driver activity pattern for a left-hand corner

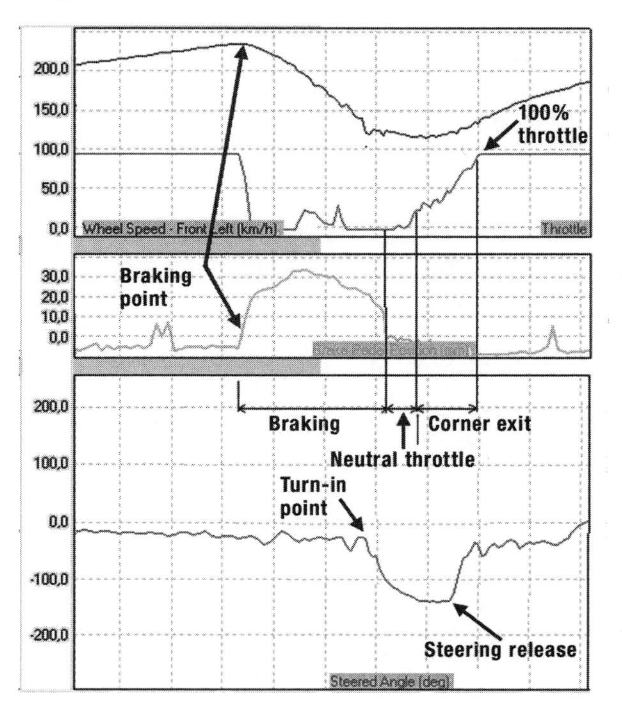

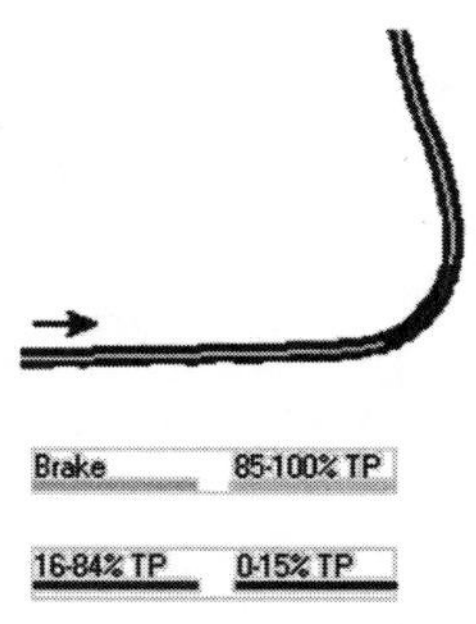

When the car develops the necessary lateral grip to get it through the corner, all four tires assume a certain degree of slip angle. The amount of slip angle for each wheel determines the lateral force that each tire develops. ***Figure 7.2*** is a plot showing the lateral force versus the slip angle of a given tire exposed to a given vertical load. Three regions can be distinguished from this graph:

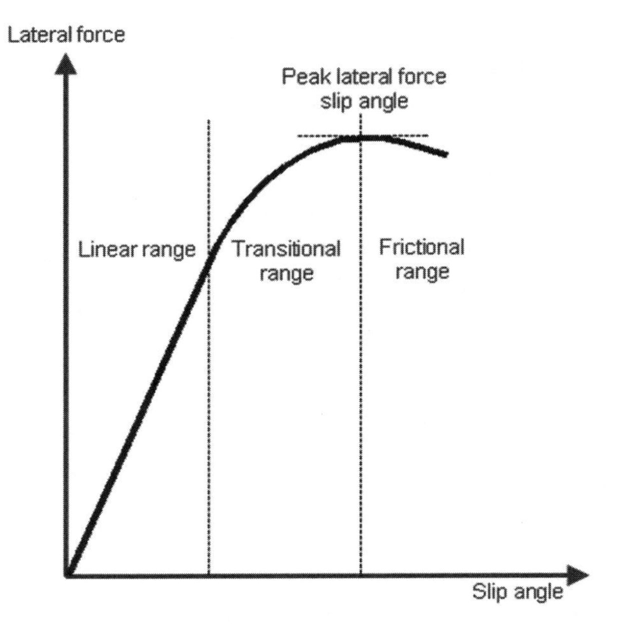

Figure 7.2 *Lateral force versus slip angle*

- The first is a linear, or elastic, range in which the developed lateral force is proportional to the amount of slip angle. The lateral force comes from the deformation in the thread surface (i.e., the tire's cornering stiffness).
- The second region is a transition range in which the relationship between the slip angle and lateral force is no longer linear. Here, the rear portion of the tire footprint begins to slide laterally along the ground.
- Third, after the maximum lateral force is reached, an increase in slip angle does not continue to result in a higher lateral force. This area is the frictional range because from this point the lateral force is merely a result of friction between the tire and the road surface.

The vehicle's balance is determined by how much lateral force one end of the vehicle develops in comparison to the other end as well as which end of the vehicle reaches the tires' maximum lateral force first. An investigation using information from the vehicle's data acquisition system occurs in the following sections.

Figure 7.3 *X-Y plot of lateral versus longitudinal acceleration illustrating the cornering potential of the vehicle*

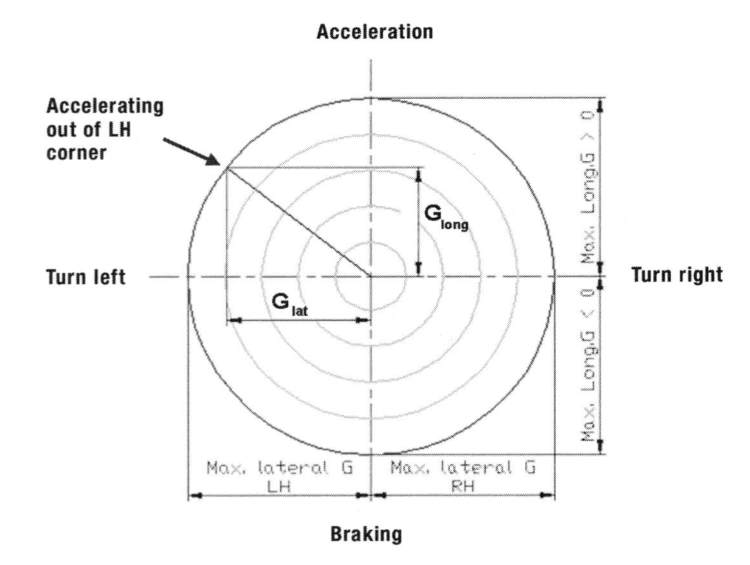

Traction Circle

A racing tire can develop approximately equal power in acceleration, braking, and cornering. Plotting the maximum forces that a tire can develop in each direction gives the traction circle of the tire. Most data acquisition packages feature an *X-Y* graph of lateral versus longitudinal acceleration, which is essentially the traction circle, or "g-g" diagram, for the entire vehicle (***Figure 7.3***). This graph represents the cornering power of the vehicle. For the traction circle to be read as illustrated in Figure 7.3, the lateral and longitudinal acceleration channels must be calibrated as indicated in Chapter 3 (in the Logging Steering Angle section).

A tire's maximum grip in any direction depends on the vertical load to which it is being subjected. This implies that the size of the traction circle is not constant. In a high-downforce corner, the circle's diameter is larger than in a slow corner.

In reality, the g-g diagram is not a circle but resembles a flattened heart shape, as maximum braking deceleration is greater than acceleration under power. The maximum longitudinal acceleration here is dictated by the traction potential of the car; this is why the upper two quadrants in the traction circle are not circular. The lower two quadrants take a circular form only if the driver and car can stay on the very limit of adhesion at every stage through a corner. This is very rare, and the lower two quadrants mostly are somewhat triangular.

Figure 7.4 illustrates acceleration data from a Dodge Viper GT car during a lap on the Circuit Zolder racetrack. The most basic information derived from this plot is the maximum acceleration in four directions, which can be read directly from the *x*- and *y*-axes.

Longitudinal acceleration under power	**0.90 G**
Longitudinal acceleration under braking	**1.70 G**
Lateral acceleration cornering left	**2.25 G**
Lateral acceleration cornering right	**2.10 G**

However, the region between the axes is more interesting. This represents the transient behavior of the car as longitudinal acceleration changes into lateral cornering acceleration. This is when the driver is braking and turning into a corner at the same time (i.e., trail braking) or getting the power down early upon exiting a corner.

The outside borderline indicates the vehicle's theoretical maximum acceleration potential in any direction, while the inside line shows how much the driver utilized this potential during that lap. Some generalities become quite clear by studying the diagram.

The area where the plot density is highest is located in the upper two quadrants, showing forward acceleration; more time was spent accelerating than braking. The scatter between 0.2 and 0.6 G in longitudinal direction represents the parts of the track where there were no cornering or brake forces. This covers the straight-line acceleration range of the vehicle.

The driver achieves limit lateral acceleration when there is very little longitudinal acceleration. As soon as there is braking power, lateral G drops significantly, so there is not much trail braking. On the corner exit, this effect reduces. Here, the driver keeps close to peak lateral G until the longitudinal G closes to 0.4 G.

Consider a random point in the plot, indicated by the small circle in the lower left quadrant. At that particular moment, the car reached a lateral deceleration of 1.10 G and a longitudinal acceleration of 0.61 G. The vector sum of these two components, the combined acceleration, can be calculated using ***Equation 7.1***.

$$G_{combined} = \sqrt{G_{lat}^2 + G_{long}^2} \qquad (Eq.\ 7.1)$$

For the indicated point, this means that the combined acceleration equals 1.25 G. There are a couple of points outside the theoretical maximum border; one is indicated by the small circle in the lower right quadrant. The combined acceleration at this point is 2.22 G, which is well within reach of the maximum lateral cornering potential. Here, for one moment, the driver is driving on the traction limit.

Combined acceleration can be calculated by the software and plotted on a time or distance graph. This graph can be particularly useful for analyzing the car's cornering potential during tran-

Figure 7.4 *Traction circle (or g-g diagram) of a lap around Zolder in a Dodge Viper GTS-R*

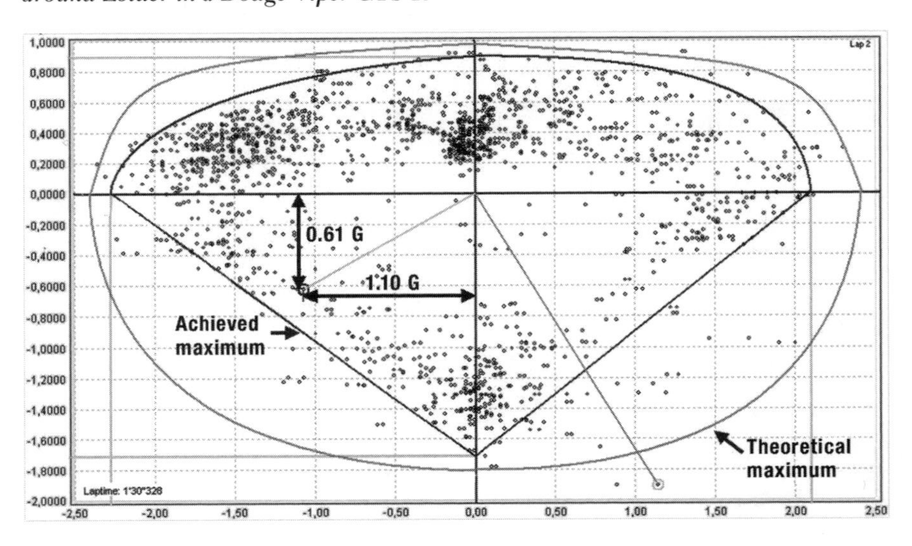

Figure 7.5 Combined acceleration

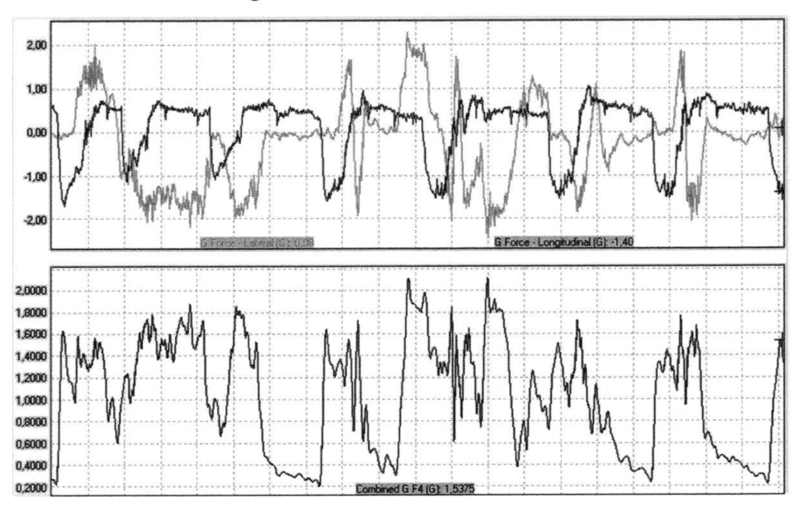

Figure 7.6 Traction circle, speed, combined acceleration, and lateral and longitudinal G graphs for one specific corner

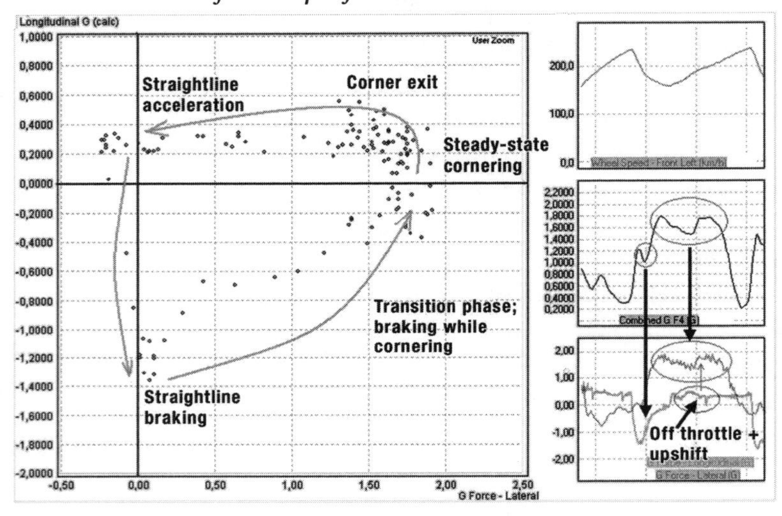

Figure 7.7 Traction circle plot for vehicle speeds below 130 km/h

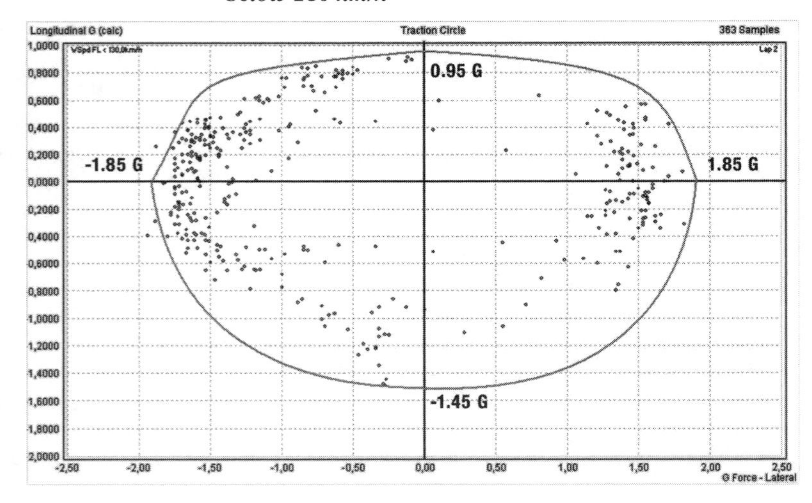

sient phases. It basically represents the radius of the vehicle's instantaneous traction circle. The example pictured in ***Figure 7.5*** is taken from the same lap as the traction circle in Figure 7.4.

The acceleration data in Figure 7.4 covers a complete lap, but separate corners also can be analyzed. ***Figure 7.6*** shows the g-g diagram of a 160-km/h sweeping corner. The arrows in the *X-Y* plot indicate the cornering phases. The driver makes all the braking effort before turning in, and most of the cornering is achieved without braking. Deceleration forces determine the first part of the combined G graph on the right. The small circle indicates a dip in the combined G graph as the driver removes his foot from the brake pedal, but momentarily there is still no cornering. The graph starts to rise again as cornering acceleration increases. In fact, at this point the lateral and combined G traces match.

As soon as acceleration becomes positive again (perhaps too quickly?), the lateral G graph takes a big dip and the car loses grip. In the traction circle, the points scatter around a lateral acceleration value of 1.5 G. The driver comes off the throttle and shifts up a gear to settle the car, after which time the lateral G increases again. As the car accelerates away and out of the corner, the lateral G drops to zero, while longitudinal G decreases slightly with increasing speed.

The traction circle and combined G are good working tools for investigating the cornering potential of a racecar. Because these tools illustrate the transient phases of the cornering sequence very well, they also can be applied for driver performance analysis.

Effects of Speed

As mentioned in the previous paragraph, the vehicle's traction circle radius is not constant. It varies with the total vertical load acting on the car's center of gravity. The most important parameter here is aerodynamic download, which is speed dependent. Therefore, a relationship exists between the vehicle's speed and its acceleration potential, both in a longitudinal and lateral sense.

Aerodynamic downforce increases proportional to the square of speed, resulting in greater cornering and braking potential. Drag has the same relationship to speed, so the more speed increases the less power remains to accelerate the vehicle.

In the following illustrations, traction circles for different speed intervals in a single lap are pictured ***(Figures 7.7–7.10)***. Speed effects also can be illustrated by plotting longitudinal and lateral acceleration against vehicle speed. ***Figures 7.11*** and ***7.12*** show an example of that for the same lap in Figures 7.8–7.10. ***Figure 7.13*** shows lateral acceleration versus vehicle speed for a Formula One car.

Throttle Histogram

A poor-handling chassis does not accept throttle input as well as a balanced one. Performing some statistical operations on the throttle data can help quantify the strength of a certain setup.

Average throttle position and the percentage of lap time spent at full throttle are statistical reference numbers for different setups. The throttle position is, however, a variable over which the driver has input. Therefore, any analysis of the throttle histogram should be undertaken with care. As drivers gain confidence, they become eager to get on the throttle earlier.

The following two laps were conducted on the Circuit Zolder racetrack with the same car and driver during the same session.

Table 7.1 *Average throttle position and full throttle percentage for two different laps around Zolder*

	Lap time	Average throttle position	Full throttle percentage
Lap 1	1'35"887	51.4%	29.8%
Lap 2	1'33"242	51.8%	30.8%

Between the two laps, a setup change that improves the car and lap time occurs. Average throttle position is improved by 0.4%. The time spent at full throttle goes up by 1%. Usually an increase in time spent at full throttle leads to an improved lap time. When this is not the case, check for a problem in straight-line acceleration (e.g., engine power, drag) or braking. The throttle histogram in ***Figure 7.14*** visually represents the throttle activities.

The throttle histogram typically includes two maximum values, at zero throttle and full throttle. Closed throttle should be approximately equal to

Figure 7.8 *Traction circle plot for vehicle speeds between 130 and 180 km/h*

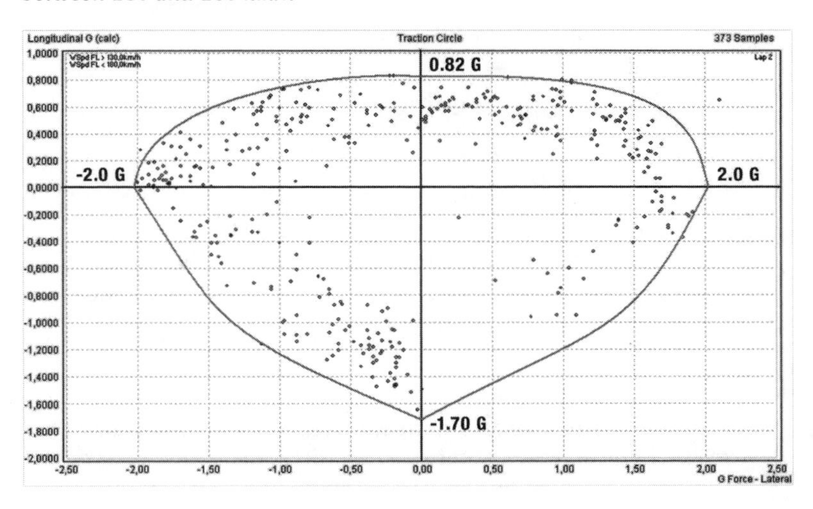

Figure 7.9 *Traction circle plot for vehicle speeds between 180 and 230 km/h*

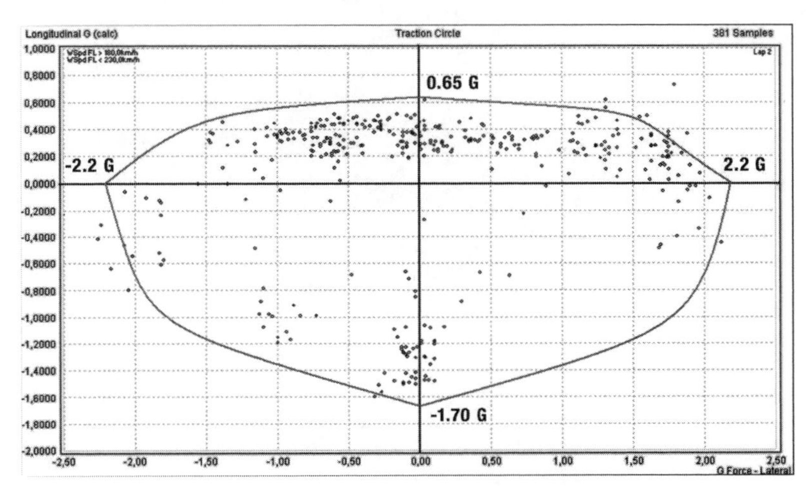

Figure 7.10 *Traction circle plot for vehicle speeds between 230 and 280 km/h*

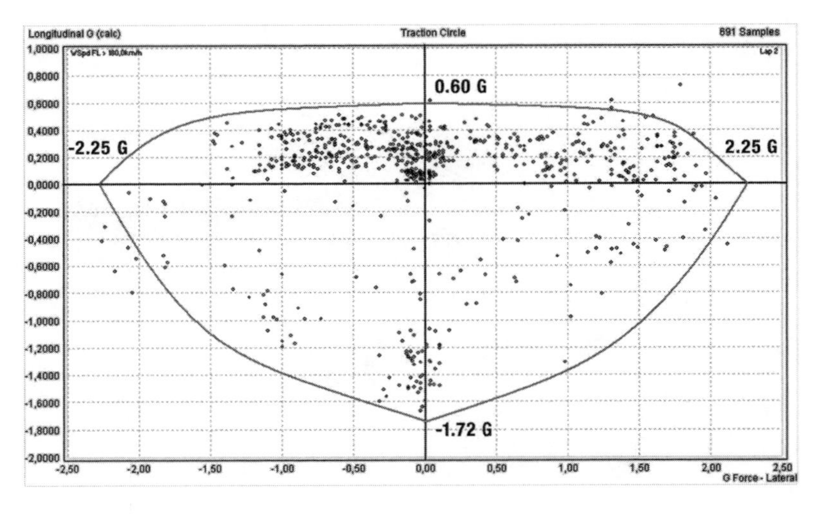

the time spent braking, unless the car is coasting. The maximum at full throttle indicates chassis or driver improvements.

Steering

Vehicle balance is commonly indicated by the terms *understeer, neutral steer,* and *oversteer.* In a simple world, understeer causes the vehicle to "push" front tires to the outside of the corner, while oversteer causes the rear axle to break out. Neutral steer is the situation where neither understeer nor oversteer are present. A more mathematical and correct definition follows in the next section.

The most common way to analyze cornering balance is to look at the input the vehicle acquires from the driver as a reaction to a handling problem. Steering movement and pedal activities can reveal much about the handling of the car. Steering angle, throttle position, and lateral acceleration are the channels to watch.

Oversteer

The driver counters oversteer by reducing the steering angle. If the oversteer is severe, this even may result in opposite lock, as demonstrated in ***Figure 7.15***. The vehicle concerned was obviously suffering from diabolical oversteer through the complete cornering sequence. The driver desperately tries to correct the rear stepping-out by jerking the steering wheel in the other direction. The gray line in the steering graph indicates approximately how the steering movement should have been addressed.

Steering corrections for oversteer very often are accompanied by little dips in the lateral G graph, parallel to the steering wheel movement. In general, lateral acceleration levels are lower than expected. Oversteer creates a rough lateral G graph as the car loses and regains grip. Variations smaller than 0.25 G and of a shorter duration than 0.3 sec are caused by irregularities in the track surface. Lateral G variations due to oversteer are confirmed in the steering angle graph.

Another oversteer indicator is when the vehicle is not willing to accept full throttle. In this example, the driver waits until the corner is completed before applying full throttle to avoid the rear stepping out.

Understeer

Understeer is a little more difficult to diagnose. Characteristic to understeer is the excessive steering angle. Often, the driver anticipates corner entry understeer by pitching the car into the corner. An oversteering moment is created around the car's center of gravity to compensate the understeer by an aggressive steering input. This pitching movement is followed by a small dip where the driver catches the rear braking out ***(Figure 7.16)***. Steady-state understeer is illustrated here by an ever-increasing steering angle and tentative throttle application.

***Figure 7.11** Lateral acceleration versus vehicle speed. Lateral grip potential increases with aerodynamic downforce.*

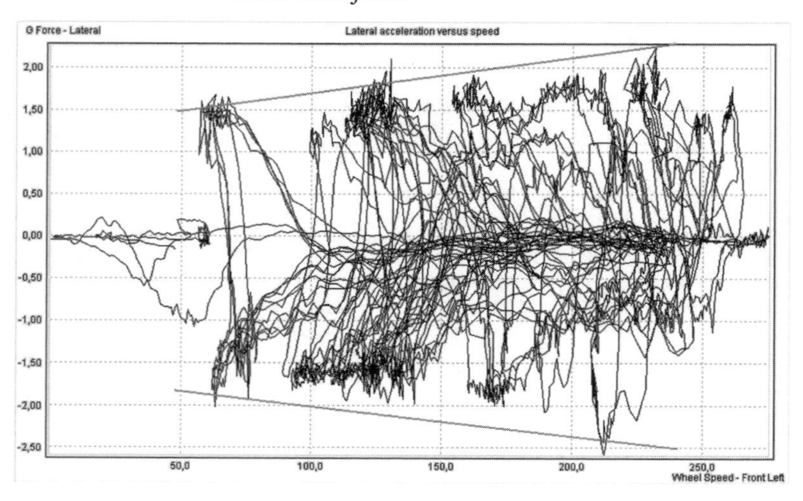

***Figure 7.12** Longitudinal acceleration versus vehicle speed. Acceleration potential under power decreases with speed due to aerodynamic drag. Deceleration potential increases with speed due to aerodynamic drag and downforce.*

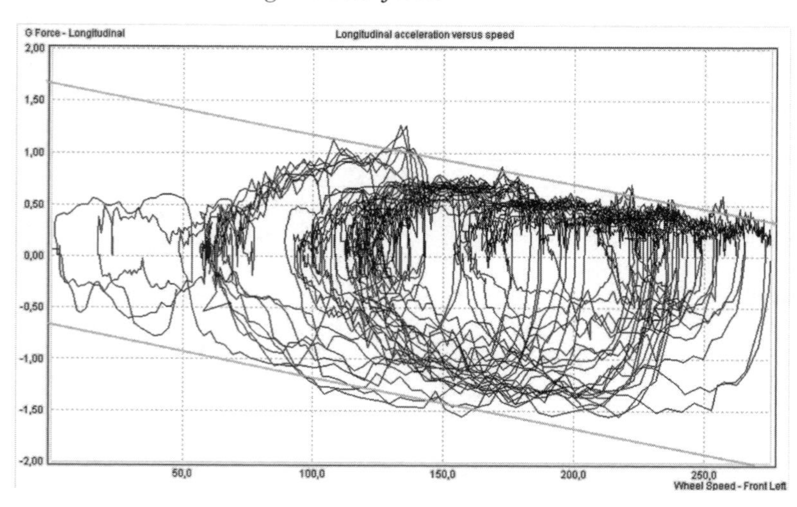

Another indication of corner entry understeer is when the steering angle peaks before the lateral Gs. As mentioned previously, when the steering angle is more than seems necessary, understeer is often the diagnosis.

The Understeer Angle

Figure 7.17 shows the bicycle model of a vehicle during cornering. The bicycle model is a mathematical model of the vehicle in which the track width is considered to be zero. It assumes a steady-state cornering situation where the vehicle takes a constant radius corner at very low speed. The steering angle is small, and front and rear slip angles are zero. R is the corner's turning radius and WB the wheelbase of the vehicle. When traveling at low speed and influences due to accelerations are negligible, the steering angle required to negotiate this corner δ_{Acker} is defined by ***Equation 7.2***.

$$\delta_{Acker} = \frac{WB}{R} \qquad (Eq.\ 7.2)$$

This angle is called the kinematic steering angle, or Ackermann steering angle (not to be confused with Ackermann steering).

As speed increases, the tires develop a slip angle, and the difference between the front and rear slip angle determines the balance of the car. This affects the required steering angle for steady-state cornering.

SAE J670[2] defines neutral steer, understeer, and oversteer as follows:

Neutral steer: A vehicle is neutral steer at a given trim if the ratio of the steering wheel angle gradient to the overall steering ratio equals the Ackermann steering gradient.

Understeer: A vehicle is understeer at a given trim if the ratio of the steering wheel angle gradient to the overall steering ratio is greater than the Ackermann steering gradient.

Oversteer: A vehicle is oversteer at a given trim if the ratio of the steering wheel angle gradient to the overall steering ratio is smaller than the Ackermann steering gradient.

By this definition, the steering wheel angle gradient is the rate of change in the steering wheel angle with respect to change in steady-state lateral acceleration. The Ackermann steering gradient is the rate of change in the Ackermann steer angle with respect to change in steady-state lateral acceleration.

The overall steering ratio is the ratio between steering wheel angle and the angle of the steered wheels.

Authors Milliken and Milliken[4] give the following definitions for neutral steer, understeer, and oversteer:

Neutral steer: Steered angle = Ackermann steering angle

Understeer: Steered angle > Ackermann steering angle

Oversteer: Steered angle < Ackermann steering angle

Figure 7.13 *Lateral acceleration versus vehicle speed for a late 1990s F1 car. This data was taken from a lap around Silverstone Circuit.*

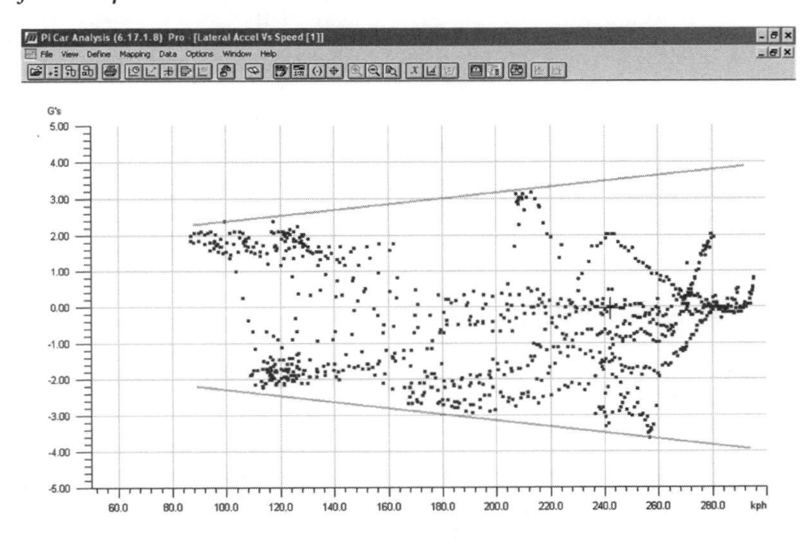

Figure 7.14 *Throttle histogram*

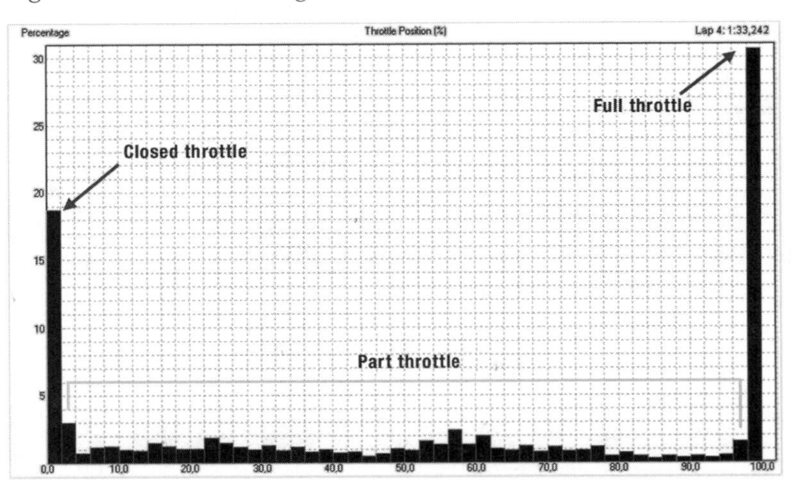

An understeer angle (δ_u) can be defined as the deviation from the Ackermann steering angle needed to follow the vehicle's intended path.

$$\delta_u = \delta - \delta_{Acker} \quad (Eq.\ 7.3)$$

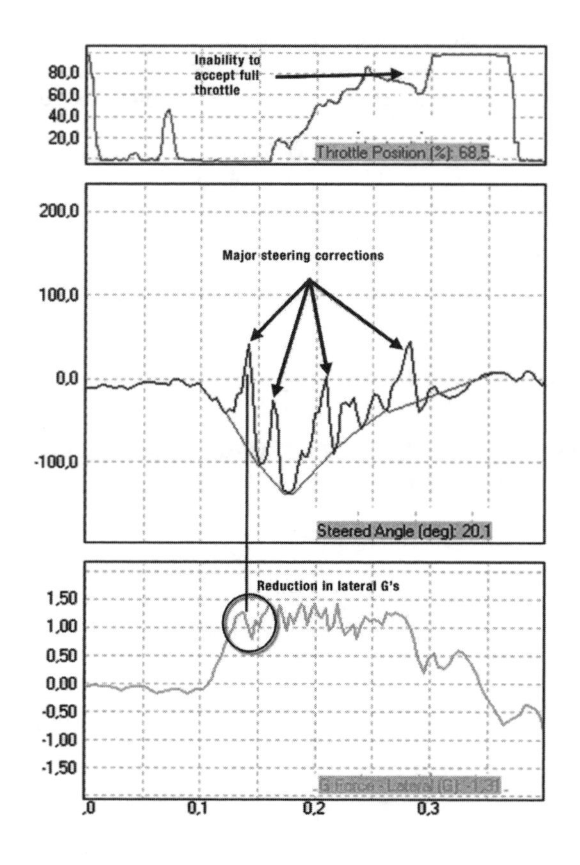

Figure 7.15 *This vehicle demonstrates the steering corrections of severe oversteer. During the corner exit phase, the driver was unable to apply full throttle.*

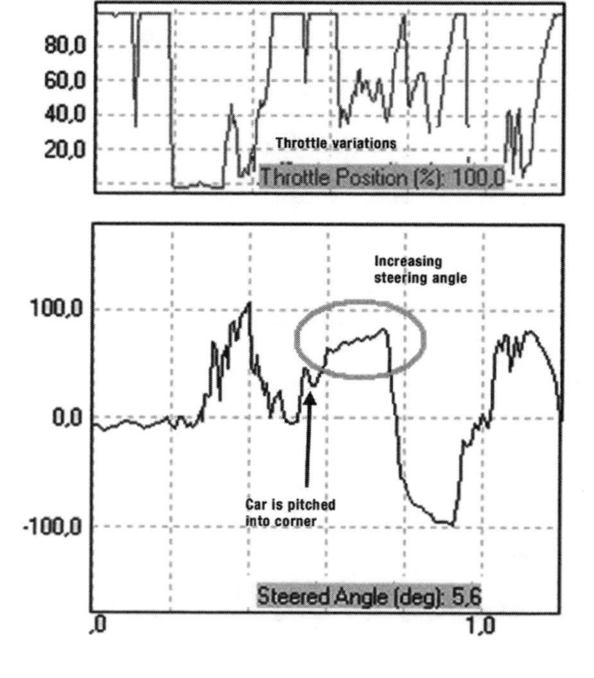

Figure 7.16 *Understeer indicators*

In this equation, δ is the actual steering angle of the front outside wheel. Cornering radius is determined by ***Equation 7.4***.

$$R = \frac{V^2}{G_{lat}} \quad (Eq.\ 7.4)$$

In this equation, V is the vehicle speed and G_{lat} is the car's lateral acceleration. Substituting this in the equation of Ackermann steering angle gives us ***Equation 7.5***.

$$\delta_{Acker} = \frac{WB \cdot G_{lat}}{V^2} \quad (Eq.\ 7.5)$$

Finally, this results in the formula to calculate the understeer angle ***(Equation 7.6)***.

$$\delta_u = \delta - \frac{WB \cdot G_{lat}}{V^2} \quad (Eq.\ 7.6)$$

A very important point to remember here is the definitions for neutral steer, understeer, and oversteer are valid only in the linear operating range of the tires and under steady-state conditions. If the equations are applied in the nonlinear operating range of the tires, their mathematical validity is lost. However, graphically represented, they can serve quite well for comparing different vehicle configurations.

The relationship between the steering wheel angle and the front wheels of the car must be determined first to create the necessary mathematical channels in the analysis software. This is best achieved with the car on turnplates from which the wheel angle can be read. A table can be created with different steering wheel angle values and the corresponding wheel angles. Always take the reading from the wheel on the outside of the corner. This table can be entered into a spreadsheet to determine a mathematical expression of wheel angle as a function of steering wheel angle ***(Figure 7.18)***, which then can be used as a mathematical channel to define δ. Depending on the vehicle's steering geometry, this can be a complex equation.

Figure 7.18 shows a steering angle measurement performed on a Corvette C5R. The plotted characteristic is nearly linear and can be approximated by ***Equation 7.7***.

$$\delta = \delta_{SW} \cdot 0.4262 \qquad (Eq.\ 7.7)$$

In this equation, δ_{SW} is the logged steering wheel angle in degrees. The next step is to create a math channel for the Ackermann steering angle. Considering that the wheelbase of the Corvette is 2.65 m, lateral acceleration is expressed in Gs, and vehicle speed is in km/h, the channel looks like ***Equation 7.8***.

$$\delta_{Acker} = \left(\frac{G_{lat} \cdot 9.81}{V^2 \cdot 0.077} \cdot 2.65 \right) \cdot 57.3 \qquad (Eq.\ 7.8)$$

In this equation, G_{lat} is converted into m/s² and speed into m/s. The 57.3 factor converts radians into degrees. Now the understeer angle (δ_u) can be calculated using ***Equation 7.9***.

$$\delta_u = |\delta| - |\delta_{Acker}| \qquad (Eq.\ 7.9)$$

The absolute values of wheel angle and Ackermann steering angle are taken to remove the sign convention between left- and right-hand corners. ***Figure 7.19*** shows the Ackermann steering angle, wheel angle, and understeer angle at the indicated corner on the Dubai Motodrom, logged on the Corvette C5R.

From the definition of understeer angle, it follows that a positive value means understeer and a negative value oversteer.

In the example proposed in Figure 7.19, this car has difficulty navigating the apex of the corner. Specifically, the understeer angle increases up on corner entry. The steering wheel movement (in this case, illustrated by the wheel angle) shows typical understeer symptoms. At mid-corner, the driver turns the steering wheel so far inward (understeer angle keeps increasing) that the understeer suddenly shifts to corner-exit oversteer. At this moment, the understeer angle turns negative. The graph starts fluctuating as the driver tries to keep the car in line by applying opposite steering lock.

The average understeer angle over a lap can be calculated to get an idea of which direction the balance of the car is going after a setup change. This is a reference number that is track dependent as it is influenced by the number of corners and the lengths of the straights.

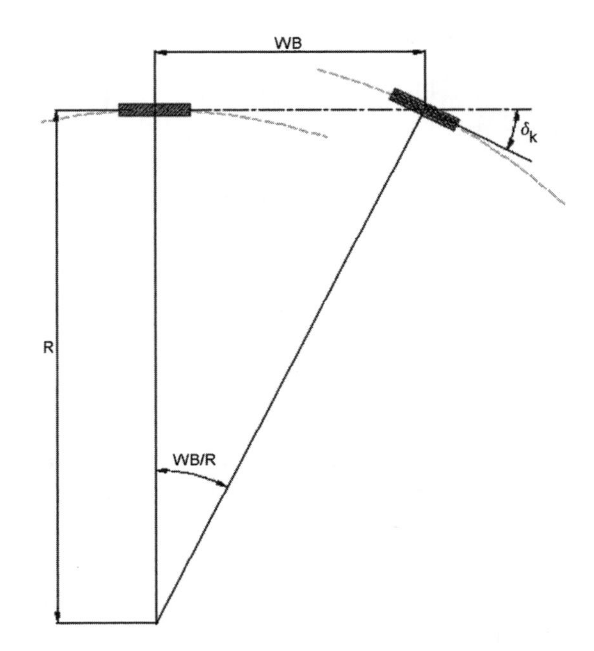

Figure 7.17
The relationship between corner radius, wheelbase, and Ackermann steering angle

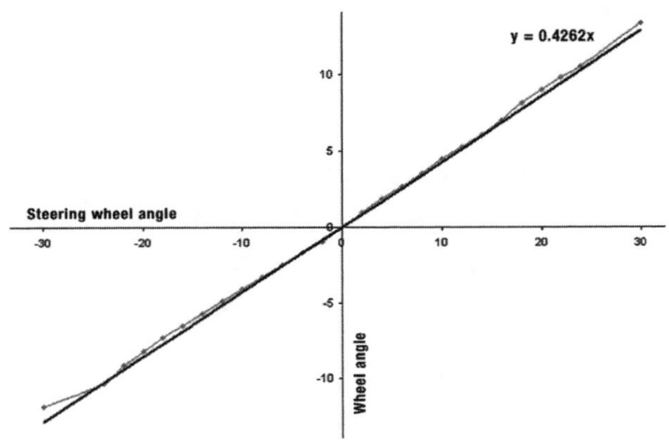

Figure 7.18
Steering wheel angle against outside corner wheel angle

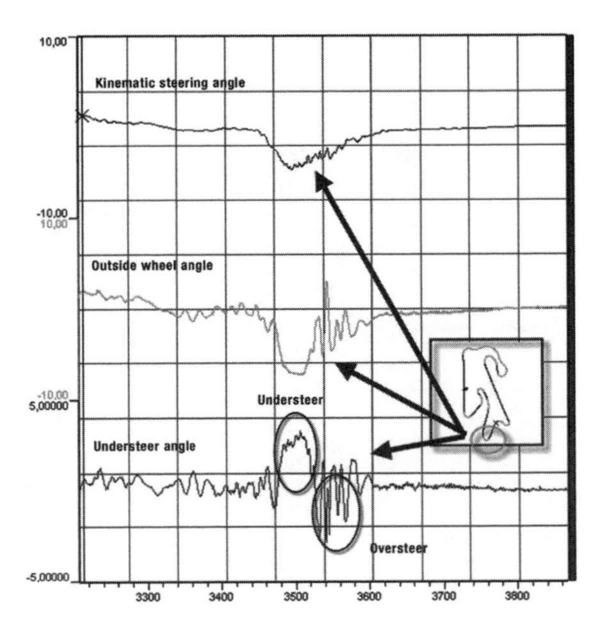

Figure 7.19
Ackermann (kinematic) steering angle, outside wheel angle, and understeer angle zoomed in at the indicated corner. Data was taken from a Corvette C5R during a test session at the Dubai Motodrom.

Attitude Velocity

The vehicle's angular rate of rotation around an axis perpendicular to the ground through the center of gravity can be measured with a gyro. This is basically the rate of change in heading or yaw angle ***(Figure 7.20)***.

In the previous section, methods for determining if a car is understeering or oversteering (i.e., to determine the balance state of the car) were reviewed. Yaw rate measurements show how this balance state is changing.

A vehicle that is beginning to oversteer experiences an increase in yaw rate, while understeering causes a decrease in yaw rate. A change in balance can be quantified by comparing the actual yaw rate to a theoretical value, which is a similar calculation (as explained previously) to determine the understeer angle.

Figure 7.20 Yaw angle

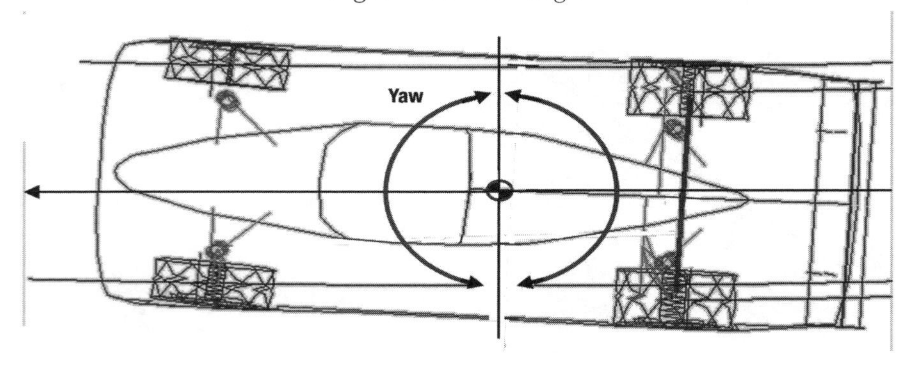

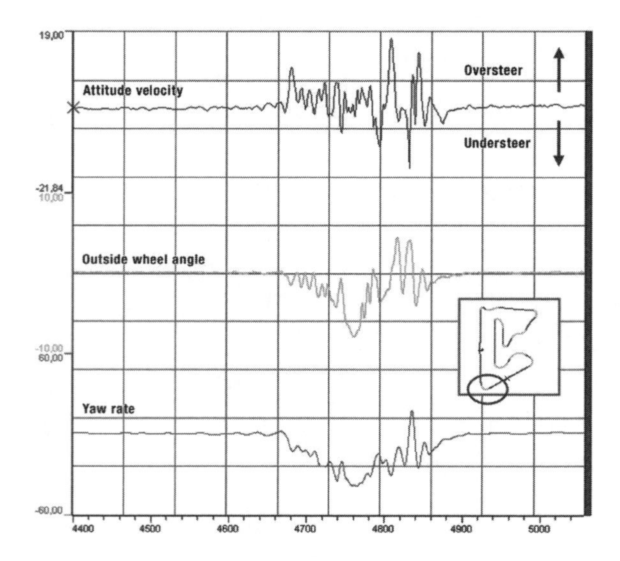

Figure 7.21 Attitude velocity

In a steady-state maneuver, a vehicle changes heading at rate defined in ***Equation 7.10***.

$$\text{angular velocity} = \frac{V}{R} \qquad (Eq.\ 7.10)$$

Substituting Equation 7.4 for corner radius R gives ***Equation 7.11***.

$$\text{angular velocity} = \frac{G_{lat}}{V} \qquad (Eq.\ 7.11)$$

Attitude velocity now can be defined as the difference between measured yaw rate (r) and angular velocity ***(Equation 7.12)***.

$$\text{attitude velocity} = r - \frac{G_{lat}}{V} \qquad (Eq.\ 7.12)$$

In ***Figure 7.21,*** a graph illustrates attitude velocity for the indicated corner on the Bahrain Grand Prix track. Equation 7.11 is modified into a mathematical channel to express angular velocity in deg/sec ***(Equation 7.13)***.

$$\text{angular velocity} = 57.3 \cdot \left(\frac{9.81 \cdot G_{lat}}{0.077 \cdot V^2} \right) \qquad (Eq.\ 7.13)$$

The result then is subtracted from the measured yaw rate to give the attitude velocity. A positive value (yaw rate > angular velocity) means that the vehicle tends to oversteer, while a negative value (yaw rate < angular velocity) indicates a vehicle with an understeering tendency.

As with the equation for determining the understeer angle, this mathematical channel does not take into account slip angles. Furthermore, to obtain an accurate result, the gyro must measure the yaw rate at the vehicle's center of gravity.

Front and Rear Lateral Acceleration

If lateral acceleration on the front and rear axles can be measured separately, the assumption can be made that the axle on which the highest value is measured will develop the highest degree of grip. The following applies:

front acceleration	>	rear acceleration	OVERSTEER
front acceleration	<	rear acceleration	UNDERSTEER

In ***Figure 7.22***, front and rear lateral acceleration traces are pictured. A mathematical channel is created in which front lateral acceleration is subtracted from rear lateral acceleration. This means that understeer shows up as a positive value, while oversteer results in a negative value.

The sensors must be placed at the location where the vehicle centerline intersects with the front and rear axle centerline. This is not always easy. If measured correctly, the comparison of front and rear lateral acceleration can be very helpful in vehicle balance analysis.

In the absence of a gyro, yaw rate can be determined from front and rear lateral acceleration channels by first calculating the yaw acceleration ***(Equation 7.14)***.

$$a_{yaw} = \frac{G_{Lat(rear)} - G_{Lat(front)}}{WB} \qquad (Eq.\ 7.14)$$

with $G_{Lat(rear)}$ = rear axle lateral acceleration
$G_{Lat(front)}$ = front axle lateral acceleration
WB = wheelbase

Then integrate this to get yaw rate ***(Equation 7.15)***.

$$r = \int a_{yaw}(t) \qquad (Eq.\ 7.15)$$

During a longer time period, this integration loses much of its accuracy because of noise in the signal and accelerometer drift. For short durations, however, it can be useful.

***Figure 7.22** Evaluation of oversteer and understeer by comparing lateral acceleration on the front and rear axles*

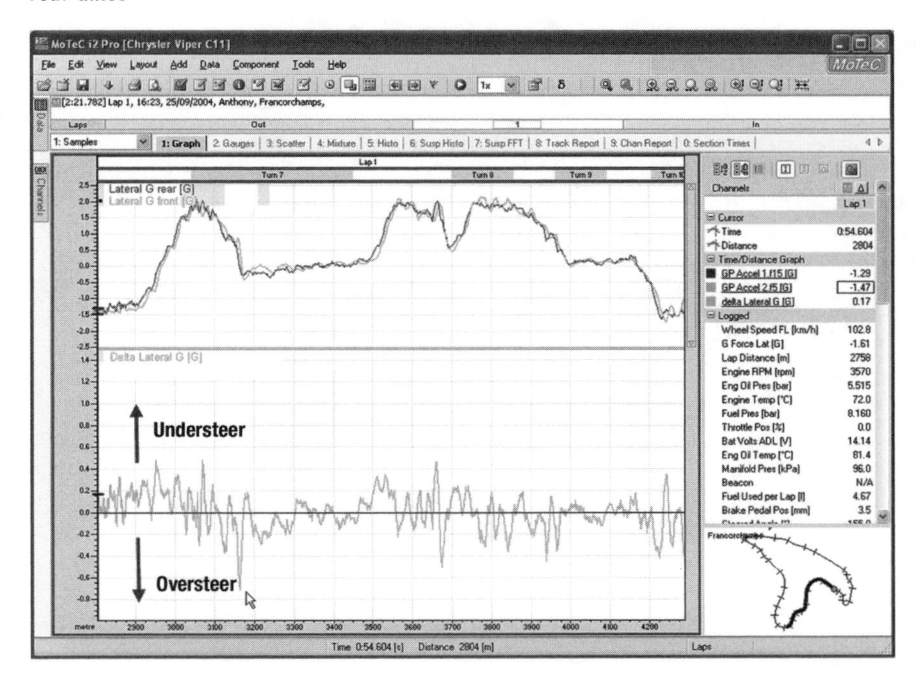

CHAPTER 8

QUANTIFYING ROLL STIFFNESS DISTRIBUTION

The car's oversteer/understeer balance during cornering is determined by the load distribution between the front and rear axles. Suspension tuning is all about influencing this balance by adjusting the vehicle's roll stiffness distribution until the desired understeer/oversteer balance is achieved. This chapter shows how the data acquisition system can assist in determining references for the suspension setup and how the effects of setup changes can be analyzed quickly.

Some setup reference numbers are discussed that can be easily calculated with the mathematical options in the data analysis software or, if possible, by exporting the required channels into a spreadsheet. They help characterize the suspension and serve as a future reference for setting up the car.

Front and Rear Roll Gradient

In vehicle dynamics, it is common to characterize the roll stiffness of a vehicle in normalized form as degrees of body roll per unit of lateral acceleration. This parameter is called the *roll gradient.* Here is a practical example of how the roll gradient is related to the car's roll stiffness. ***Table 8.1*** provides an overview of the dynamic parameters of a GT-type racecar. This data was calculated by SusProg3D, a software package used for suspension geometry analysis.

Equation 8.1 is the basic definition of roll gradient with α_{roll} being the vehicle roll angle and G_{lat} the lateral acceleration acting on the center of gravity.

$$RG = \frac{\alpha_{roll}}{G_{lat}} \qquad (Eq.\ 8.1)$$

From the data in Table 8.1, the roll moment (M_{roll}) at 1-G lateral acceleration is calculated. M_{roll} is defined by ***Equation 8.2***.

$$M_{roll} = h_{roll} \cdot \left(W_{sF} + W_{sR}\right) \qquad (Eq.\ 8.2)$$

with h_{roll} = the distance between the car's center of gravity and the roll axis (the line connecting front and rear roll centers)

W_{sF} = static sprung weight front axle

W_{sR} = static sprung weight rear axle

Figure 8.1 illustrates the parameters.

In Table 8.1, the following information is provided:

$W_{sF} = 636 - 61 = 575$ kg

$W_{sR} = 739.5 - 98 = 641.5$ kg

h_{roll} = sprung C of G height above roll axis = 323.61 mm

Entering this information into Equation 8.2 gives us ***Equation 8.3***.

$$M_{roll} = 32.361\ \text{cm} \cdot 1216\ \text{kg} = 39351\ \text{kgcm} \qquad (Eq.\ 8.3)$$

The roll angle then can be expressed as a function of roll moment ***(Equation 8.4)***.

$$\alpha_{roll} = \frac{M_{roll}}{K_{rolltot}} \qquad (Eq.\ 8.4)$$

$K_{rolltot}$ is the total roll stiffness of the car, which is 83166 kgcm/deg (Table 8.1). Because the roll moment is calculated at a lateral acceleration of 1 G, Equation 8.3 equals the roll gradient (RG).

The data in Table 8.1 is calculated for a lateral acceleration of 1 G. The total roll angle of the vehicle in this table is 0.5 deg, which equals the roll gradient calculated in ***Equation 8.5***.

$$RG = \frac{39351\ \text{kgcm}}{83166\ \text{kgcm/deg}} = 0.473\ \text{deg/G} \qquad (Eq.\ 8.5)$$

Note that this calculation uses the total roll stiffness, including the stiffness of the tires. The method explained later in this section calculates roll gradients using the data from the suspension

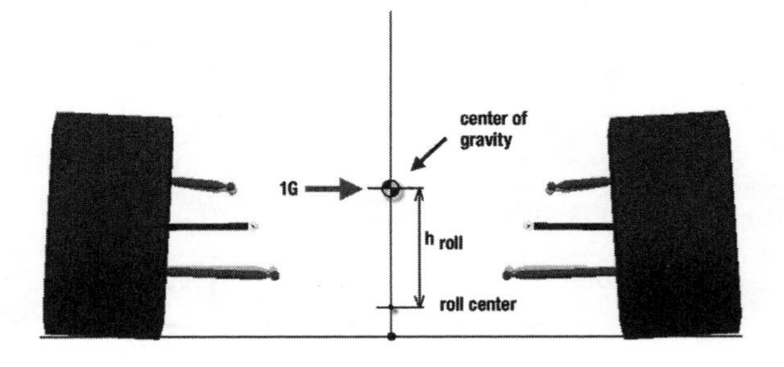

Figure 8.1 Distance between center of gravity and roll center

Table 8.1 *SusProg3D calculation of vehicle parameters from a GT racecar*

SusProg3D GT2-2002.s3dDynamic			
Roll center height	front/rear	46.49 mm	51.03 mm
Track	front/rear	1649.54 mm	1684.19 mm
Wheelbase		2402.87 mm	
Vehicle weight	front/rear	636.00 kg	739.50 kg
Unsprung weight	front/rear	61.00 kg	98.00 kg
Lateral acceleration		1.00 G	
Vehicle C of G	from ground		369.00 mm
	from front axle cl		1291.84 mm
Sprung C of G	from ground		372.50 mm
	from front axle cl		1267.11 mm
	above roll axis		323.61 mm
Antiroll stiffness	front	tires	97368.66 kgcm/deg
		spring	40752.86 kgcm/deg
		antiroll bar	64312.73 kgcm/deg
		total	50535.40 kgcm/deg
Antiroll stiffness	rear	tires	112147.73 kgcm/deg
		spring	23765.06 kgcm/deg
		antiroll bar	22255.07 kgcm/deg
		total	32630.23 kgcm/deg
Antiroll stiffness	car	tires	209516.38 kgcm/deg
		spring	64517.92 kgcm/deg
		antiroll bar	86567.80 kgcm/deg
		total	83165.62 kgcm/deg
Roll angle	front	tires	0.28 deg
		suspension	0.22 deg
		total	0.50 deg
	rear	tires	0.19 deg
		suspension	0.31 deg
		total	0.50 deg

displacement channels, which do not measure deflection of the tires. If the elasticity of the tires is not taken into account in the calculation, RG becomes as shown in ***Equation 8.6*** with $K_{rolltot}$ = 64518 + 86568 = 151086 kgcm/deg (because antirollbar and suspension springs are parallel springs, these can be added together).

$$RG = \frac{39351 \text{ kgcm}}{151086 \text{ kgcm/deg}} = 0.260 \text{ deg/G} \qquad (Eq.\ 8.6)$$

Table 8.2 states typical numbers for roll gradient.

Table 8.2 *Typical roll gradients*[4]

Very soft—Economy and basic family transportation, pre-1975	8.5 deg/G
Soft—Basic family transportation, after 1975	7.5 deg/G
Semisoft—Contemporary middle market sedans	7.0 deg/G
Semifirm—Sport sedans	6.0 deg/G
Firm—Sport sedans	5.0 deg/G
Very firm—High performance (e.g., Camaro Z-28, Firebird Trans Am)	4.2 deg/G
Extremely firm—Contemporary very high-performance sports (Corvette), street cars extensively modified to increase roll stiffness	3.0 deg/G
Hard—Racing cars only	1.5 deg/G
Active suspension, servo-controlled roll stiffness, roll-in, zero-roll, and roll out all possible	————

Table 8.3 gives the experience numbers for roll gradient.

Table 8.3 *Typical roll gradients*[9, 10]

Stiff high-downforce racecars	0.2–0.7 deg/ G
Low downforce sedans	1.0–1.8 deg/ G

Both references state that racecars with any amount of downforce have a roll gradient lower than 1.5 deg/G. ***Table 8.4*** shows some measured examples of various racecars. Keep in mind these numbers were derived from suspension potentiometer data and that tire spring rates were not taken into account. When the roll angle of the tires is added to the measured suspension roll angle, the roll gradients in Table 8.4 are greater.

Table 8.4 *Typical roll gradients for various types of racecars*

Dodge Viper GTS-R racecar	0.44–0.55 deg/G
Corvette C6R racecar	0.25–0.35 deg/G
Corvette C5R racecar	0.20–0.40 deg/G
2002 Formula One car	0.03–0.10 deg/G
2001 Indycar	0.10–0.20 deg/G
2005 Daytona prototype	0.35–0.55 deg/G

By measuring the lateral acceleration and the movement of the suspension, the data analysis software can calculate the roll gradient of the car. From the suspension data, the overall roll angle of the car can be calculated by creating the mathematical channel in ***Equation 8.7***. This equation results in a positive roll angle when the chassis rolls clockwise and vice versa, facing forward along the vehicle's longitudinal centerline. This is in accordance with SAE's Vehicle Axis System.[2]

If the analysis software package allows the use of *X-Y* graphs, a chart can be created illustrating the roll angle versus lateral acceleration as shown in ***Figure 8.2***. The advantage of putting this information in a graph is that in one view shows the vehicle's maximum roll angle and the acceleration at which this angle is reached. It shows this for left- and right-hand cornering. The relationship between roll angle and lateral acceleration is linear in most cases (although not in the case where progressive suspension elements are used), and the slope of the graph should equal the roll gradient. In the illustration, the equation for a linear trend line is calculated. The roll gradient, or the slope of the trend line, is 0.48 deg/G in this case.

Note further that this calculation is made taking the roll angle of the suspension springs and antiroll bars into account. This does not include the roll angle resulting from tire deflection.

The *X-Y* plot in Figure 8.2 shows a certain degree of scatter in the data points, which can be explained by the following:

- the damping of the system momentarily changing the roll stiffness distribution of the car;
- chassis torsion, especially when driving over big bumps or curbs; and
- accuracy of measuring the suspension positions and proper zeroing of sensors.

$$\alpha_{roll} = \arctan\left(\frac{(x_{suspensionLF} - x_{suspensionRF}) \cdot MR_F + (x_{suspensionLR} - x_{suspensionRR}) \cdot MR_R}{T_F + T_R}\right) \cdot 57.3 \qquad (Eq.\ 8.7)$$

with
α_{roll} = total suspension roll angle (deg)
$x_{suspensionLF}$ = left-front suspension potentiometer travel (mm)
$x_{suspensionRF}$ = right-front suspension potentiometer travel (mm)
$x_{suspensionLR}$ = left-rear suspension potentiometer travel (mm)
$x_{suspensionRR}$ = right-rear suspension potentiometer travel (mm)
MR_F = front suspension motion ratio
MR_R = rear suspension motion ratio
T_F = front track width (mm)
T_R = rear track width (mm)

Figure 8.2 *X-Y graph of overall roll angle versus lateral G, taken from a lap around Nürburgring by Bert Longin in a Dodge Viper GTS-R*

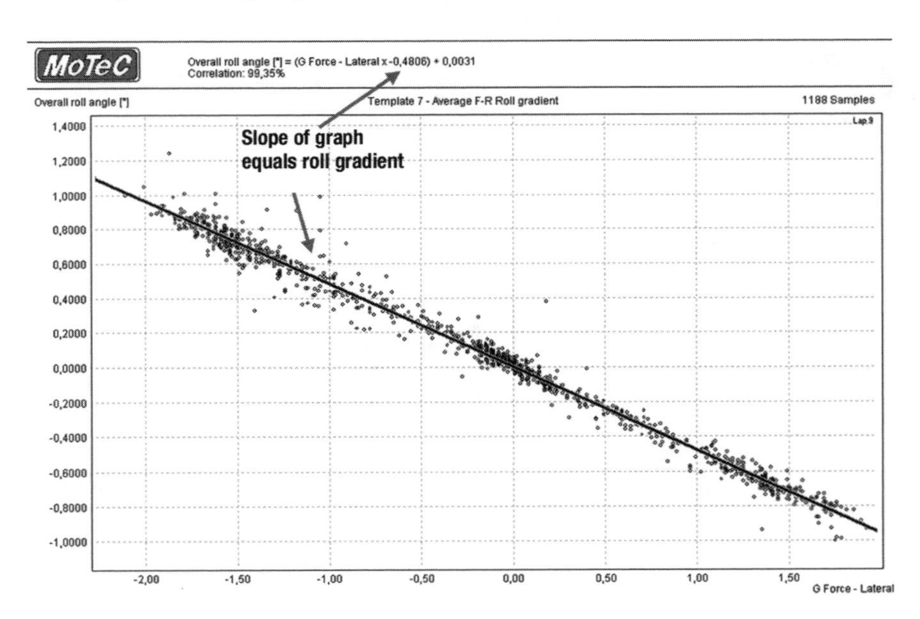

Each axle also can be viewed separately and roll gradients calculated for the front and rear suspension. For this, the overall roll angle in the *X-Y* graph is replaced by the roll angle of the front and rear suspension, respectively. From the same session covered by Figure 8.2, the front and rear roll gradients are shown in ***Figures 8.3*** and ***8.4***.

In Equation 8.7, the overall roll gradient is calculated from the average roll angle between the front and rear suspension, while the lateral force remains the same. This means that the overall roll gradient is the average between the front and rear roll gradient.

Figure 8.3 Front roll gradient

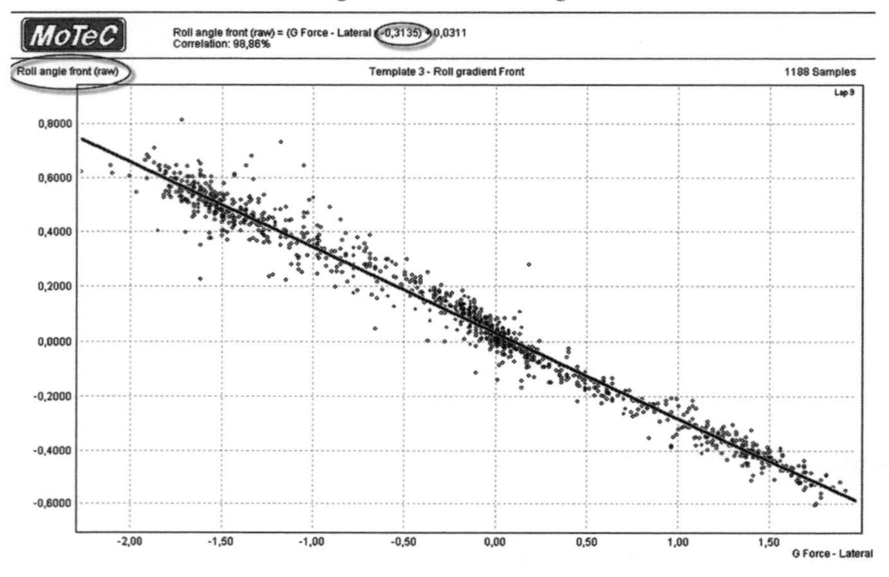

Figure 8.4 Rear roll gradient

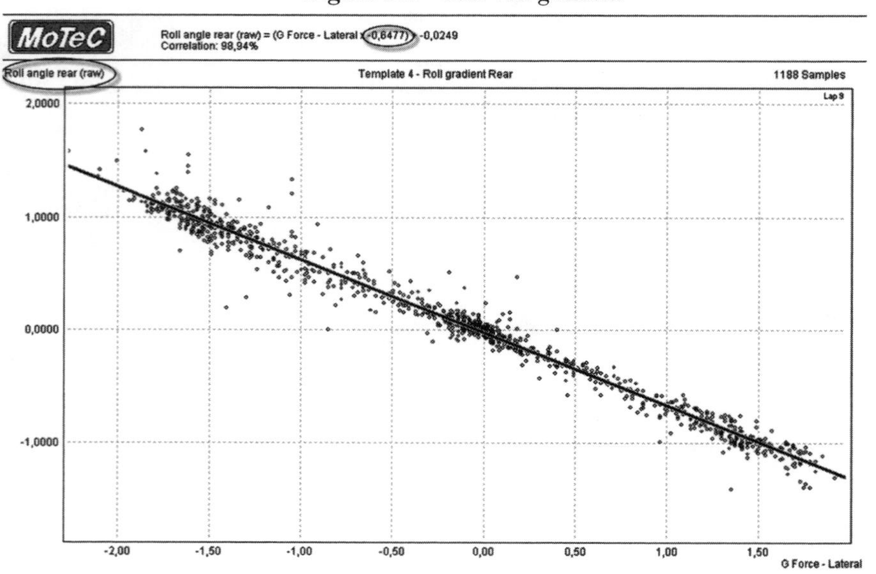

The advantage of calculating front and rear roll gradients separately is the information obtained about the roll stiffness distribution of the vehicle. The lower the roll gradient is on one side of the vehicle, the higher the resistance is to roll movement on that side, and vice versa.

From these roll gradients, the actual roll stiffness of the front and rear suspension can be calculated. However, for analysis purposes, merely observing roll gradients provides the engineer with a good idea about the roll stiffness distribution.

If the driver is pleased with the car's handling during a track session, the roll gradients can be recorded as a future reference for that particular racetrack. If in a following test or race on this track the handling is not as good as before, the engineer can try to restore the same roll gradients with the proper setup changes.

Using Roll Gradients as a Setup Tool

How useful roll gradients can be is illustrated with an example. In Figures 8.3 and 8.4, the analysis software calculated these numbers:

$$RG_F = 0.3135 \text{ deg/G}$$
$$RG_R = 0.6477 \text{ deg/ G}$$

This means that the overall roll gradient (RG) is 0.4806 deg/G. The car in question has the following properties:

front axle weight	=	636 kg
rear axle weight	=	739 kg
front unsprung weight	=	61 kg
rear unsprung weight	=	98 kg
front roll center height (h_{RCf})	=	46.5 mm
rear roll center height (h_{RCr})	=	51 mm
front spring rate (per wheel) (SR_f)	=	320 kg/cm
rear spring rate (per wheel) (SR_f)	=	280 kg/cm
front suspension motion ratio (MR_F)	=	1.373
rear suspension motion ratio (MR_R)	=	1.725
front antiroll bar motion ratio (MR_{rollF})	=	1.495
rear antiroll bar motion ratio (MR_{rollR})	=	1.550
front track width (T_F)	=	1,650 mm
rear track width (T_R)	=	1,685 mm
wheelbase (WB)	=	2,403 mm
height sprung center of gravity from ground (h_{CoG})	=	372 mm

First h_{roll} is calculated:

front sprung weight (W_{sF}) = 636 – 61 = 575 kg
rear sprung weight (W_{sR}) = 739 – 98 = 641 kg

Using basic statics and trigonometry, the equation for h_{roll} becomes ***Equations 8.8*** and ***8.9***.

$$h_{roll} = h_{CoG} - \left(h_{RCf} + \frac{(h_{RCr} - h_{RCr}) \cdot W_{sR}}{W_{sF} + W_{sR}} \right)$$

(Eq. 8.8)

$$h_{roll} = 372 - \left(46.5 + \frac{(51 - 46.5) \cdot 641}{575 + 641} \right) = 323 \text{ mm}$$

(Eq. 8.9)

Entering this into Equation 8.2 gives the roll moment at 1-G lateral acceleration ***(Equation 8.10)***.

$$M_{roll} = 32.3 \text{ cm} \cdot 1216 \text{ kg} = 39277 \text{ kgcm} \quad \textit{(Eq. 8.10)}$$

The total roll stiffness then becomes ***Equation 8.11***.

$$K_{rolltot} = \frac{M_{roll}}{RG} = \frac{39277}{0.4806} = 81725 \text{ kgcm/deg}$$

(Eq. 8.11)

With the front and rear roll gradient, it can be determined how this total roll stiffness is distributed over the front and rear axles ***(Equations 8.12*** and ***8.13)***.

$$K_{rollf} = K_{rolltot} \cdot \frac{RG_R}{RG_F + RG_R} \quad \textit{(Eq. 8.12)}$$

$$K_{rollr} = K_{rolltot} - K_{rollf} \quad \textit{(Eq. 8.13)}$$

with K_{rollf} = roll stiffness front axle
K_{rollr} = roll stiffness rear axle

So, K_{rollf} = 55083 kgcm/deg and K_{rollr} = 26642 kgcm/deg. The roll stiffness distribution is biased 67.4% to the front.

Because suspension springs and antiroll bars are parallel springs, ***Equations 8.14*** and ***8.15*** apply.

$$K_{rollr} = K_{rollrSPRINGS} + K_{rollrARB} \quad \textit{(Eq. 8.14)}$$

$$K_{rollr} = K_{rollrSPRINGS} + K_{rollrARB} \quad \textit{(Eq. 8.15)}$$

with $K_{rollfSPRINGS}$ = roll stiffness front axle due to suspension springs
$K_{rollfARB}$ = roll stiffness front axle due to antiroll bar
$K_{rollrSPRINGS}$ = roll stiffness rear axle due to suspension springs
$K_{rollrARB}$ = roll stiffness rear axle due to antiroll bar

The springs now on the car result in the following wheel rates ***(Equations 8.16–8.18)***.

$$WR = \frac{SR}{MR^2} \quad \textit{(Eq. 8.16)}$$

with WR = wheel rate
SR = spring rate
MR = suspension motion ratio

$$WR_f = \frac{320}{1.373^2} = 170 \text{ kg/cm} \quad \textit{(Eq. 8.17)}$$

$$WR_r = \frac{280}{1.725^2} = 95 \text{ kg/cm} \quad \textit{(Eq. 8.18)}$$

The roll stiffness produced by these wheel rates can be determined using ***Equations 8.19–8.22***.

$$K_{rollfSPRING} = \frac{T_F^2 \cdot WR_f}{4} \cdot \frac{\pi}{180} \quad \textit{(Eq. 8.19)}$$

$$K_{rollfSPRING} = \frac{\pi \cdot 165^2 \cdot 170}{720} = 20194 \text{ kgcm/deg}$$

(Eq. 8.20)

$$K_{rollrSPRING} = \frac{T_R^2 \cdot WR_r}{4} \cdot \frac{\pi}{180} \quad \textit{(Eq. 8.21)}$$

$$K_{rollrSPRING} = \frac{\pi \cdot 168.5^2 \cdot 95}{720} = 11767 \text{ kgcm/deg}$$

(Eq. 8.22)

Using Equations 8.14 and 8.15, the roll stiffness (measured at the wheel) produced by the antiroll bars becomes as given in ***Equations 8.23*** and ***8.24***.

$$K_{rollfARB} = K_{rollf} - K_{rollfSPRINGS} = 55083 - 20194 = 34844 \text{ kgcm/deg}$$

(Eq. 8.23)

$$K_{rollrARB} = K_{rollr} - K_{rollrSPRINGS} = 26642 - 11767 = 14875 \text{ kgcm/deg}$$

(Eq. 8.24)

Equations 8.25 and ***8.26*** show the actual antiroll bar rates.

As an example, an attempt assumedly is made to resolve a steady-state understeer problem with this car, but the overall roll gradient needs to remain the same. To do this, the front axle weight transfer could be decreased (i.e., the front roll stiffness decreased). This requires a higher front roll gradient. The problem is solved by mounting different springs with which the engineer would like to obtain a roll stiffness bias of 65% on the front axle. To maintain the same overall roll gradient, the front and rear springs must be changed. What spring rates must be put on the car?

First, the required roll gradients are determined ***(Equations 8.27*** and ***8.28)***.

$$SR_{rollf} = MR_{rollf}^{\ 2} \cdot K_{rollfARB} = 1.495^2 \cdot 34844 = 77877 \text{ kgcm/deg} \quad (Eq. 8.25)$$

$$SR_{rollR} = MR_{rollR}^{\ 2} \cdot K_{rollrARB} = 1.550^2 \cdot 14875 = 35375 \text{ kgcm/deg} \quad (Eq. 8.26)$$

$$RG_{F(required)} = 0.35 \cdot 2 \cdot 0.4806 = 0.336 \text{ deg/G} \quad (Eq. 8.27)$$

$$RG_{R(required)} = 0.65 \cdot 2 \cdot 0.4806 = 0.625 \text{ deg/G} \quad (Eq. 8.28)$$

$$K_{rollf(required)} = 81725 \cdot \frac{0.625}{0.9612} = 53140 \text{ kgcm/deg} \quad (Eq. 8.29)$$

$$K_{rollr(required)} = 28585 \text{ kgcm/deg} \quad (Eq. 8.30)$$

$$K_{rollfSPRING(req.)} = 53140 - 34844 = 18296 \text{ kgcm/deg} \quad (Eq. 8.31)$$

$$WR_{f(req.)} = 162.4 \text{ kg/cm} \quad (Eq. 8.32)$$

$$SR_{f(req.)} = 162.4 \cdot 1.373^2 = 306 \text{ kg/cm} \quad (Eq. 8.33)$$

$$K_{rollrSPRING(req.)} = 28585 - 14875 = 12710 \text{ kgcm/deg} \quad (Eq. 8.34)$$

$$WR_{r(req.)} = 102.6 \text{ kg/cm} \quad (Eq. 8.35)$$

$$SR_{r(req.)} = 102.6 \cdot 1.725^2 = 305 \text{ kg/cm} \quad (Eq. 8.36)$$

The total roll stiffness remains the same. The required distribution between front and rear axle is as shown in ***Equations 8.29*** and ***8.30***.

Assuming that no changes are made to the antiroll bars, the required spring rates can be calculated ***(Equations 8.31–8.36)***.

In this example, the roll stiffness distribution is shifted forward by changing the front and rear spring rates, which is required to achieve the desired roll gradients. In addition, the stiffness of the antiroll bars is measured. Different setup changes require small variations in the previous calculations.

Front and Rear Roll Angle Ratio

Another reference number that tells something about the roll stiffness distribution is the roll ratio (ζ) between the rear and front roll angle ***(Equation 8.37)***.

$$\varsigma = \frac{\alpha_{rollR}}{\alpha_{rollF}} \quad (Eq. 8.37)$$

with α_{rollR} = rear suspension roll angle
α_{rollF} = front suspension roll angle

This ratio also can be calculated from the front and rear roll gradients using ***Equation 8.38***.

$$\varsigma = \frac{RG_R}{RG_F} \quad (Eq. 8.38)$$

If the software allows, the relationship between the front and rear roll angle also can be illustrated in an *X-Y* graph as shown in ***Figure 8.5***. Note that this is usually a linear relationship (as the roll ratio is constant), except for progressive suspension systems. The graph shows the same pattern of scatter as the roll gradient *X-Y* graphs for the same reasons.

In Figure 8.5, the ratio between the rear and front roll angle is 2.02. This means that in this particular situation the front roll angle is about half the rear roll angle. Remember that *suspension roll angle* is being addressed. The total vehicle roll angle is the sum of the suspension roll angle and the roll angle induced by the tire spring rates.

Assuming an infinitely stiff chassis, ***Equation 8.39*** is true, with α_{rollF} and α_{rollR} being the front and rear suspension roll angle, respectively, and $\alpha_{rolltiresF}$ and $\alpha_{rolltiresR}$ the roll angles induced by the front and rear pairs of tires, respectively.

$$\alpha_{rollR} + \alpha_{rolltiresF} = \alpha_{rollR} + \alpha_{rolltiresR} \qquad \textit{(Eq. 8.39)}$$

Expressing the rear suspension roll angle as a function of the front suspension roll angle produces ***Equation 8.40***.

$$\alpha_{rollF} + \alpha_{rolltiresF} = \varsigma \cdot \alpha_{rollF} + \alpha_{rolltiresR} \qquad \textit{(Eq. 8.40)}$$

From this another expression for ζ follows ***(Equation 8.41)***.

$$\varsigma = \left(\frac{\alpha_{rolltiresF} - \alpha_{rolltiresR}}{\alpha_{rollF}} \right) + 1 \qquad \textit{(Eq. 8.41)}$$

With this expression, the effect of roll stiffness variations on one axle on ζ can be investigated. The possibilities are summarized in ***Table 8.5***.

Table 8.5 *Changing the roll stiffness on one vehicle axle, either on the suspension or the tires, and its effects on* ζ

	INCREASE	DECREASE
Front roll stiffness		
Suspension	ζ increases	ζ decreases
Tires	ζ decreases	ζ increases
Rear roll stiffness		
Suspension	ζ decreases	ζ increases
Tires	ζ increases	ζ decreases

From Equations 8.38 and 8.41, it can be concluded that when the tire spring rates change (due to a different construction or running a different tire pressure), the suspension roll gradients change also.

For example, Table 8.1 had the following roll angles:

Front	Suspension	0.22 deg
	Tires	0.28 deg
	Total	0.50 deg
Rear	Suspension	0.32 deg
	Tires	0.19 deg
	Total	0.50 deg

So ζ = 0.32 / 0.22 = 1.45.

Now, the rear tire spring rate is increased by 10% so instead of 453 kg/cm, this result becomes 498 kg/cm. Running the calculation again using the SusProg3D software offers the following results:

Front	Suspension	0.21 deg
	Tires	0.28 deg
	Total	0.49 deg
Rear	Suspension	0.32 deg
	Tires	0.17 deg
	Total	0.49 deg

ζ becomes 0.32 / 0.21 = 1.52 and has increased by 4.8%.

To correct this back to the old roll ratio by modifying the spring rate, the rear wheel rate needs to be increased by 25% (to 120 kg/cm) to approach the old roll ratio.

Front	Suspension	0.21 deg
	Tires	0.28 deg
	Total	0.49 deg
Rear	Suspension	0.30 deg
	Tires	0.20 deg
	Total	0.49 deg

ζ becomes 0.30 / 0.21 = 1.43.

In this particular example, a 10% change to the tire spring rates equals a 25% change to the car's spring rate!

Figure 8.5 *Front roll angle versus rear roll angle*

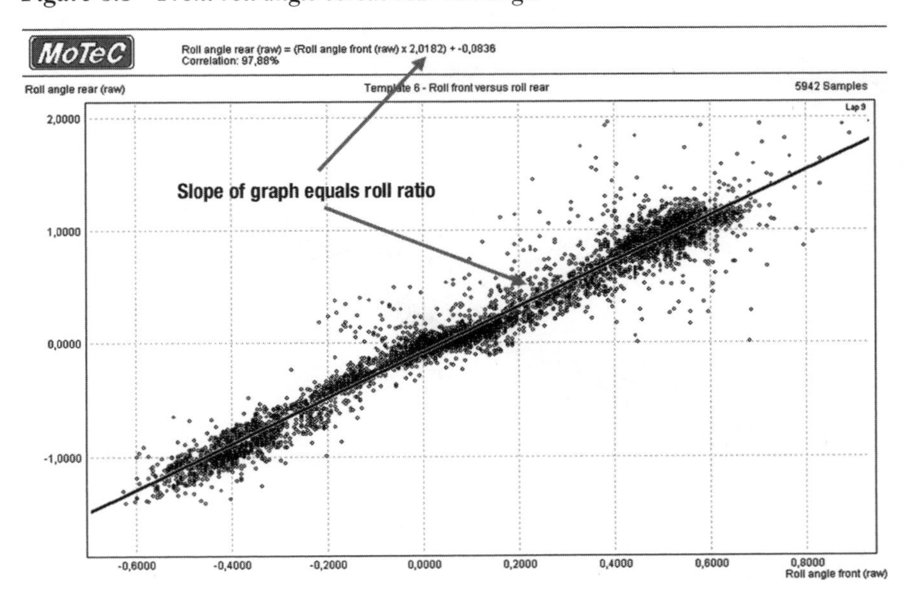

Using the Roll Ratio as a Setup Tool

The practical application of the roll angle ratio is illustrated using a real-world example. ***Figures 8.6*** and ***8.7*** give the roll ratio (ζ) before and after a change to the front antiroll bar setting. The first graph shows a lap done by Bert Longin in a Dodge Viper on the Circuit Zolder racetrack. In this situation, the front antiroll bar was in full soft position. In the second graph, this antiroll bar was adjusted to full hard, changing the vehicle roll stiffness distribution by quite a margin.

The roll gradients are not illustrated in a graph, but ***Table 8.6*** summarizes the three calculated stiffness characteristics.

Figure 8.6 Front versus rear roll angle during a lap around Zolder in a Dodge Viper GTS-R, driven by Bert Longin; front antiroll bar in softest position

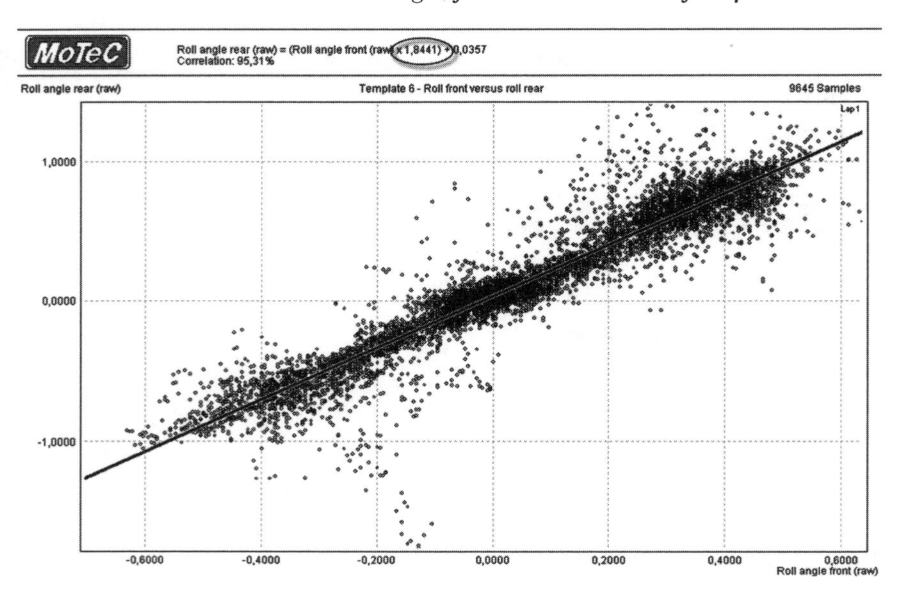

Figure 8.7 Same car, same driver, and same set of tires, now with the front antiroll bar in its hardest position

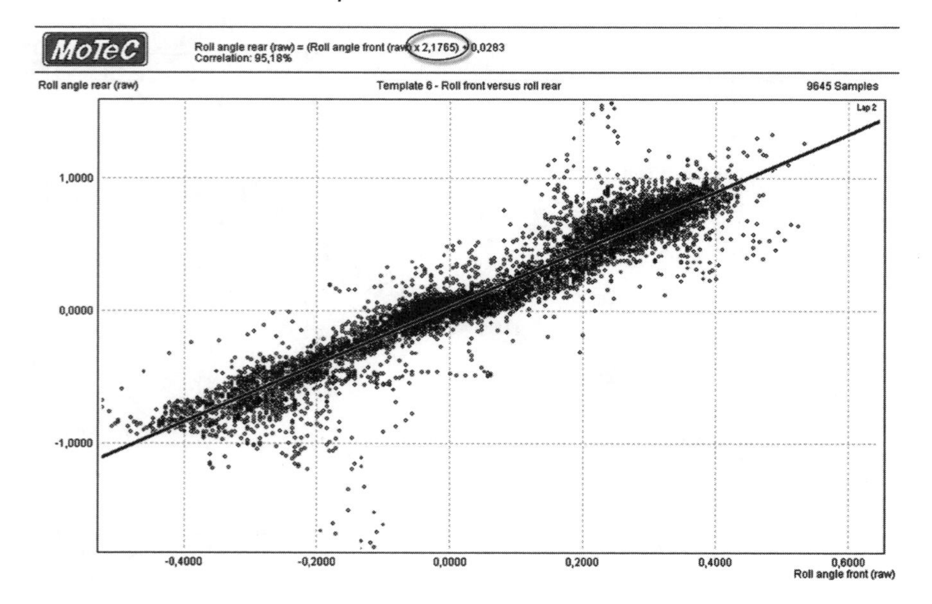

Table 8.6 *Measured roll gradients and roll ratio before and after the setup change*

	FARB soft	FARB hard
Front roll gradient (RG_F)	0.2754	0.2253
Rear roll gradient (RG_R)	0.5386	0.5137
Roll ratio (ζ)	1.8441	2.1765

The charts in Figures 8.6 and 8.7 also show that by changing the front antiroll bar stiffness from its minimum to maximum the front maximum roll angle decreases approximately 0.1 deg, while the rear maximum roll angle remains the same.

All the equations in the previous paragraphs are valid when chassis torsion rates are not taken into consideration. In practice, the chassis spring rate may not be high enough to be of no importance in these calculations. Chassis torsional spring rates influence the roll stiffness distribution of the vehicle and, therefore, the roll gradients change as well. However, when comparing setups within one vehicle, the chassis spring rate can be ignored because normally it is a parameter in the vehicle configuration that does not change.

Suspension Troubleshooting

In the previous sections, linear functions were created from some suspension parameters. In the case where a progressive suspension is applied, these functions are not linear. Where a linear suspension is applied and roll gradients or roll angle ratio are not constants, something is wrong. ***Figure 8.8*** shows an example of this situation.

In this example, there is obviously a problem with the suspension. The rear roll angle remains zero until the front angle reaches 0.1 deg. In this particular example, the rear antiroll bar blade was broken.

Pitch Gradient

Similar to the roll gradient, it is possible to calculate a gradient for the vehicle's pitch movement. The pitch angle can be calculated from the suspension position channels and expressed as a function of longitudinal *g*-force with β_{pitch} being the vehicle's pitch angle measured at the chassis centerline and G_{long} being the longitudinal acceleration acting on the car's center of gravity ***(Equation 8.42)***.

$$PG = \frac{\beta_{pitch}}{G_{long}} \qquad (Eq.\ 8.42)$$

Unlike the roll gradient, the pitch gradient is not dependent on chassis torsion (the pitch angle being determined at the vehicle's longitudinal centerline). Pitch gradient also is not commonly a linear function when antisquat and antidive suspension geometry is applied. It is, however, a good way to visualize different setups graphically and characterize a suspension. ***Figure 8.9*** shows an *X-Y* plot of the pitch angle against longitudinal acceleration that clearly illustrates this.

Figure 8.8 *Roll angle ratio affected by a broken rear antiroll bar blade*

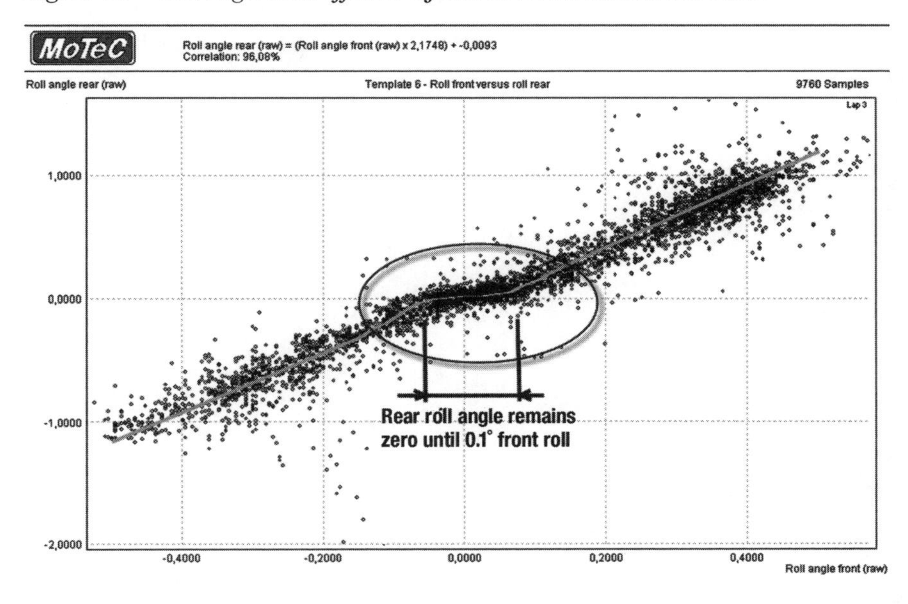

Figure 8.9 *Pitch gradient*

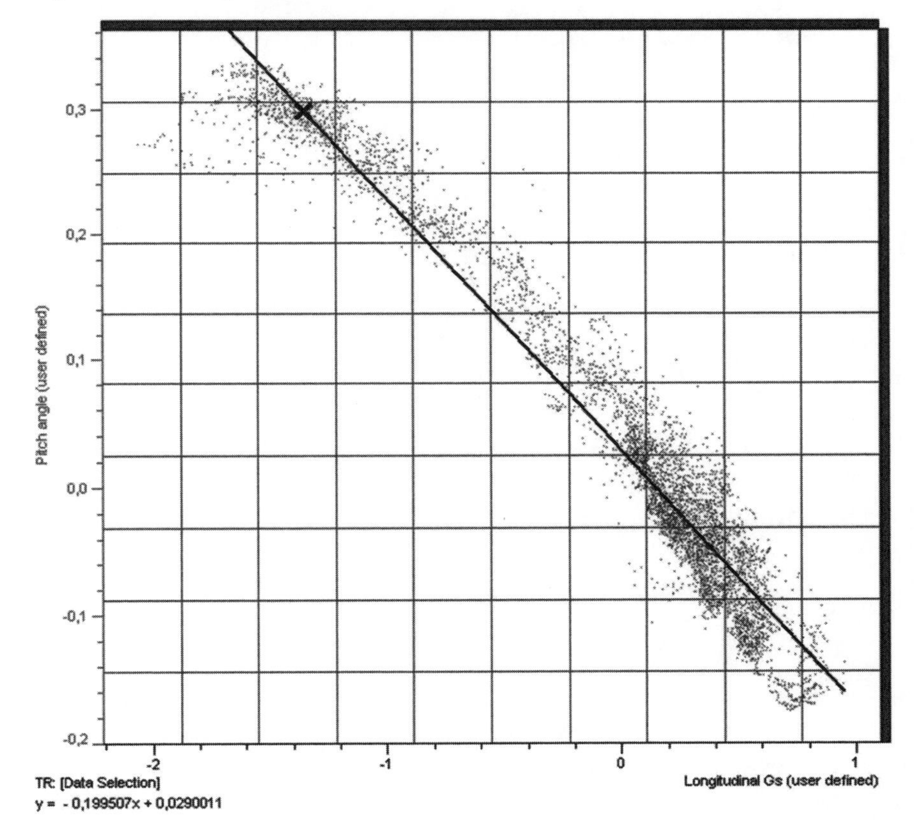

CHAPTER 9

WHEEL LOADS AND WEIGHT TRANSFER

The loads on each wheel determine a vehicle's maximum cornering capability. This chapter shows how to measure and calculate the dynamic wheel loads resulting from these forces.

Lateral Weight Transfer

During cornering, a car accelerates toward the center of the corner by the tire contact patch forces. The vehicle mass resists this acceleration with a force acting at its center of gravity. This inertial (or centrifugal) force generates lateral weight transfer. This situation is illustrated in ***Figure 9.1*** for a right-hand turn, where the cornering forces (F_L and F_R) result in a lateral acceleration (G_{Lat}). The reacting inertial force is calculated as ***Equation 9.1***.

$$F = W \cdot G_{Lat} \qquad \textit{(Eq. 9.1)}$$

By taking the moments about one of the wheel centers, the weight transfer due to cornering is found ***(Equation 9.2)***.

$$\Delta W_{Lat} = \frac{W \cdot G_{Lat} \cdot h}{T} \qquad \textit{(Eq. 9.2)}$$

where
- W = vehicle weight
- ΔW_{Lat} = increase in left side load and decrease in right side load
- h = center of gravity height from ground
- T = track width
- G_{Lat} = lateral acceleration at center of gravity

Figure 9.1 ***Lateral weight transfer (right-hand turn)***

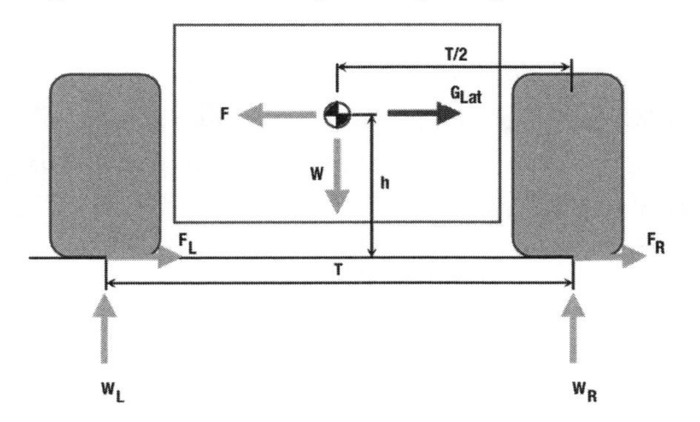

Equation 9.2 indicates that the total lateral weight transfer is proportional to the lateral acceleration and the center of gravity height and inversely proportional to the track width. This means that weight transfer increases with the weight of the car and the height of the center of gravity and decreases with a greater track width.

The center of gravity height can be determined by first weighing the car on a level surface and then raising its rear end and weighing the front end again ***(Figure 9.2)***. The following procedure should be observed:

- Each shock absorber should be replaced by a solid link to eliminate any suspension travel. The length of these links should be dimensioned carefully to put the car at its exact static ride height. Inaccuracy here can influence greatly the scale readings.
- The tires should be overinflated as much as possible to limit sidewall flexing.
- All fluids should be at the right level.
- A driver should be in the car (or at least an object that is equivalent to the driver's weight).
- The total weight on the front axle should be measured while the car is on a level surface.
- The rear end of the car should be raised as much as possible. The calculation is based on the change in weight on the front wheels

Figure 9.2 ***Determining the center of gravity height requires the rear end of the car to be raised by a substantial amount.***

in relation to the angle to which the car is raised. The higher the distance the car can be raised, the bigger the shift in front wheel weight and the more accurate the height calculation.

- The height of the center of gravity above the axle centerlines now can be calculated using ***Equation 9.3***. ***(Figure 9.3)***

$$h = \frac{WB \cdot \Delta W}{W \cdot \tan\alpha} \qquad \textit{(Eq. 9.3)}$$

with WB = wheelbase
W = total vehicle weight
$\Delta W = W_{f2} - W_{f1}$
W_{f1} = front weight measured on level surface
W_{f2} = front weight measured with raised rear axle
$\tan\alpha = B/A$

$$A = \sqrt{WB^2 - B^2}$$

- To obtain the height of the center of gravity above the ground plane, the tire radius should be added to the calculated height above axle centerlines (h).
- Often the center of gravity is close to the wheel center height, and the change in front weight as the car is jacked up is small. To get the best results, measure at different jacking heights and average the results.

Equation 9.2 defines the total amount of lateral weight transfer of a car during cornering.

Figure 9.3 Center of gravity height calculation

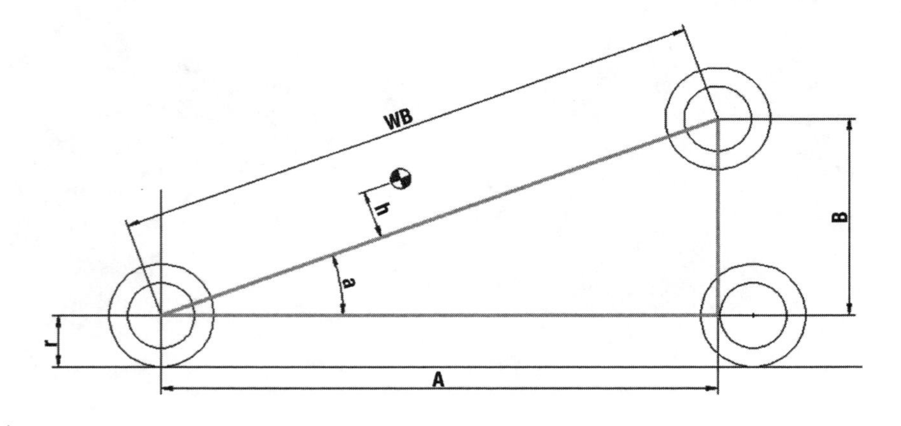

There are three mechanisms that distribute the total lateral weight transfer over the front and rear axle.

- unsprung weight transfer (ΔW_{uF} and ΔW_{uR})
- geometric weight transfer (ΔW_{gF} and ΔW_{gR})
- sprung weight transfer (ΔW_{sF} and ΔW_{sR})

The suspension splits the total weight of the vehicle into sprung weight (chassis, bodywork, and driveline) and unsprung weight (tires, wheels, brakes, and about half of the suspension links and driveshafts). Because the weight transfer of the unsprung mass is not influenced by the roll stiffness of the suspension, it is calculated separately.

Understand that this portion of the total lateral weight transfer is *not* measured by any suspension load cells nor can it be calculated from the suspension potentiometer readings. It is, however, calculable using ***Equations 9.4*** and ***9.5***.

$$\Delta W_{uF} = \frac{W_{uF} \cdot G_{Lat} \cdot h_F}{T_F} \qquad \textit{(Eq. 9.4)}$$

$$\Delta W_{uR} = \frac{W_{uR} \cdot G_{Lat} \cdot h_R}{T_R} \qquad \textit{(Eq. 9.5)}$$

with W_{uF} = front unsprung weight
W_{uR} = rear unsprung weight
h_F = front unsprung weight center of gravity height
h_R = rear unsprung weight center of gravity height
T_F = front track width
T_R = rear track width

Geometric weight transfer results from a direct application of the tire forces to the chassis through the front and rear roll centers. In addition, this part of the lateral weight transfer cannot be measured with the suspension load cells or potentiometers. However, it can be calculated if the front and rear roll center locations are known ***(Equations 9.6*** and ***9.7)***.

$$\Delta W_{gF} = \frac{W_{sF} \cdot G_{Lat} \cdot \left[\frac{a}{WB}\right] \cdot h_{RCf}}{T_F} \qquad \textit{(Eq. 9.6)}$$

$$\Delta W_{gR} = \frac{W_{sR} \cdot G_{Lat} \cdot \left[\frac{WB - a}{WB}\right] \cdot h_{RCr}}{T_R} \quad (Eq.\ 9.7)$$

with W_{sF} = front sprung weight
W_{sR} = rear sprung weight
h_{RCf} = front roll center height from ground plane
h_{RCr} = rear roll center height from ground plane
WB = wheelbase
a = distance between rear axle centerline and sprung mass center of gravity

Sprung weight transfer makes the chassis roll during cornering. It is derived from the suspension springs, shock absorbers, and antiroll bars, whereas the geometric weight transfer is obtained from the suspension links. Sprung weight transfer can be measured with suspension load cells or calculated from the suspension potentiometer signals.

The total sprung weight transfer can be determined with ***Equation 9.8***.

$$\Delta W_s = \frac{W_s \cdot G_{Lat} \cdot h_{roll}}{T} \quad (Eq.\ 9.8)$$

where W_s = total sprung weight = $W_{sF} + W_{sR}$
h_{roll} = distance between roll axis and sprung center of gravity

The distribution between the front and rear axles depends on the roll stiffness distribution of the suspension, so front and rear lateral sprung weight transfer is as shown ***Equations 9.9–9.11***.

$$\Delta W_{sF} = \frac{W_s \cdot G_{Lat} \cdot h_{roll}}{T_F} \cdot q \quad (Eq.\ 9.9)$$

$$\Delta W_{sR} = \frac{W_s \cdot G_{Lat} \cdot h_{roll}}{T_R} \cdot (1 - q) \quad (Eq.\ 9.10)$$

$$q = \frac{K_{rollf}}{K_{rollf} + K_{rollr}} = \frac{RG_R}{RG_F + RG_R} \quad (Eq.\ 9.11)$$

with K_{rollf} = roll stiffness front axle (at the wheels)
K_{rollr} = roll stiffness rear axle (at the wheels)
RG_F = front roll gradient
RG_R = rear roll gradient

Finally, the total lateral weight transfer is determined by adding all the portions together ***(Equation 9.12)***.

$$\Delta W_{Lat} = \Delta W_{uF} + \Delta W_{uR} + \Delta W_{gF} + \Delta W_{gR} + \Delta W_{sF} + \Delta W_{sR} \quad (Eq.\ 9.12)$$

Longitudinal Weight Transfer

Similar to the centrifugal force that occurs when cornering, an inertial reaction force develops at the center of gravity when the car is subjected to longitudinal acceleration. This inertial force creates a longitudinal weight transfer equal to ***Equation 9.13***.

$$\Delta W_{Long} = \frac{W \cdot G_{Long} \cdot h}{WB} \quad (Eq.\ 9.13)$$

ΔW_{Long} is the decrease in front axle weight and the correspondent increase in rear axle weight.

Banking and Grade Effects

As discussed earlier in this chapter, the effects of lateral weight transfer were investigated on a flat track surface. The picture changes somewhat when the car negotiates a banked corner, as illustrated in ***Figure 9.4***. The gravitational force (W) can be resolved into a force parallel to the road surface ($W \cdot \sin \alpha$) and perpendicular to it ($W \cdot \cos \alpha$). The centrifugal force (here named $G_{Lat\alpha}$ to indicate the banking angle) that exists during cornering on a banked corner is no different than what occurs when the car is on a horizontal surface. However, this is not the lateral acceleration measured by the accelerometer in the vehicle. There is also a component of the centrifugal force along an axis perpendicular to the road surface ***(Equations 9.14*** and ***9.15)***.

$$G_{LatMeasured} = G_{Lat\alpha} \cdot \cos \alpha \quad (Eq.\ 9.14)$$

$$G_{Vert} = G_{Lat\alpha} \cdot \sin \alpha \quad (Eq.\ 9.15)$$

where $G_{LatMeasured}$ = lateral acceleration measured by vehicle accelerometer
G_{Vert} = vertical acceleration measured by vehicle accelerometer
α = banking angle

This means that the wheel loads are influenced in two ways. First, the weight force perpendicular to the road is ***Equation 9.16***.

$$W \cdot \left(G_{Lat\alpha} \cdot \sin\alpha + \cos\alpha\right) \qquad \textit{(Eq. 9.16)}$$

Second, the inertial force creating the lateral weight transfer is ***Equation 9.17***.

$$W \cdot \left(G_{Lat\alpha} \cdot \cos\alpha + \sin\alpha\right) \qquad \textit{(Eq. 9.17)}$$

The banking angle can be determined by ***Equation 9.18***.

$$\alpha = \text{Arc}\tan\left[\frac{G_{Vert} - 1}{G_{Lat}}\right] \qquad \textit{(Eq. 9.18)}$$

Figure 9.4
The effects of banking

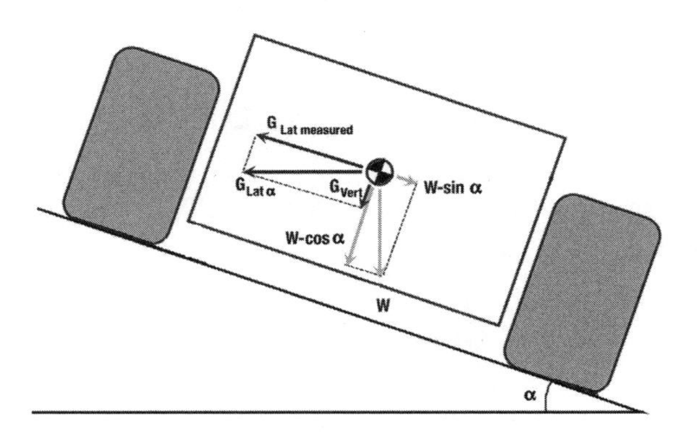

***Figure 9.5** A lap around Phoenix International Raceway (PIN) by a 2001 Indycar (Courtesy of Pi Research, England)*

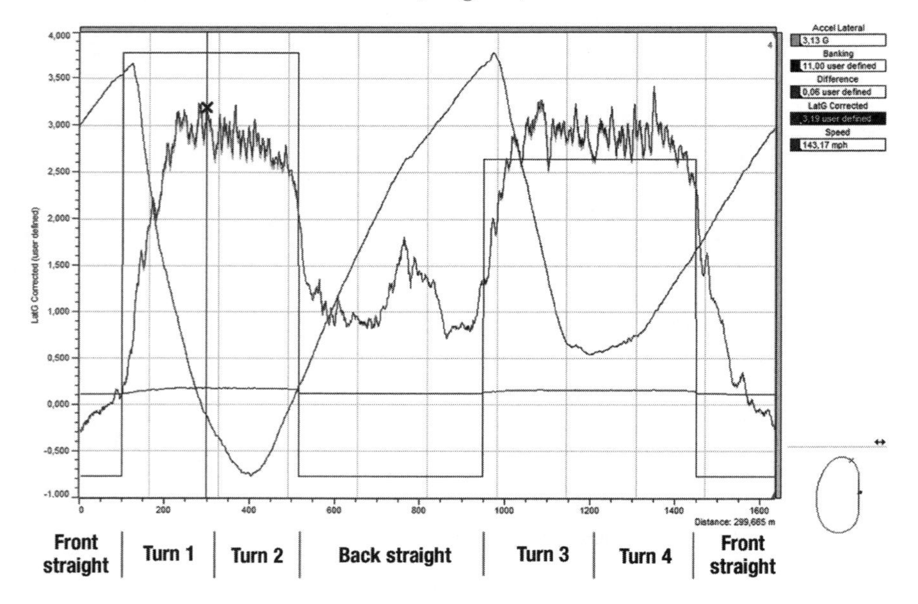

Figure 9.5 shows an example of a 2001 Indycar taking a lap around Phoenix International Raceway. The following channels are indicated:

- Speed
- Banking—This block-shaped trace defines the banking of the track as a function of distance. The banking figures were taken from ***Figure 9.6***.
- Accel Lateral—This is the measured lateral acceleration of the vehicle.
- LatG Corrected—This is a mathematical channel that defines the real lateral acceleration experienced by the vehicle by dividing the measured acceleration by the cosine of the banking angle.
- Difference—This mathematical channel calculates the difference between actual and measured lateral acceleration.

At the cursor point in ***Figure 9.5,*** the measured lateral acceleration is 3.13 G. At this point on the track, the banking angle is 11 deg. ***Equation 9.19*** shows the real lateral acceleration experienced by the vehicle.

$$G_{Lat\alpha} = \frac{G_{Latmeasured}}{\cos(\alpha)} = \frac{3.13}{\cos(11°)} = 3.19\ G \qquad \textit{(Eq. 9.19)}$$

A similar effect takes place when the vehicle runs up or down a sloped road, as pictured in ***Figure 9.7***. Depending on the size of the slope angle θ, the total normal force is reduced to W·cos θ. In addition, because of a weight component parallel to the road surface, the slope angle creates a

***Figure 9.6** PIN turn nomenclature and banking angles*

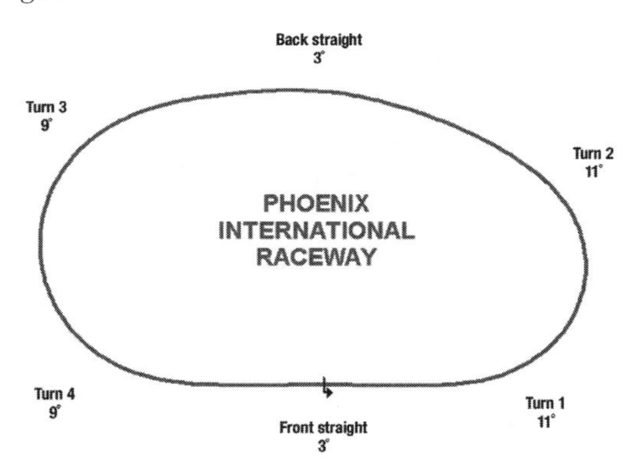

longitudinal weight transfer. The front and rear axle weights (W_F and W_R) are as given in ***Equations 9.20*** and ***9.21***.

$$W_F = W \cdot \frac{a}{WB} \cdot \cos\theta - W \cdot \frac{h}{WB} \cdot \left(G_{Long} + \sin\theta\right) \quad \text{(Eq. 9.20)}$$

$$W_R = W \cdot \frac{WB - b}{WB} \cdot \cos\theta + W \cdot \frac{h}{WB} \cdot \left(G_{Long} + \sin\theta\right) \quad \text{(Eq. 9.21)}$$

with a = distance between rear axle centerline and center of gravity
WB = wheelbase
G_{Long} = longitudinal acceleration

To calculate the slope angle, ***Equation 9.22*** applies.

$$\theta = \text{Arc}\tan\left[\frac{G_{Vert} - 1}{G_{Long}}\right] \quad \text{(Eq. 9.22)}$$

Total Wheel Loads

To calculate the individual loads on each wheel, the effect of each following situation must be determined:

- static weight distribution,
- lateral weight transfer,
- longitudinal weight transfer,
- banking effects,
- track slope effects, and
- aerodynamic forces.

The total load on each wheel is the sum of all the above-mentioned effects. To illustrate this, consider a vehicle with the properties given in ***Table 9.1***.

Figure 9.8 shows the speed and the lateral and longitudinal acceleration of this car exiting a slow corner. The wheel loads for the acceleration values at the cursor point are calculated. The aerodynamic forces are not considered in this example. More on aerodynamics follows in Chapter 11.

Use ***Equations 9.23*** and ***9.24*** for unsprung weight transfer. Geometric weight transfer can be calculated with ***Equations 9.25–9.29***. Sprung weight transfer is shown in ***Equations 9.30–9.32***. ***Equation 9.33*** is longitudinal weight transfer (per wheel = 55.95 kg). ***Table 9.2*** summarizes the results.

Figure 9.7 *The effects of track slope*

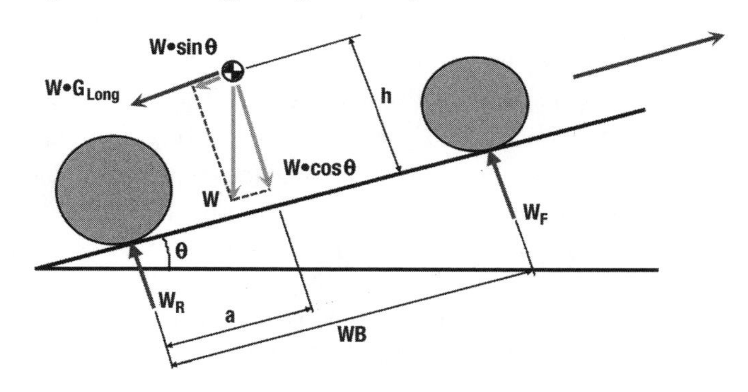

Table 9.1 *Vehicle properties*

Vehicle weight	W = 1375 kg	W_f = 636 kg	W_r = 739 kg
Unsprung mass	W_u = 159 kg	W_{uF} = 61 kg	W_{uR} = 98 kg
Track width	T_F = 1.649 mm	T_R = 1684 mm	
Wheelbase	WB = 2403 mm		
Center of gravity height	h = 369 mm		
Roll center height	h_{RCf} = 46.5 mm	h_{RCr} = 51.0 mm	h_{roll} = 323.6 mm
Roll gradient	RG_F = 0.313 deg/G	RG_R = 0.647 deg/G	
Banking angle	0 deg		
Track slope	0 deg		

Figure 9.8 *Speed and lateral and longitudinal Gs during a corner exit*

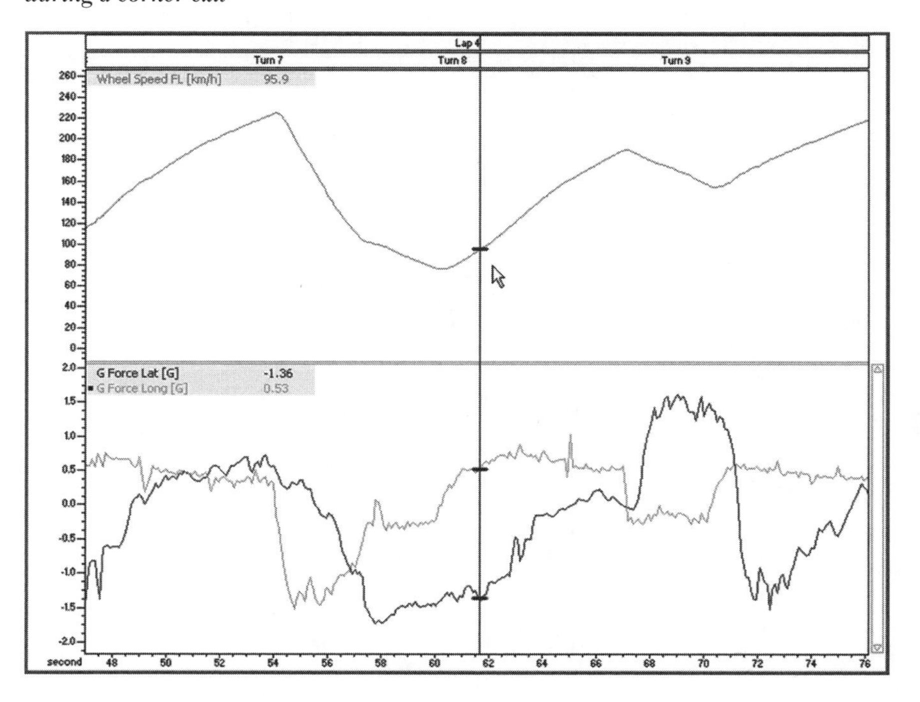

These equations can be used to create a mathematical channel in the data analysis software for each separate wheel load. To ensure sufficient accuracy, have detailed knowledge of the necessary vehicle parameters. For a more accurate picture of how the wheel loads are developed, some effects in Table 9.2 should be measured directly. How this is achieved follows in the next sections.

Determining Wheel Loads with Modal Analysis

Without considering camber change, toe change, and other wheel movements, an individual suspension corner basically lets the wheel go up and down. A 4-wheel suspension system can have combinations of compression and extension at either wheel corner. Any position of or motion in this system can be described in millimeters of one or more of four fundamental modes—heave, roll, pitch, and warp.[11] Although closely related, the suspension system modes do not describe the motions of the suspended mass, which has six degrees of freedom (heave, pitch, roll, yaw, sideslip, and forward and rearward motion). If the chassis is not considered infinitely stiff, warp is the seventh degree of freedom (chassis torsion). This can be envisioned as the tire contact patches moving in heave, roll, pitch, and warp as the vehicle moves over the track. The chassis also moves in these four modes, while suspension motions make up the difference.

$$\Delta W_{uF} = \frac{W_{uF} \cdot G_{Lat} \cdot h_F}{T_F} = \frac{61 \cdot (-1.36) \cdot 325}{1649} = -16.35 \text{ kg} \quad \textit{(Eq. 9.23)}$$

$$\Delta W_{uR} = \frac{W_{uR} \cdot G_{Lat} \cdot h_R}{T_R} = \frac{98 \cdot (-1.36) \cdot 325}{1649} = -26.27 \text{ kg} \quad \textit{(Eq. 9.24)}$$

$$W_{SF} = W_f - w_{UF} = 575 \text{ kg} \quad \textit{(Eq. 9.25)}$$

$$W_{sR} = W_r - W_{uR} = 641 \text{ kg} \quad \textit{(Eq. 9.26)}$$

$$a = WB \cdot \frac{W_r}{W} = 1292 \text{ mm} \quad \textit{(Eq. 9.27)}$$

$$\Delta W_{gF} = \frac{W_{sF} \cdot G_{Lat} \cdot \left[\frac{a}{WB}\right] \cdot h_{RCf}}{T_F} = \frac{575 \cdot (-1.36) \cdot \left[\frac{1292}{2403}\right] \cdot 46.5}{1649} = -11.85 \text{ kg} \quad \textit{(Eq. 9.28)}$$

$$\Delta W_{gR} = \frac{W_{sR} \cdot G_{Lat} \cdot \left[\frac{WB - a}{WB}\right] \cdot h_{RCr}}{T_R} = \frac{641 \cdot (-1.36) \cdot \left[\frac{2403 - 1292}{2403}\right] \cdot 51.0}{1684} = -12.21 \text{ kg} \quad \textit{(Eq. 9.29)}$$

$$q = \frac{RG_R}{RG_F + RG_R} = \frac{0.647}{0.313 + 0.647} = 0.674 \quad \textit{(Eq. 9.30)}$$

$$\Delta W_{sF} = \frac{W_s \cdot G_{Lat} \cdot h_{roll}}{T_F} \cdot q = \frac{1216 \cdot (-1.36) \cdot 323.6}{1649} \cdot 0.674 = -218.7 \text{ kg} \quad \textit{(Eq. 9.31)}$$

$$\Delta W_{sR} = \frac{W_s \cdot G_{Lat} \cdot h_{roll}}{T_R} \cdot (1 - q) = \frac{1216 \cdot (-1.36) \cdot 323.6}{1684} \cdot (1 - 0.674) = -103.6 \text{ kg} \quad \textit{(Eq. 9.32)}$$

$$\Delta W_{Long} = \frac{W \cdot G_{Long} \cdot h}{WB} = \frac{1375 \cdot 0.53 \cdot 369}{2403} = 111.9 \text{ kg} \quad \textit{(Eq. 9.33)}$$

Each mode has a spring rate and a damping rate, depending on the suspension system's components. The car springs provide the basic wheel rates for the four modes. Roll bars, third springs, and the chassis add to the wheel rate in some modes while leaving others unaffected. These rates are not necessarily linear, but for simplification they are considered to be so in the rest of this chapter. The shock absorbers provide the damping rates for all four modes. In the case of a conventional suspension with four shock absorbers, the damping rate for all four modes is equal.

Heave is the synchronous motion of all four wheels, all in the same direction ***(Figure 9.9)***. The total force developed by heave motion is the sum of the vertical forces acting on each wheel as a result of the sum of each wheel displacement. For this motion, an elasticity constant (K_H) can be defined that relates the heave force to the wheel displacements ***(Equation 9.34)***.

Pitch is a synchronous motion where front and rear wheel pairs move in opposite direction ***(Figure 9.10)***. In this case, the elasticity constant (K_P) relates the total pitch force to the pitch displacement ***(Equation 9.35)***.

Roll is also a synchronous oppositional motion but between the left and right wheel pairs ***(Figure 9.11)***. The roll rate (K_R) is the ratio between the total roll force and the roll displacement ***(Equation 9.36)***.

Finally, warp is the synchronous motion of diagonal wheel pairs in opposite directions ***(Figure 9.12)***. The warp rate (K_W) establishes the relationship as shown in ***Equation 9.37***.

Illustrating how the four modes can be composed from the individual wheel movements, ***Figure 9.13*** features four recorded wheel-travel signals of a vehicle going through a right-hand corner. In this case, compression is indicated with a positive sign, while extension is negative. The cursor is pointed at one particular point where the driver is starting to accelerate out of a corner. The car is simultaneously running off a curb with the left-rear wheel. The wheel positions are as follows:

left front x_{LF}	9.12-mm compression
right front x_{RF}	3.49-mm extension
left rear x_{LR}	37.29-mm compression
right rear x_{RR}	6.46-mm compression

Table 9.2 *Wheel load summary*

	Left-front	Right-front	Left-rear	Right-rear
Static weight	318	318	369.5	369.5
Unsprung weight transfer	16.3	–16.3	26.3	–26.3
Geometric weight transfer	11.9	–11.9	12.21	–12.21
Sprung weight transfer	218.7	–218.7	103.6	–103.6
Longitudinal weight transfer	–55.9	–55.9	55.9	55.9
Banking	0	0	0	0
Track slope	0	0	0	0
Total	508.7 kg	15.1 kg	567.5 kg	283.6 kg

$$F_{LF} + F_{RF} + F_{LR} + F_{RR} = K_H \cdot (x_{LF} + x_{RF} + x_{LR} + x_{RR}) \quad \textit{(Eq. 9.34)}$$

$$F_{LF} + F_{RF} - F_{LR} - F_{RR} = K_P \cdot (x_{LF} + x_{RF} - x_{LR} - x_{RR}) \quad \textit{(Eq. 9.35)}$$

$$F_{LF} - F_{RF} + F_{LR} - F_{RR} = K_R \cdot (x_{LF} - x_{RF} + x_{LR} - x_{RR}) \quad \textit{(Eq. 9.36)}$$

$$F_{LF} - F_{RF} - F_{LR} + F_{RR} = K_W \cdot (x_{LF} - x_{RF} - x_{LR} + x_{RR}) \quad \textit{(Eq. 9.37)}$$

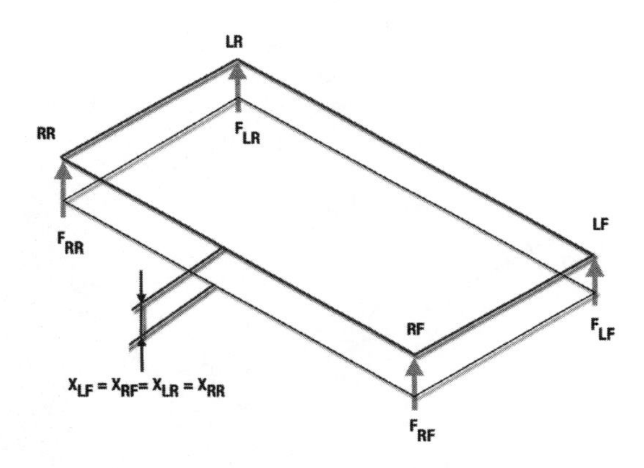

Figure 9.9
Heave

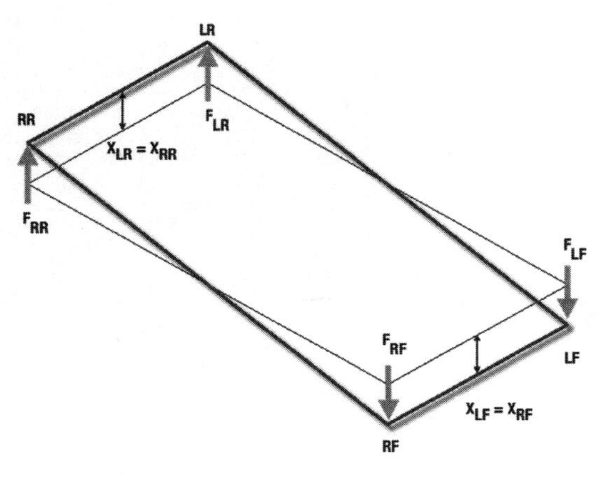

Figure 9.10
Pitch

These are graphically indicated in ***Figure 9.14***. The four-mode displacements are calculated as follows:

Heave	**9.12 + (–3.49) + 37.29 + 6.46 = 49.38 mm**
Pitch	**9.12 + (–3.49) – 37.29 – 6.46 = –38.12 mm**
Roll	**9.12 – (–3.49) + 37.29 – 6.46 = 43.44 mm**
Warp	**9.12 – (–3.49) – 37.29 + 6.46 = –18.12 mm**

The negative sign in front of the amount of pitch indicates that the body is moving rearward (i.e., suspension compression is greater on the rear axle). The warp is also negative, which indicates that the cumulative wheel travel on the right front/left rear diagonal is higher than that on the left front/right rear.

The plane formed in Figure 9.14 by the four suspension corners does not indicate the chassis attitude. If the track surface is completely flat (which in this example it clearly is not), the sprung mass experiences 12.3 mm of heave (total suspension heave divided over four corners). It pitches 19.1 mm downward on the rear axle and the same amount upward on the front axle. The sprung mass rolls 21.7 mm, and the chassis twists 9.1 mm.

Figure 9.15 shows the four suspension system modes, calculated from the wheel-travel signals. At the point indicated by the cursor, the following values are given:

heave	**9.52 mm**
pitch	**–7.33 mm**
roll	**36.21 mm**
warp	**–13.75 mm**

To calculate the corresponding wheel positions, ***Equations 9.38–9.41*** apply.

$$x_{LF} = \frac{\text{Heave} + \text{Pitch} + \text{Roll} + \text{Warp}}{4} = 6.16 \text{ mm} \quad \textit{(Eq. 9.38)}$$

$$x_{RF} = \frac{\text{Heave} + \text{Pitch} - \text{Roll} - \text{Warp}}{4} = -5.06 \text{ mm} \quad \textit{(Eq. 9.39)}$$

$$x_{LR} = \frac{\text{Heave - Pitch} + \text{Roll} - \text{Warp}}{4} = 16.70 \text{ mm} \quad \textit{(Eq. 9.40)}$$

Figure 9.11 *Roll*

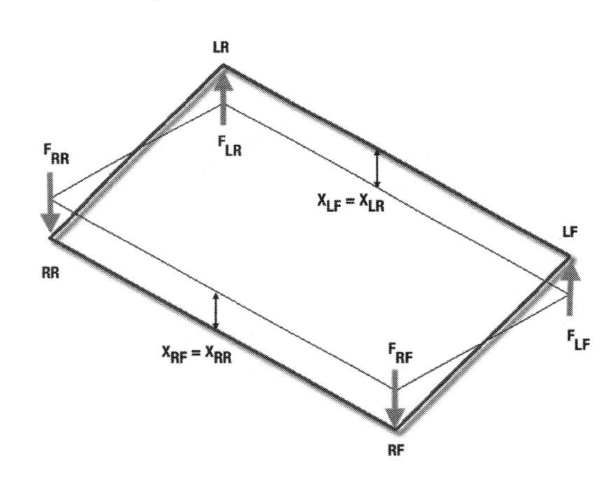

Figure 9.12 *Warp*

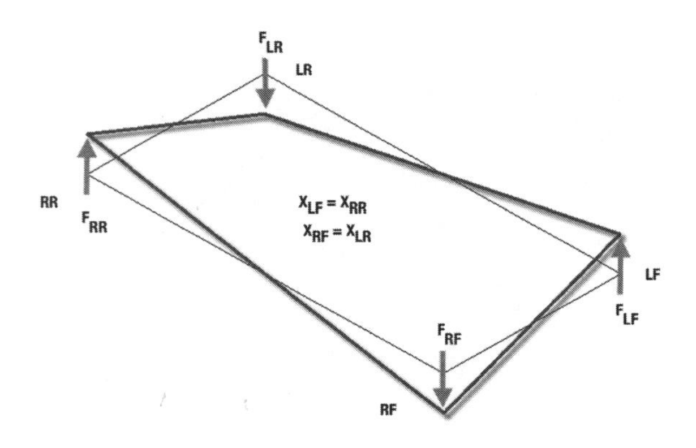

Figure 9.13 ***Wheel travel signals of a vehicle in steady-state cornering and running over a curb***

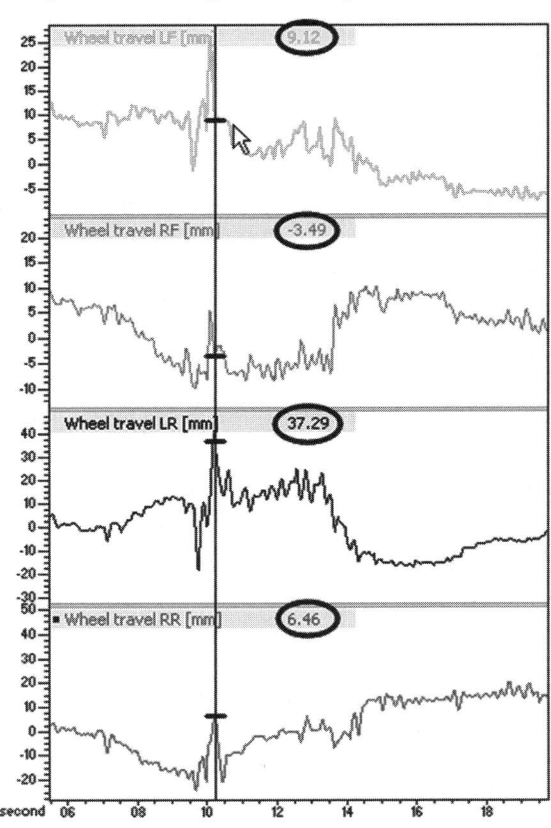

$$x_{RR} = \frac{\text{Heave} - \text{Pitch} - \text{Roll} + \text{Warp}}{4} = -8.27 \text{ mm}$$

(Eq. 9.41)

The values of the four modes can be used to estimate how loads are being transferred between the four wheel corners. Therefore, the wheel rate of each mode must be determined. For a conventional suspension system with four springs and two antiroll bars, the wheel rates for heave and pitch are equal ***(Equation 9.42)***.

$$K_H = K_P = \frac{1}{2} \cdot \left(WR_{SpringF} + WR_{SpringR}\right)$$ *(Eq. 9.42)*

with $WR_{SpringF}$ = wheel rate of front suspension springs

$WR_{SpringR}$ = wheel rate of rear suspension springs

Similarly, if a chassis of infinite stiffness is assumed, the wheel rates for roll and warp are equal as well. In ***Equation 9.43,*** WR_{RollF} = wheel rate of front antiroll bar and WR_{RollR} = wheel rate of rear antiroll bar.

Often the suspension setup features different spring rates and antiroll bar rates on the front and rear axles, and the motion ratio for these components can vary from one axle to another. To calculate the dynamic load distribution over the four suspension corners, the asymmetries between axles must be considered because there is a difference in load transfer between them when roll and warp wheel rates at the front and rear axles are not equal. This effect is one of the most important ways to influence the vehicle's understeer/oversteer balance ***(Equation 9.44)***.

Using either ***Equations 9.45*** and ***9.46*** or the suspension roll gradients mentioned earlier in this chapter (Equation 9.11), the weight transfer bias for roll and warp between the front and rear axles now can be determined.

As previously indicated, on a conventional suspension system, *q* always equals *w* as long as the chassis stiffness is considered infinite. There are, however, suspension systems providing warp rates that are softer than the roll rate.

As an exercise, plug in the values from Table 9.3a and follow the progression through ***Equations 9.47–9.54***.

$$K_R = K_W = \frac{1}{2} \cdot \left(WR_{Spring\,F} + 2 \cdot WR_{RollF}\right) + \frac{1}{2} \cdot \left(WR_{Spring\,R} + 2 \cdot WR_{RollR}\right)$$

(Eq. 9.43)

$$a = b = \frac{1}{2} \cdot \left(WR_{SpringF} + 2 \cdot WR_{RollF}\right) - \frac{1}{2} \cdot \left(WR_{Spring\,R} + 2 \cdot WR_{RollR}\right)$$

(Eq. 9.44)

with a = front-to-rear asymmetry between front and rear axle for roll

b = front-to-rear asymmetry between front and rear axle for pitch

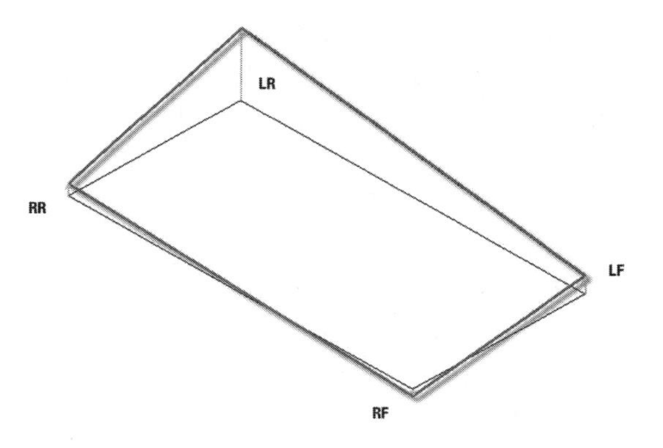

Figure 9.14
Wheel positions at the indicated cursor point in Figure 9.13

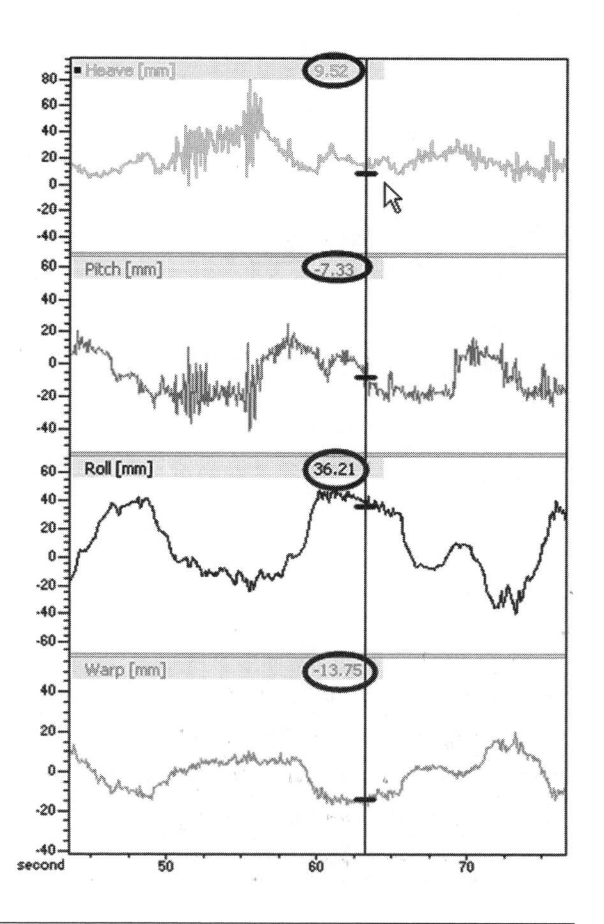

Figure 9.15
Mathematical channels of the four suspension system modes

From ***Equations 9.34–9.37,*** the definition of the wheel load on each suspension corner can be derived as ***Equations 9.55–9.58***.

The right-hand side of these equations can be considered as a constant. The modal displacements can be measured and the wheel rate of each mode calculated following the method covered previously ***(Equations 9.59–9.62)***. From this follows ***Equations 9.63–9.66***.

Solving these equations for F_{LF}, F_{RF}, F_{LR}, and F_{RR} gives ***Equations 9.67–9.70***.

When Equations 9.59–9.62 are substituted in the above, the expressions for each wheel load become ***Equations 9.71–9.74***.

Finally, the asymmetries between the front and rear axles for roll and warp must be taken into account and the modal spring forces added to the static wheel load ***(Equations 9.75–9.78)***.

These equations can be entered as mathematical channels into the data analysis software. The result of this is given in ***Figure 9.16,*** where a lap around Nurburgring is illustrated. For ease, use an interactive setup sheet in a spreadsheet that is linked to the data analysis software (Chapter 2). In this spreadsheet, the modal wheel rates can be calculated and used by the analysis software as session constants. This software option is provided in most popular data acquisition packages. Because setup changes can be stored in the setup sheet, the analysis software can link the right constants to each data file. Keep in mind that the wheel loads calculated in Figure 9.16 represent only portions of the load the wheel actually experiences. Only the forces acting on the wheels that are distributed to the chassis through the springs are determined with the previous equations. Geometric and unsprung weight transfer are not taken into account by this calculation, but they can be calculated and added to the total wheel loads separately using the methods outlined earlier in this chapter. ***Table 9.4*** summarizes the loads taken into account in the wheel load traces in Figure 9.16.

Another way to figure the individual wheel loads using the suspension potentiometer signals is to calculate the forces developed by the suspension springs and antiroll bars and resolve them with their respective motion ratios to forces at the wheels. The advantage of analyzing the vehicle's modal responses is that the four suspension corners are seen as one system and the effect of each suspension parameter on the complete system can be investigated.

$$q = \frac{K_R + a}{K_R - a} - 1 = \frac{WR_{SpringF} + 2 \cdot WR_{RollF}}{WR_{SpringR} + 2 \cdot WR_{RollR}} - 1 \quad (Eq.\ 9.45)$$

$$w = \frac{K_W + b}{K_W - b} - 1 \quad (Eq.\ 9.46)$$

$$WR_{SpringF} = \frac{SR_F}{MR_F^2} = 14.9 \text{ kg/mm} \quad (Eq.\ 9.47)$$

$$WR_{SpringR} = \frac{SR_R}{MR_R^2} = 10.8 \text{ kg/mm} \quad (Eq.\ 9.48)$$

$$WR_{RollF} = \frac{SR_{RollF}}{MR_{RollF}^2} = 26.8 \text{ kg/mm} \quad (Eq.\ 9.49)$$

$$WR_{RollR} = \frac{SR_{RollR}}{MR_{RollR}^2} = 16.3 \text{ kg/mm} \quad (Eq.\ 9.50)$$

Table 9.3a

Front spring rate SR_F = 28 kg/mm	**Rear spring rate SR_R = 32 kg/mm**
Front antiroll bar rate SR_{rollF} = 60 kg/mm	**Rear antiroll bar rate SR_{rollR} = 40 kg/mm**
Front spring motion ratio MR_F = 1.373	**Rear spring motion ratio MR_R = 1.725**
Front antiroll bar motion ratio MR_{rollF} = 1.495	**Rear antiroll bar motion ratio MR_{rollR} = 1.566**
Left-front corner weight W_{LF} = 318 kg	**Left-rear corner weight W_{LR} = 369 kg**
Right-front corner weight W_{RF} = 318 kg	**Right-rear corner weight W_{RR} = 369 kg**

$$K_H = K_P = \frac{1}{2} \cdot \left(WR_{SpringF} + WR_{SpringR}\right) = 12.85 \text{ kg/mm} \qquad (Eq.\ 9.51)$$

$$K_R = K_W = \frac{1}{2} \cdot \left(WR_{SpringF} + 2 \cdot WR_{RollF}\right) + \frac{1}{2} \cdot \left(WR_{SpringF} + 2 \cdot WR_{RollR}\right) = 55.96 \text{ kg/mm} \qquad (Eq.\ 9.52)$$

$$a = b = \frac{1}{2} \cdot \left(WR_{SpringF} + 2 \cdot WR_{RollF}\right) - \frac{1}{2} \cdot \left(WR_{SpringF} + 2 \cdot WR_{RollF}\right) = 12.58 \text{ kg/mm} \qquad (Eq.\ 9.53)$$

$$q = w = \frac{K_R + a}{K_R - a} - 1 = 0.58 \qquad (Eq.\ 9.54)$$

$$F_{LF} + F_{RF} + F_{LR} + F_{RR} = K_{Heave} \cdot \left(x_{LF} + x_{RF} + x_{LR} + x_{RR}\right) \qquad (Eq.\ 9.55)$$

$$F_{LF} + F_{RF} - F_{LR} - F_{RR} = K_{Pitch} \cdot \left(x_{LF} + x_{RF} - x_{LR} - x_{RR}\right) \qquad (Eq.\ 9.56)$$

$$F_{LF} - F_{RF} + F_{LR} - F_{RR} = K_{Roll} \cdot \left(x_{LF} - x_{RF} + x_{LR} - x_{RR}\right) \qquad (Eq.\ 9.57)$$

$$F_{LF} - F_{RF} - F_{LR} + F_{RR} = K_{Warp} \cdot \left(x_{LF} - x_{RF} - x_{LR} + x_{RR}\right) \qquad (Eq.\ 9.58)$$

$$F_{LF} + F_{RF} + F_{LR} + F_{RR} = X \qquad (Eq.\ 9.59)$$

$$F_{LF} + F_{RF} - F_{LR} - F_{RR} = Y \qquad (Eq.\ 9.60)$$

$$F_{LF} - F_{RF} + F_{LR} - F_{RR} = Z \qquad (Eq.\ 9.61)$$

$$F_{LF} - F_{RF} - F_{LR} + F_{RR} = T \qquad (Eq.\ 9.62)$$

$$F_{LF} + F_{RF} = \frac{X + Y}{2} \qquad (Eq.\ 9.63)$$

$$F_{LR} + F_{RR} = \frac{X - Y}{2} \qquad (Eq.\ 9.64)$$

$$F_{LF} - F_{RF} = \frac{Z + T}{2} \qquad (Eq.\ 9.65)$$

$$F_{LR} - F_{RR} = \frac{Z - T}{2} \qquad (Eq.\ 9.66)$$

$$F_{LF} = \frac{1}{4} \cdot (Z + T) + \frac{1}{4} \cdot (X + Y) \qquad (Eq.\ 9.67)$$

$$F_{RF} = \frac{1}{4} \cdot (X + Y) + \frac{1}{4} \cdot (Z + T) \qquad (Eq.\ 9.68)$$

$$F_{LR} = \frac{1}{4} \cdot (Z - T) + \frac{1}{4} \cdot (X - Y) \qquad (Eq.\ 9.69)$$

$$F_{RR} = \frac{1}{4} \cdot (X - Y) - \frac{1}{4} \cdot (Z - T) \qquad (Eq.\ 9.70)$$

Figure 9.16 Calculated wheel loads from modal displacements and wheel rates

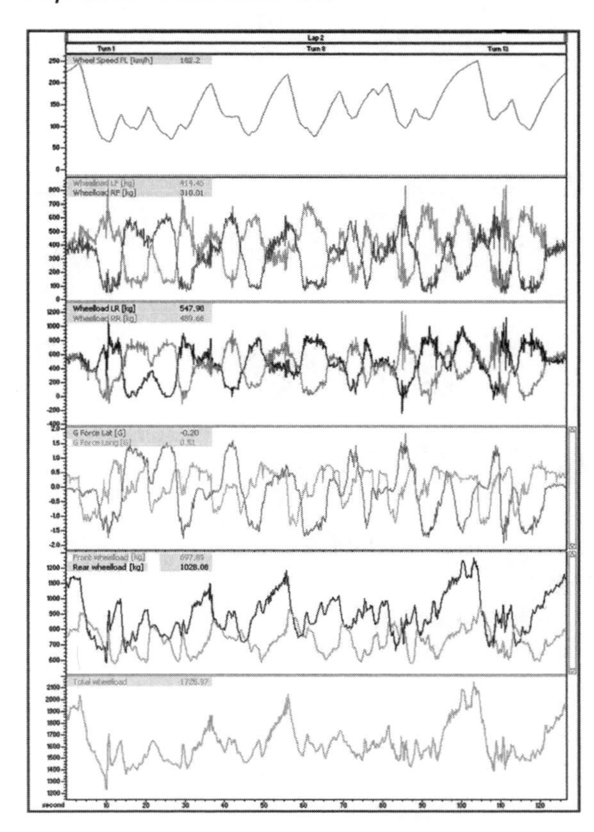

$$F_{LF} = \frac{K_H \cdot (x_{LF} + x_{RF} + x_{LR} + x_{RR}) + K_P \cdot (x_{LF} + x_{RF} - x_{LR} - x_{RR}) + K_R \cdot (x_{LF} - x_{RF} + x_{LR} - x_{RR}) + K_W \cdot (x_{LF} - x_{RF} - x_{LR} + x_{RR})}{4}$$

(Eq. 9.71)

$$F_{RF} = \frac{K_H \cdot (x_{LF} + x_{RF} + x_{LR} + x_{RR}) + K_P \cdot (x_{LF} + x_{RF} - x_{LR} - x_{RR}) - K_R \cdot (x_{LF} - x_{RF} + x_{LR} - x_{RR}) - K_W \cdot (x_{LF} - x_{RF} - x_{LR} + x_{RR})}{4}$$

(Eq. 9.72)

$$F_{LR} = \frac{K_H \cdot (x_{LF} + x_{RF} + x_{LR} + x_{RR}) - K_P \cdot (x_{LF} + x_{RF} - x_{LR} - x_{RR}) + K_R \cdot (x_{LF} - x_{RF} + x_{LR} - x_{RR}) - K_W \cdot (x_{LF} - x_{RF} - x_{LR} + x_{RR})}{4}$$

(Eq. 9.73)

$$F_{RR} = \frac{K_H \cdot (x_{LF} + x_{RF} + x_{LR} + x_{RR}) - K_P \cdot (x_{LF} + x_{RF} - x_{LR} - x_{RR}) - K_R \cdot (x_{LF} - x_{RF} + x_{LR} - x_{RR}) + K_W \cdot (x_{LF} - x_{RF} - x_{LR} + x_{RR})}{4}$$

(Eq. 9.74)

$$F_{LF} = W_{LF} + \frac{K_H \cdot (x_{LF} + x_{RF} + x_{LR} + x_{RR}) + K_P \cdot (x_{LF} + x_{RF} - x_{LR} - x_{RR}) + K_R \cdot (x_{LF} - x_{RF} + x_{LR} - x_{RR}) \cdot q}{4} + \frac{K_W \cdot (x_{LF} - x_{RF} - x_{LR} + x_{RR}) \cdot (1 - w)}{4}$$

(Eq. 9.75)

$$F_{RF} = W_{RF} + \frac{K_H \cdot (x_{LF} + x_{RF} + x_{LR} + x_{RR}) + K_P \cdot (x_{LF} + x_{RF} - x_{LR} - x_{RR}) - K_R \cdot (x_{LF} - x_{RF} + x_{LR} - x_{RR}) \cdot q}{4} - \frac{K_W \cdot (x_{LF} - x_{RF} - x_{LR} + x_{RR}) \cdot (1 - w)}{4}$$

(Eq. 9.76)

$$F_{LR} = W_{LR} + \frac{K_H \cdot (x_{LF} + x_{RF} + x_{LR} + x_{RR}) - K_P \cdot (x_{LF} + x_{RF} - x_{LR} - x_{RR}) + K_R \cdot (x_{LF} - x_{RF} + x_{LR} - x_{RR}) \cdot (1 - q)}{4} - \frac{K_W \cdot (x_{LF} - x_{RF} - x_{LR} + x_{RR}) \cdot w}{4}$$

(Eq. 9.77)

$$F_{RR} = W_{RR} + \frac{K_H \cdot (x_{LF} + x_{RF} + x_{LR} + x_{RR}) - K_P \cdot (x_{LF} + x_{RF} - x_{LR} - x_{RR}) - K_R \cdot (x_{LF} - x_{RF} + x_{LR} - x_{RR}) \cdot (1 - q)}{4} + \frac{K_W \cdot (x_{LF} - x_{RF} - x_{LR} + x_{RR}) \cdot w}{4}$$

(Eq. 9.78)

with W_{LF} = left-front static corner weight force
W_{RF} = right-front static corner weight force
W_{LR} = left-rear static corner weight force
W_{RR} = right-rear static corner weight force

Table 9.4 Breakdown of the individual wheel loads

Change in wheel load due to:	Included in modal analysis?
Static weight	Yes
Unsprung weight transfer	No
Geometric weight transfer	No
Sprung weight transfer	Yes
Longitudinal weight transfer	Yes
Banking	Yes
Track slope	Yes
Aerodynamic forces	Yes
Bumps, road surface irregularities	Yes

Measuring Wheel Loads with Suspension Load Cells

The accuracy of the wheel load calculations can be improved greatly by measuring the strain in the suspension members with load cells and calibrating these so that they output the vertical loads acting on the wheels. The greatest accuracy improvement is the sensor resolution. By measuring the load directly, the wheel loads do not need calculation through spring and wheel rates; those variables do not come into play. Remember that Table 9.2 also applies here. Suspension load cells do not measure unsprung or geometric weight transfer.

Figure 9.17 shows the four dynamic (i.e., static weight not included) wheel loads measured on an early 1990s Formula One car during a lap around Hockenheim Ring. These four traces clearly show the change in load at each wheel as the car accelerates, brakes, and negotiates corners. The increase in downforce as speed increases is also apparent. By manipulating these signals and using them in mathematical channels, the different components making up the total wheel loads can be investigated.

Adding the four wheel loads creates a channel that gives the total download on the car. Illustrated in ***Figure 9.18***, this channel clearly shows the amount of downforce that develops—the trace is almost identical to the speed trace. This is not the case with a low-downforce car. This channel also can be used to investigate the effect of banked corners and track slopes. These track properties change the measured vertical load.

Lateral sprung weight transfer can be determined by creating the following math channels:

front weight transfer = load LF – load RF

rear weight transfer = load LR – load RR

The bias between front and rear axle is as given in ***Equation 9.79***. These channels are shown in ***Figure 9.19***.

The channel showing the lateral weight transfer bias between the front and rear axles is particularly useful to investigate transient effects at corner entry and exit. The shock absorbers temporarily change the lateral weight transfer distribution to a value different than that of steady-state cornering.

$$\text{Weight transfer bias} = \frac{\text{Front weight transfer}}{\text{Front weight transfer} + \text{Rear weight transfer}} \cdot 100\% \qquad (Eq.\ 9.79)$$

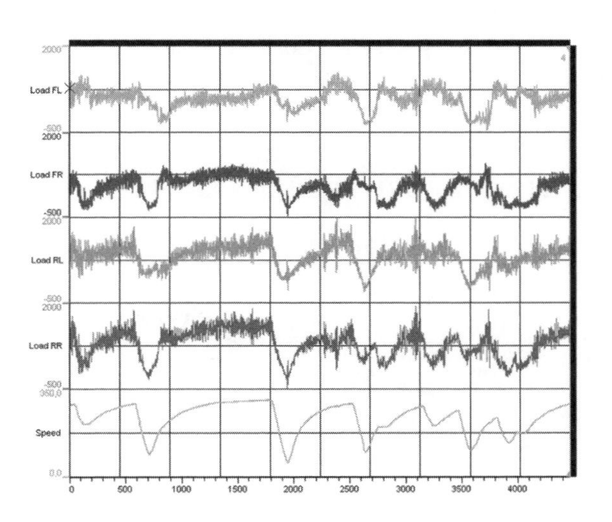

Figure 9.17 *Wheel loads determined with strain gage measurement (Courtesy of Pi Research, England)*

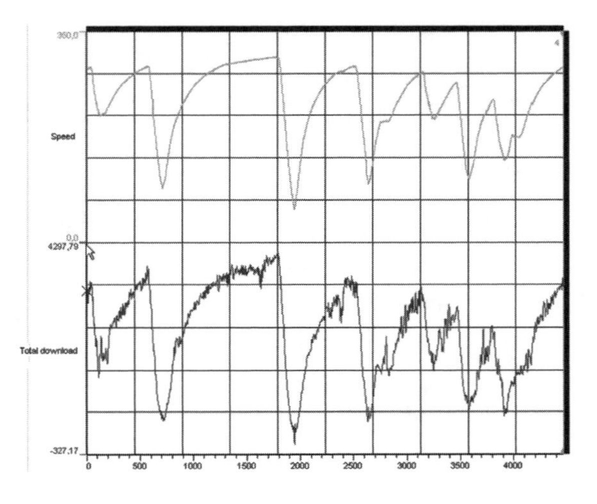

Figure 9.18 *Total download (Courtesy of Pi Research, England)*

For a quick analysis of steady-state weight transfer (e.g., after a spring or antiroll bar change), the *X-Y* chart in ***Figure 9.20*** is helpful. The slope of this graph gives the ratio between rear and front lateral weight transfer.

Similarly, the longitudinal weight transfer can be determined using ***Equation 9.80*** as a math channel ***(Figure 9.21)***. In addition, the percentage of load on the front axle is given in ***Equation 9.81***.

The longitudinal weight bias graph clearly shows how weight is shifted forward during braking and rearward during acceleration. On the straights, it is possible to determine the shift in aerodynamic center of pressure with increasing speed.

Tire Spring Rates

Until now the suspension system was considered as having four springs and two antiroll bars, but the tires represent four more spring rates that also influence the lateral weight transfer distribution. The proposed method for calculating the wheel loads from the suspension modes did not take the tire spring rates into account. However, when the wheel loads are measured directly by suspension load cells, the measured load distribution incorporates the lateral weight transfer bias through the tire spring rates.

Each tire spring works in series with the corresponding suspension spring and parallel to the antiroll bar on that axle. Therefore, the roll rate bias (q) (Equation 9.45) can be rewritten to take into account the tire spring rates ***(Equation 9.82)***.

This equation can be used only if the tire spring rates are known, which often is not the case. The best way to assess the lateral weight transfer bias is to measure the wheel loads directly.

$$\text{Longitudinal weight transfer} = \text{Load LR} + \text{Load RR} - \text{Load LF} - \text{Load RF} \qquad (Eq.\ 9.80)$$

$$\text{Longitudinal weight bias} = \frac{\text{Load LF} + \text{Load RF}}{\text{Load LF} + \text{Load RF} + \text{Load LR} + \text{Load RR}} \cdot 100\% \qquad (Eq.\ 9.81)$$

$$q = \frac{\left[\dfrac{WR_{SpringF} \cdot SR_{TireF}}{WR_{SpringF} + SR_{TireF}}\right] + 2 \cdot WR_{RollF}}{\left[\dfrac{WR_{SpringR} \cdot SR_{TireR}}{WR_{SpringR} + SR_{TireR}}\right] + 2 \cdot WR_{RollR}} - 1 \qquad (Eq.\ 9.82)$$

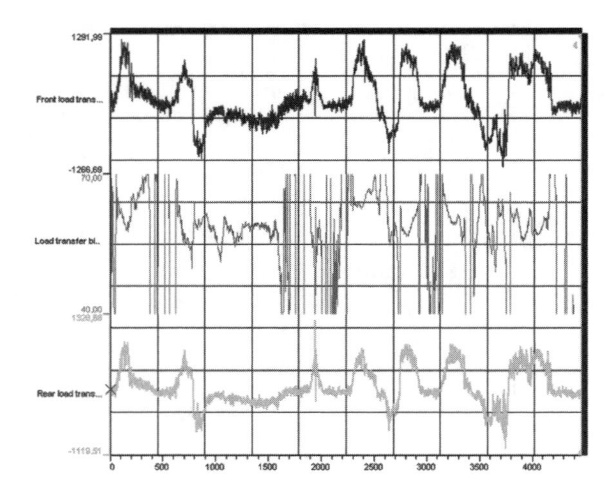

***Figure 9.19** Lateral (sprung) weight transfer (Courtesy of Pi Research, England)*

Chassis Torsion

When calculating lateral weight transfer distribution on the front and rear axles, it has been always assumed that the chassis stiffness is of such magnitude that it can be ignored. This may not be necessarily the case. A flexible chassis could be considered as an extra (torsional) spring in the suspension system. Calculated weight transfers may not represent the real situation when ignoring the chassis stiffness.

Chassis stiffness does not only imply the rigidness of the frame but also the compliance in the suspension pick-up points, bodywork attachments to the chassis, and engine and driveline support points on the chassis.

The nonlinear behavior of the racecar tire provides the means of tuning the vehicle's handling balance by changing the amount of weight transfer on one axle. For an understeering car, the engineer normally tries to decrease the weight transfer on the front axle or increase it at the rear, and vice versa for an oversteering car. However, the car handling can be influenced only in this way if the chassis serves as a platform to feed the involved torques through.

In the following discussion, distinguish between the different components that make up the overall vehicle roll angle—suspension roll, tire roll, and chassis torsion. As an example, the following suspension roll angles were measured on a racecar:

front suspension roll angle = 16.61 mm

rear suspension roll angle = 30.71 mm

This vehicle's respective front and rear roll (wheel) rates are 78.54 kg/mm and 23.38 kg/mm The tire spring rates are 50 kg/mm (front) and 55 kg/mm (rear).

If a chassis with a finite torsional stiffness is considered, ***Equation 9.83*** applies with α_{rollF} = front suspension roll angle, $\alpha_{\text{rolltiresF}}$ = front tires roll angle, α_{rollR} = rear suspension roll angle, $\alpha_{\text{rolltiresR}}$ = rear tires roll angle, and α_{torsion} = chassis torsion angle.

$$\alpha_{\text{rollF}} + \alpha_{\text{rolltiresF}} = \alpha_{\text{rollR}} + \alpha_{\text{rolltiresR}} + \alpha_{\text{torsion}} \qquad \text{(Eq. 9.83)}$$

This equation differs from the definition given in the previous chapter by the addition of a torsion component (α_{torsion}).

Multiplying front and rear roll angles with the corresponding roll rates gives the lateral weight transfer on each axle ***(Equations 9.84*** and ***9.85)***.

$$\Delta W_{sF} = 16.61 \cdot 78.54 = 1304.5 \text{ kg} \qquad \text{(Eq. 9.84)}$$

$$\Delta W_{sR} = 30.71 \cdot 23.38 = 718.0 \text{ kg} \qquad \text{(Eq. 9.85)}$$

The tire roll angles ($\alpha_{\text{rolltiresF}}$ and $\alpha_{\text{rolltiresR}}$) are given in ***Equations 9.86*** and ***9.87***.

$$\alpha_{\text{rolltiresF}} = \frac{1304.5}{2 \cdot 50} = 13.0 \text{ mm} \qquad \text{(Eq. 9.86)}$$

$$\alpha_{\text{rolltiresR}} = \frac{718.0}{2 \cdot 55} = 6.5 \text{ mm} \qquad \text{(Eq. 9.87)}$$

Entering these values in Equation 9.82 finally gives the chassis torsion angle α_{torsion} = 7.6 mm. When this is related to the overall lateral weight transfer, the torsional stiffness of the chassis (SR_{Chassis}) is known ***(Equation 9.88)***.

$$SR_{\text{Chassis}} = \frac{\frac{1}{2} \cdot \left(\Delta W_{sF} + \Delta W_{sR} \right)}{\alpha_{\text{Torsion}}} = 133.4 \text{ kg/mm} \qquad \text{(Eq. 9.88)}$$

Compared to the front roll stiffness, the magnitude of chassis torsion in this example certainly affects the lateral weight transfer distribution between the front and rear axles. In this case, do not ignore the torsional stiffness of the chassis.

***Figure 9.20** X-Y chart of rear versus front lateral weight transfer (Courtesy of Pi Research, England)*

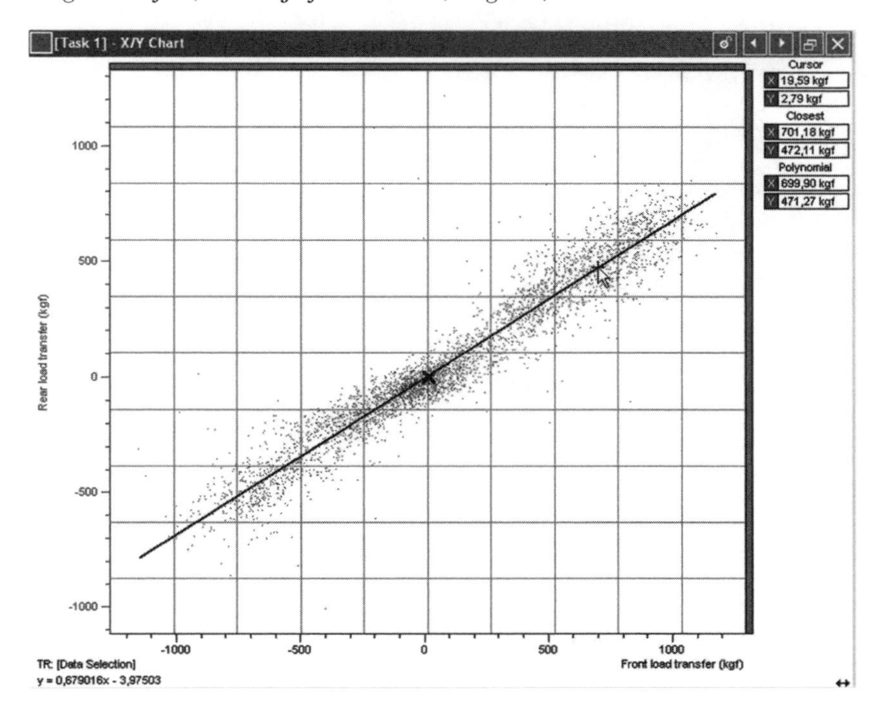

***Figure 9.21** Longitudinal weight transfer (Courtesy of Pi Research, England)*

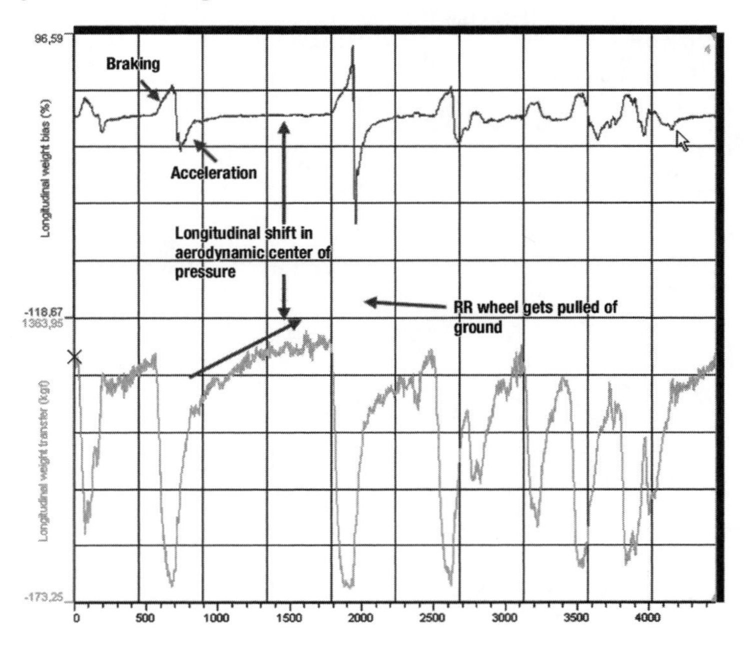

CHAPTER 10

FREQUENCIES AND SHOCK ABSORBERS

Cars are equipped with shock absorbers to minimize variation in contact between tires' contact patches and the track surface and to control transient chassis movements due to lateral, longitudinal, and vertical acceleration. The influence of shock absorbers on racecar dynamics is investigated in the first half of this chapter.

The second half introduces frequency analysis, a method used to optimize the interface between the vehicle and the road surface. This part of the chapter is written by Josep Fontdecaba I. Buj, engineering director at Creuat S.L., a Spanish company specializing in interconnected suspension systems. Given his expertise in suspension optimization, it is more appropriate that he write this part.

Damping Analysis

The force developed by a spring is proportional to its compression ***(Equation 10.1)***.

$$F_{spring} = SR \cdot s \qquad \text{(Eq. 10.1)}$$

with F_{spring} = spring force (N)
SR = spring rate (N/mm)
s = spring compression (mm)

Shock absorbers are speed sensitive. They develop a force proportional to the speed at which they compress or extend ***(Equation 10.2)***.

$$F_{shock} = C \cdot v \qquad \text{(Eq. 10.2)}$$

with F_{shock} = shock absorber force (N)
C = damping coefficient (Ns/mm)
v = shock absorber velocity (mm/s)

Basically this means that a spring develops its highest force at maximum deflection, whereas a shock absorber reaches maximum force at maximum shaft velocity. Shock absorber speed is an important parameter to measure to understand shock absorber performance. Modern racecar shock absorbers are often adjustable, and adjustments apply to different velocity ranges. By identifying the speed of the shock absorber in problem areas, setup adjustments may be assessed better.

Figure 10.1 illustrates shock velocity and spring travel for a single wheel traveling over an irregularity in the road surface. The shock absorber speed channel was created by differentiating the shock travel signal.

The shock travel trace shows the shock absorber remaining in extension (or rebound) initially, at which point the shaft speed of the shock is zero. As the shaft changes direction from extension into compression (or bump), the shock speed increases to its maximum at the point halfway through the compression motion. At this point, the shock absorber develops a force proportional to this maximum speed and the applied damping coefficient. Shock speed reaches zero again, where the spring compression is at its maximum (and spring force subsequently reaches a local maximum).

Also worth noting in Figure 10.1 is the difference in duration between extension and compression. The extension movement takes longer than the subsequent compression movement for two reasons:

- The damping coefficient for rebound is greater than that of bump.

Figure 10.1 *The shock absorber velocity and travel for a single wheel going over a bump*

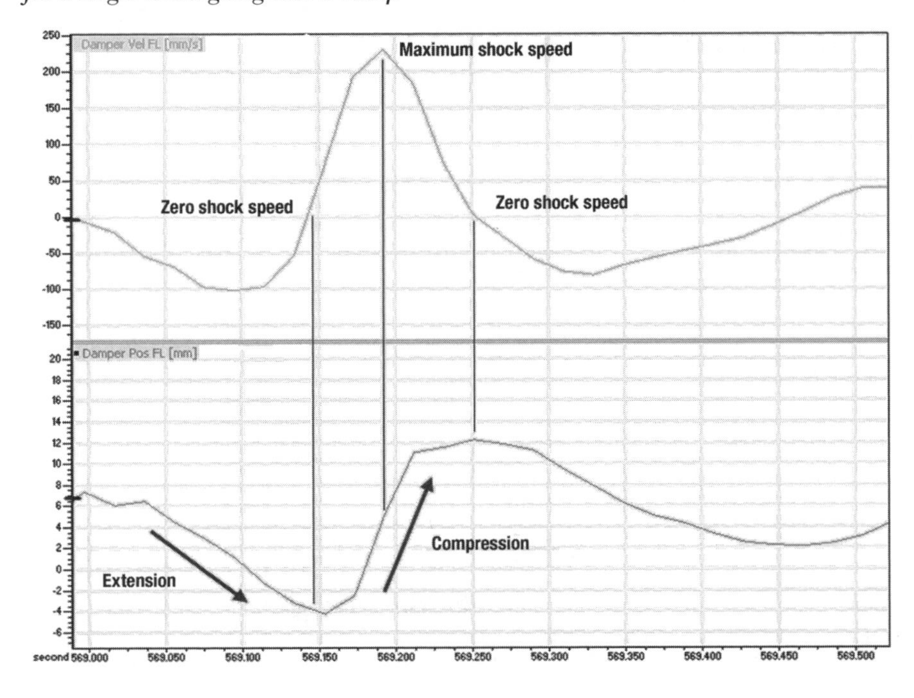

- The road irregularity itself introduces a considerable amount of kinetic energy into the suspension during the compression phase. During the extension phase, the suspension returns to its normal position without the presence of this kinetic energy (the bump already has been transmitted into the chassis).

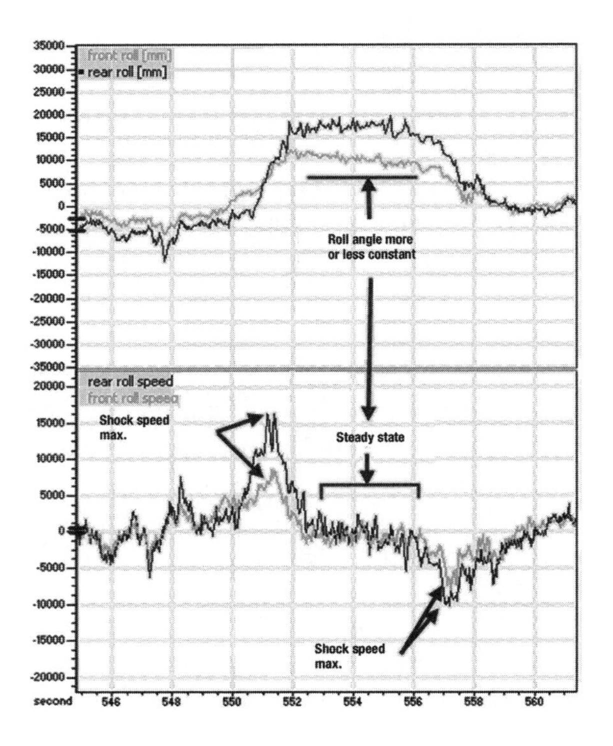

***Figure 10.2** The roll angle and speed of a car negotiating a corner*

Shock absorbers deal with road irregularities at higher shaft velocities. Transient effects due to lateral and longitudinal load transfer occur at low speeds, but also here the shock absorber develops its maximum force there where it reaches its highest speed. ***Figure 10.2*** indicates what happens when a car goes through a corner. The upper portion shows the front and rear roll angle (expressed in millimeters), which was calculated by subtracting the left and right shock travel signals from one another. The lower graph shows the front and rear roll speeds, which were calculated by differentiating the roll angle channels. These roll speed channels indicate that the shock absorbers add force to the wheels at the moment the roll movement is initiated (at the corner entry) and it is at that point that the roll angle returns to zero again (at the corner exit). The area in between measures a roll speed of nearly zero. This is called steady-state cornering, where the vehicle balance is influenced mainly by springs and antiroll bars.

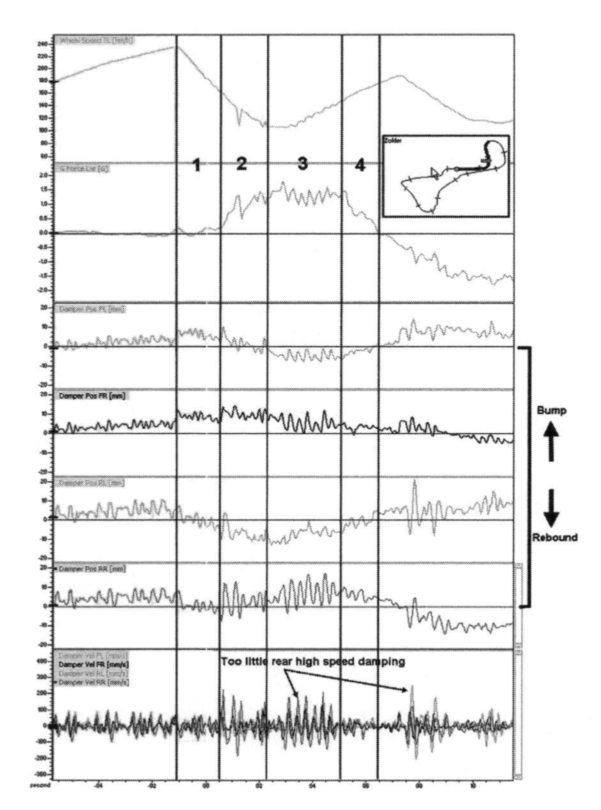

***Figure 10.3** The first lefthander after the start/finish on the Zolder. Displayed channels are wheel speed, lateral G, shock travel, and velocity.*

Most racing shock absorbers can be adjusted separately, in bump and rebound. The position in the corner where the shocks influence the cornering balance of the car has been identified, but to tune this balance with shock settings the phase of the corner in which the handling problem occurs needs determining. Next, the wheel in bump or rebound motion must be identified. ***Figure 10.3*** shows wheel speed, lateral acceleration, and the four shock displacement and velocity channels for a car negotiating the first lefthander after the start/finish line on the Circuit Zolder. The corner is divided into four separate sections:

Section 1: Straight-line Braking

The speed trace shows where braking begins. Both front wheels suddenly go into bump and remain relatively constant until cornering commences. Both rear wheels extend more gradually into rebound.

Section 2: Initial Cornering While Continuing to Brake

Here the left-front (on the inside of the corner) shock absorber extends again (rebound), while on the right front wheel it continues to fluctuate around a stable average. The left-rear wheel goes even more into rebound as the right rear decreases its compression (or goes into rebound movement).

Section 3: Steady-state Cornering Phase, Followed by Throttle Application (While Still Cornering)

During this phase, the front wheels remain in constant compression, while the rear wheels go into bump upon throttle application. Note that despite the two rear wheels being in bump movement, the left-rear shock absorber is still in extension (i.e., longer than its static length.) The right-rear shock absorber already was compressed and now compresses even further.

Section 4: Corner Exit, Steering Wheel Unwinds

As the vehicle's roll angle decreases, the right-hand wheels go into rebound and the left-hand wheels into bump.

These observations are summarized in ***Table 10.1*** and can be used to suggest changes in shock absorber settings. For instance, in the case of understeer during initial cornering, front rebound could be decreased to improve the compliance between the left-front tire and the track surface. Furthermore, a low-speed damping change on the front axle would have no influence on handling during a steady-state cornering phase. Apart from the high-speed movement caused by track irregularities, there is no substantial movement in the low-speed region of the front shock absorbers in this corner section. Tables like this tend to suggest asymmetric shock absorber settings, because often during cornering the wheels on the opposite side on an axle experience an opposite movement. Asymmetry, however, creates different handling characteristics in left- and right-hand corners. Although in some isolated cases asymmetric damping can improve the car's balance, concentrate on the wheel creating the least amount of grip (which is usually the wheel on the inside of the corner) and keep the changes in damping the same on the left- and right-hand sides.

The shock absorber speed traces in Figure 10.3 provide a good indication of the high-speed damping characteristics of the vehicle. One conclusion drawn from this data is that the rear axle has too little high-speed damping in bump and rebound. During the steady-state cornering phase in this left-hand corner, the right-rear shock absorber experiences speeds far greater than those of the other three wheels. This is confirmed in the next (right-hand) corner where the same occurs on the left-rear shock absorber. The rear axle does not handle road irregularities as well as the front axle, thereby increasing the rear tires' contact patch load variation and therefore decreasing the maximum grip level in the tires.

Shock Speed Histogram

As mentioned earlier in this chapter, damping adjustments in the low-speed range influence the transient handling of the car, while the high-speed range takes care of road input. To put some figures on it, the shock absorber speed range is divided into the sections in ***Table 10.2***.

Table 10.1 *Shock movement and wheel position summary for traces in Figure 10.3*

	Section 1		Section 2		Section 3		Section 4	
	Straight-line Braking		Initial Cornering + Braking		Steady-state Cornering		Corner Exit Acceleration	
	Wheel Position	Shock Movement	Wheel Position	Shock Movement	Wheel Position	Shock Movement	Wheel Position	Shock Movement
LF	Compression	Bump	Compression	Rebound	Extension	———	Extension	Bump
RF	Compression	Bump	Compression	———	Compression	———	Compression	Rebound
LR	Extension	Rebound	Extension	Rebound	Extension	Bump	Compression	Bump
RR	Extension	Rebound	Compression	Bump	Compression	Bump	Compression	Rebound

The first speed interval (below 5 mm/s) is dominated primarily by friction in the suspension system. Most racecars have a considerable amount of sliding contact pivots (spherical bearings), but also the contact surface between the shock absorber shaft and the seals in its housing can account for a significant portion of the total suspension friction. Other than mechanically trying to minimize friction, not much damping tuning occurs in this speed range. However, this does not mean that suspension friction should be ignored.

Generally called the low-speed area, the second speed interval (5–25 mm/s) is where the shock absorber responds to chassis motions resulting from braking, accelerating, and cornering. This is also the area with the greatest influence on how the driver feels the car.

Table 10.2 *Shock absorber speed range*

Speed range	Influence
Below 5 mm/s	Friction (damper shaft and seals, and suspension joints)
5–25 mm/s	Inertial chassis motion (roll, pitch, and heave)
25–200 mm/s	Road input (bumps)
Plus 200 mm/s	Curbs

Figure 10.4 *Shock absorber speed ranges illustrated in a shock speed histogram*

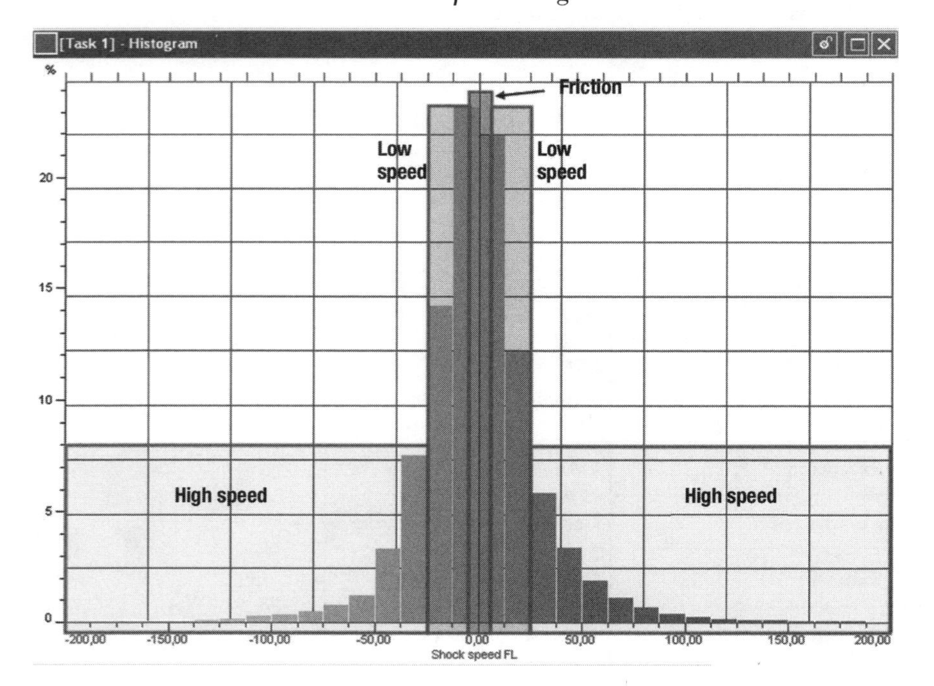

Road input results in shock absorber speeds above 25 mm/s. Damping in this area should be optimized to minimize tire contact patch load variation. The ability of a suspension to cope with road input rarely is assessed reliably by a driver. However, some techniques are available to analyze this using a data acquisition system. This topic is discussed later in this chapter. Finally, curbs typically result in shock absorber speeds greater than 200 mm/s.

To visualize how much time a shock absorber is spending in each time interval, one can prepare a histogram of its shock speed channel ***(Figure 10.4)***. Depending on the calibration of the linear potentiometers measuring shock travel, the negative values in the histogram represent rebound travel and positive bump, or vice versa. The shock speed histogram of the vehicle's shock absorbers can be used as a tuning tool, a method pioneered by Claude Rouelle in his data acquisition seminars.[10]

The shock speed histogram is essentially a characterization of the shock absorber while it is on the car. To maximize the performance of the shock absorber, setup adjustments should be implemented to make the shock speed histogram as symmetrical as possible. Ideally, the histogram resembles a Gaussian distribution (also known as normal distribution or bell curve). The goal here is to maintain a tire contact patch load that is as constant as possible.

When looking at a wheel passing over a single bump, initially there is a positive shock speed when the bump is hit and then a negative speed as the wheel passes over it. To maintain the balance of the chassis, positive and negative velocities should be as close as possible to each other in magnitude and duration. Now, extrapolate this picture to a complete lap around the track, where each shock movement applies to this condition and the result is a perfectly symmetrical shock speed histogram. In other words, an ideal shock absorber setup dissipates equal amounts of energy into the shock absorber in bump and rebound movements. ***Figure 10.5*** shows histograms of the four corners of a racecar. The data was taken from the same lap as in Figure 10.3. The darker color bars represent the low-speed range of the shock absorber (Lo% 0–25 mm/s), while the lighter color bars indicate how much time is spent in the high-speed range (Hi% >25 mm/s). The total percentage of time spent in each of these ranges is indicated in each histogram for bump and rebound travel. Additionally, the average speed in bump and rebound is indicated in the graphs.

Table 10.3 summarizes the percentage of time spent in each speed range. The third column illustrates the difference between the bump and rebound durations. The fourth column shows the average between bump and rebound to offer an indication of how much time is spent at low and high speed, regardless of the direction of damper travel. From this, one can conclude that in the high-speed range there is not a great deal of asymmetry between bump and rebound, the maximum difference being 1.6% for the right-rear corner. The low-speed range can use some tweaking. The left-front shock absorber spends 3.5% more time in rebound, while the right rear is 5.1% more in bump. Therefore, the left-front shock absorber could use a bit more rebound damping and the right rear more bump damping.

Although there is reasonable symmetry in the high-speed range, Figure 10.3 shows that there is not enough rear high-speed damping. This is confirmed in the shock speed histograms. There is a difference of approximately 6% between the front and rear high-speed ranges. To obtain a more even distribution between the front and rear, the rear high-speed damping should be increased in bump and rebound.

The shock speed histograms in Figure 10.5 were created from data covering an entire lap. Particular corners can be magnified. However, in this case, attention should be paid to the beginning and ending point of the magnified area. These points should have approximately the same shock potentiometer value. If this is not the case, one shock absorber spent more time in the bump or rebound phase, which creates an offset in the shock speed histograms.

The general tendencies of the shock speed histogram can be determined by investigating the behavior of a simple mass-spring-shock absorber system. ***Figure 10.6*** shows three different situations. The upper graphs are the time histories of the shock absorber speeds created by a single stepped input, and the lower graphs are the frequency distributions of the respective shock speeds.

The first situation illustrates the reaction of an undamped mass-spring-shock absorber system (damping factor = 0). Here, the histograms shows two peaks on the left and right side equal to the system's maximum speed in bump and rebound. The values in between are less frequent.

In the second situation, the system damping factor is increased to 0.1. The effect of this is a histogram showing that the shock absorber spends more time at lower speeds.

Figure 10.5 *Shock speed histograms for the four corners of a racecar*

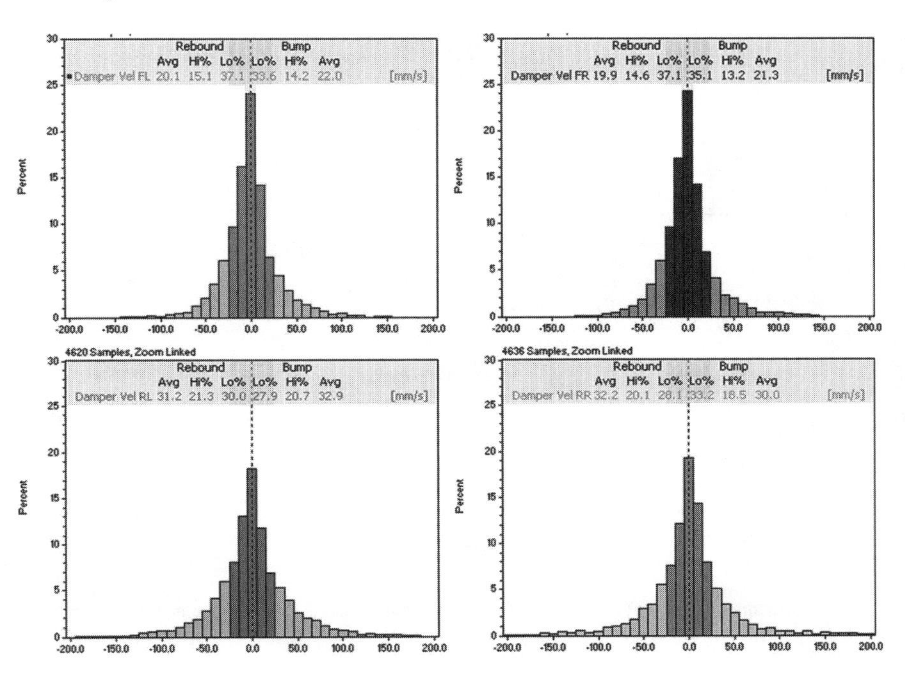

Table 10.3 *Shock speed asymmetries between bump and rebound*

		Bump	Rebound	Difference	Average
Left Front	Lo%	33.6	37.1	3.5	35.4
	Hi%	14.2	15.1	0.9	14.7
Right Front	Lo%	35.1	37.1	2.0	36.1
	Hi%	13.2	14.6	1.4	13.9
Left Rear	Lo%	27.9	30.0	2.1	28.9
	Hi%	20.7	21.3	0.6	21.0
Right Rear	Lo%	33.2	28.1	–5.1	30.7
	Hi%	18.5	20.1	1.6	19.3

Finally, when the damping factor is further increased to 1 (critical damping), the histogram shows one pronounced peak at zero shock speed.

These results indicate the following:

- The more damping is present in the shock absorber, the narrower the shock speed histograms become.
- The time a shock absorber spends in one speed interval can be decreased or increased by increasing or decreasing the damping in that respective interval.

Except for the average shock speed percentages in bump and rebound in the high- or low-speed ranges as calculated in the previous example, other statistical parameters can be employed to provide more information about the histogram shape. These parameters also make it easier to compare different histograms. Often the statistical functions discussed below are not available in the data acquisition analysis software. Therefore, the data in the next example is imported into a spreadsheet. This spreadsheet, including the sample data, can be downloaded from http://jorge.segers.googlepages.com/technical. ***Figure 10.7*** shows the output of this spreadsheet. As indicated previously, the ideal shock speed histogram resembles a normal distribution curve. Therefore, a measured histogram is parameterized by determining how much it deviates from a normal distribution.

Figure 10.6 *Shock speed histogram tendencies for different damping values*

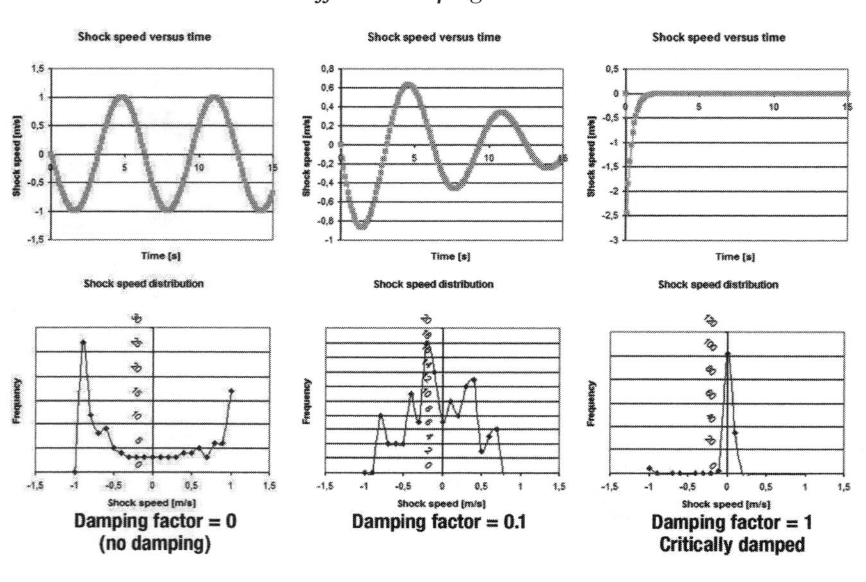

Figure 10.7 *Shock speed histogram statistics*

Laptime	seconds			
Number of samples	Samples			
Sampling frequency	Hz			

Statistics	LF	RF	LR	RR
Average shock speed [mm/s]	-0,01	0,01	-0,01	0,00
Median [mm/s]	-1,76	-2,24	-0,40	-0,27
Variance	804	768	1373	1277
Standard deviation	28,3	27,7	37,0	35,7
Skewness (assymmetry)	0,368	0,441	-0,059	-0,162
Kurtosis (peakedness)	4,51	4,85	3,73	3,70

	LF	RF	LR	RR
Max shock speed in BUMP [mm/s]	182,2	196,4	202,8	164,2
Max shock speed in REBOUND [mm/s]	-198,1	-243,1	-265,0	-269,1
Suspension friction [%]	19,4%	19,7%	16,0%	16,9%
Kerbstones [%]	0,000%	0,016%	0,064%	0,048%

	LF	RF	LR	RR
Number of samples in BUMP	2870	2816	3055	3076
Percentage	46,2%	45,3%	49,1%	49,5%
Avg shock speed in BUMP [mm/s]	21,58	21,58	26,62	25,54
LOW SPEED BUMP	1989	1955	1852	1919
Percentage	32,0%	31,4%	29,8%	30,9%
HIGH SPEED BUMP	881	861	1203	1157
Percentage	14,2%	13,8%	19,4%	18,6%

	LF	RF	LR	RR
Number of samples in REBOUND	3347	3401	3162	3141
Percentage	53,8%	54,7%	50,9%	50,5%
Avg shock speed in REBOUND [mm/s]	-18,52	-17,86	-25,73	-25,00
LOW SPEED REBOUND	2543	2614	1985	2009
Percentage	40,9%	42,0%	31,9%	32,3%
HIGH SPEED REBOUND	804	787	1177	1132
Percentage	12,9%	12,7%	18,9%	18,2%

Initially, the average shock speed should be zero. If this is not the case, there is probably something wrong with the suspension travel sensors or the math channel that calculates the shock speed. One property of a normal distribution is that the median value $\mu_{1/2}$ (the middle value of an array) equals the average value of that distribution. For instance, in the array 5, 7, 8, 10, 15, the average is 9 and the median 8. In the case of the shock speed histogram, the median should be as close to zero as possible.

A median greater than zero indicates that there are more measured samples in the bump range, while a negative median signifies a greater number of samples in the rebound range. In other words, $\mu_{1/2}$ is a first measure of the asymmetry of the shock speed histogram between bump and rebound.

Variance and standard deviation are two statistical measures with the same meaning; they measure dispersion (i.e., the scattering of measured values around their average). Variance σ^2 is the average distance of each data point from the average value of all the samples. For a discrete collection of samples, it can be expressed mathematically as ***Equation 10.3***.

$$\sigma^2 = \frac{1}{N} \cdot \sum_{i=1}^{N} (x_i - \mu)^2 \qquad \textit{(Eq. 10.3)}$$

where N = total number of samples

μ = average value of all samples

Standard deviation σ is simply the square root of the variance ***(Equation 10.4)***.

$$\sigma = \sqrt{\frac{1}{N} \cdot \sum_{i=1}^{N} (x_i - \mu)^2} \qquad (Eq.\ 10.4)$$

Standard deviation is expressed in the same units as the measurement; in the case of this example, it is in mm/s. A small σ indicates that the data points are clustered closely to the histogram's average (which indicates a lot of damping). Meanwhile, a large σ indicates that they are distributed a significant distance from the average. Essentially this is a measure of the width of the shock speed histogram. A large standard distribution indicates that the shock absorber sees more high-speed movement, and vice versa. The rear shock absorbers in Figure 10.7 have a greater standard deviation than the front ones, so by comparison they experience more high-speed movement.

Pay attention to the fact that the standard deviation does not take into account any asymmetry between the bump and rebound side of the histogram.

Skewness is a measure of the asymmetry of the shock speed histogram. A histogram biased to the bump side has negative skew. A positive skew means the histogram is biased to the rebound side ***(Figure 10.8)***. Mathematically, skewness A can be expressed as ***Equation 10.5***.

$$A = \frac{N}{(N-1) \cdot (N-2)} \cdot \sum_{i=1}^{N} \left(\frac{x_i - \mu}{\sigma} \right)^3 \qquad (Eq.\ 10.5)$$

where N = total number of samples
μ = average value of all samples
σ = standard deviation of all samples

A normal distribution has a skewness equal to zero. If the shock speed average and median are not equal, skewness is not zero.

Kurtosis is the degree of peakedness of a distribution. A higher kurtosis histogram has a sharper peak and fatter tails, while a lower kurtosis histogram has a more rounded peak with wider shoulders ***(Figure 10.9)***. ***Equation 10.6*** determines the kurtosis of a distribution.

Kurtosis indicates that the collection of samples is spread in a wider fashion than the normal distribution entails. A normal distribution has a kurtosis of zero (mesokurtic). A distribution with positive kurtosis is called *leptokurtic* and one with a negative kurtosis *platykurtic*. In Figure 10.7, all histograms are leptokurtic, but the front shock absorbers have a higher kurtosis than the rear ones. This means that more movement on the front shock absorbers is concentrated in the low-speed range.

If a histogram has a higher kurtosis, this does not mean necessarily that it has a lower standard deviation. The distributions pictured in Figure 10.9 may have the same standard deviation.

Often, shock absorbers are not used only to handle the transient characteristics of the vehicle and the irregularities in the track surface. In some cases, they are used to control the attitude of the chassis. An example of this is a racecar possessing considerable aerodynamic downforce with excessive front rebound damping applied to jack the car's nose down to improve the airflow under the car. In this case, the front downforce pushes the front of the car down, creating a greater rake angle (the longitudinal angle of inclination of the vehicle floor). A high rebound damping helps keep it there.

$$G_2 = \left[\frac{N \cdot (N+1)}{(N-1) \cdot (N-2) \cdot (N-3)} \cdot \sum_{i=1}^{N} \left(\frac{x_i - \mu}{\sigma} \right)^4 \right] - \frac{3 \cdot (N-1)^2}{(N-2) \cdot (N-3)} \qquad (Eq.\ 10.6)$$

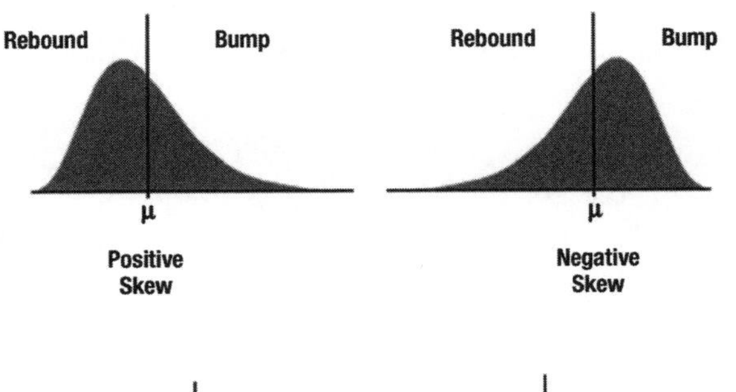

Figure 10.8
Skewness

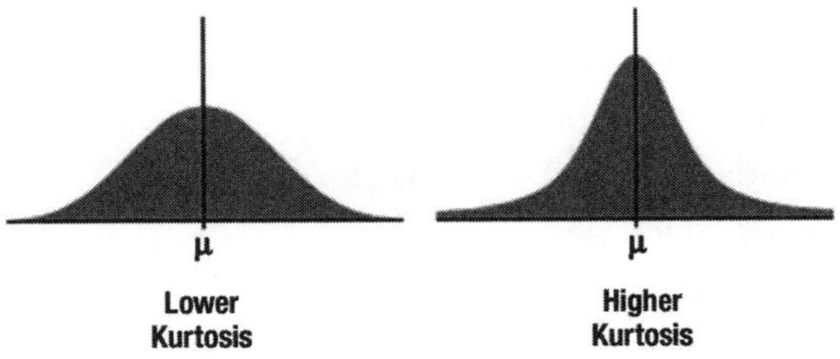

μ
Lower Kurtosis
μ
Higher Kurtosis

Figure 10.9
Kurtosis

By doing this, some mechanical grip is sacrificed to improve aerodynamic performance. A deliberate asymmetry is created in the shock speed histograms, and there is an offset to the rebound side.

Shock speed histograms are not affected only by changes in the damping characteristics of the vehicle. Springs, antiroll bars, aerodynamics, tires, tire pressures, and many other factors influence the histogram shape. The shock absorbers always should be matched to the rest of the suspension to obtain their best performance. The shock speed histogram is a proven tool to extract the maximum mechanical grip from a tire.

Introducing Frequency Analysis

Analyzing vehicle suspension is undertaken to understand and optimize the interface between the vehicle and the road. In this section, some analysis tools and concepts are proposed that may be useful for this purpose. These methods imply some complex calculations that possibly are not within the scope of a user's data acquisition analysis software. Therefore, this type of analysis is conducted primarily by exporting the logged data into a mathematical package (e.g., Matlab, MathCAD, or a spreadsheet) or into customized software packages.

The suspension of a four-wheeled vehicle can be defined as a combination of rigid and elastic components that link each wheel to the vehicle body. In effect, the suspension is designed to behave as rigidly as possible in response to lateral and longitudinal forces while allowing a softer response to vertical movements to cope with road irregularities.

Although springs and dampers define the main properties of the vehicle suspension, one must consider the suspension's geometrical properties, the distribution of the vehicle's mass, and the tire properties to prepare a useful model.

In addition to the model complexity, the suspension analysis must take into account the random nature of the input, which is a combination of road irregularities and dynamic loads induced by the driving conditions.

The nature of this problem is very complex. This text focuses on the effect of springs and dampers because they are responsible for the vertical movements of the wheels (i.e., the vertical movements of the vehicle body with respect to the road).

The frequency analysis is one of the best analysis tools because it can handle efficiently the road input, cover the entire range of body movements, and characterize suspension behavior.

In terms of functionality, the purpose of the suspension is twofold. First, one needs it to isolate the vehicle body movements from the dynamic forces generated by the vehicle driving conditions and the irregularities of the road. Second, one needs it to help the tires follow the terrain to maximize their grip capacity.

From this, it follows that the suspension must minimize the following:

- the body movements induced by driving and the road input, and
- the tire load fluctuations induced by driving and the road input.

The first requirement represents a contradiction. Body movements generated by driving dynamics are minimized with stiff suspension components, while road input is isolated better with soft settings.

The second requirement is easier to fulfill with soft settings. Nevertheless, the suspension needs to distribute tire loads in a particular way to keep the desired car balance, so in this case there are other considerations.

Given these requirements, it is not easy to optimize the suspension. Any configuration is partially optimal, making a compromise necessary. This creates the need for powerful analysis tools that make it easier to select the correct compromises.

Those trying to improve the response to vertical road input could easily achieve this with soft springs and dampers. Soft settings also guarantee minimum tire load fluctuations as the tires follow the road with smaller changes in suspension forces.

However, soft settings cannot handle the dynamic forces induced by driving. The suspension influences other vehicle properties such as steering response time, balance, transient lateral stability, driver feel, traction, and (in some cases) aerodynamic issues. The pitch movement at braking and the roll movement in cornering should be controlled because these negatively influence the suspension geometry and the optimal tire contact patch. These movements can be reduced with the appropriate geometry adjustments, but this approach has many side effects. Most racecars make very limited use of them.

Basically, the suspension needs to be as soft as possible for the vertical movement (and the warp movement as well) but stiffer for pitch and roll movements.

Pitch and roll movements are influenced greatly by the driving conditions, meaning the input frequencies indicate a different spectrum than the road input. This issue alone justifies performing the frequency analysis.

Grip is a concept related to how the tire adheres to the road. Such adherence can be increased with the appropriate material (i.e., rubber compounds) and with the load on the tire. Once the tires are selected, the suspension must maintain the load on the tire and keep it as high as possible. Intuitively, the tire should follow the terrain, so the irregularities do not translate into tire deflections. This intuitive reasoning is called *minimizing tire load fluctuations*.

In addition, the nonlinear behavior of grip forces versus tire loads ***(Figure 10.10)*** is ultimately responsible for the car's balance. The suspension is designed usually to take advantage of balance by transferring weight differently on the front and rear axles during roll. This normally is achieved using different settings for roll stiffness on the front and rear, so the suspension settings can make the car oversteer or understeer, depending on the nonlinear characteristics of the tires.

For racecars, tires play a significant role in the suspension analysis and optimization. Compound characteristics are important when combined with the vehicle's suspension because of the low-damping capabilities of the tires. There is also the need to work the tires so they quickly reach operating temperatures. It is fundamental to consider the tire parameters carefully to analyze their influence on grip.

Weight transfers are induced in the suspension because of lateral and longitudinal acceleration. Although the total magnitude of these weight transfers cannot be altered by the suspension, they can be influenced by it in the following ways:

- The suspension can distribute the total amount of weight transfer over four wheels.
- The suspension can minimize weight transfer fluctuations to avoid the lower tire load values that cause the start of breakaway under limited maneuver conditions.

Therefore, if load fluctuations are the key issue to consider when optimizing the suspension, frequency analysis is the tool of choice.

One can attempt to measure suspension movements and loads with the aim of characterizing vehicle response to road input. Ultimately, judgments must be made based on the results obtained. Many racecars install position and load sensors in some key suspension elements. In some cases, accelerometers are installed not only on the car body but also in the wheel hubs.

Measurements are required because not all data is known. Normally appropriate laboratory measuring devices can characterize spring rates and damper rates quite accurately. However, the entire suspension system includes many unknown parameters, and nonlinear elements and sometimes becomes too complex. Therefore, measurements must be taken under operating conditions.

Nevertheless, there are important limitations to measuring data on track. Competition vehicles have nonconstant and nonlinear characteristics that are very difficult to model. A real circuit is the only place to record data for optimizing the suspension settings, but crucial data items such as tire contact patch load are impossible to measure because there is no sensor for a rolling tire. Other inputs, such as

Figure 10.10 *Nonlinear tire characteristics*

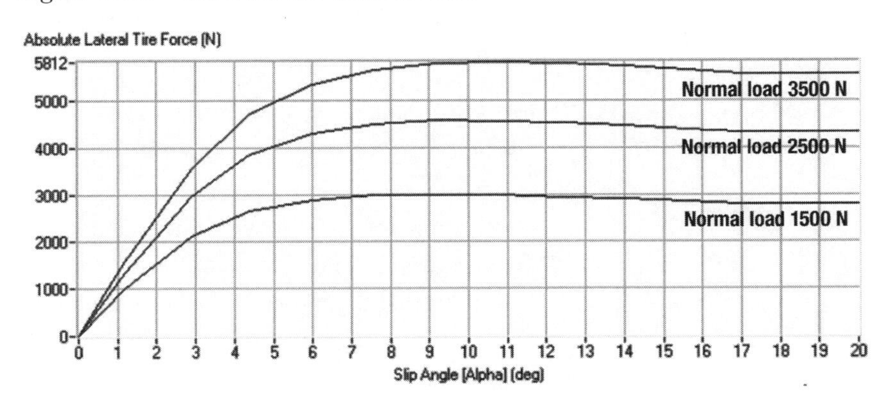

road actual position, are also difficult to measure accurately. In general, the following are limitations for an appropriate analysis:

- Logged data is limited to the available (and possible) sensors,
- Logged data is limited to the logging capabilities (i.e., resolution and frequency),
- Logged data is lap-, circuit-, and weather-dependent.
- On-circuit tests are expensive, particularly when considering the previously mentioned limitations.

The four-post rig is an alternative to some limitations of on-track logged data. This approach offers a more structured test bed that can address the specific measurements and test conditions, therefore providing more concise answers to the characterization of the suspension system.

The four-post rig positions the vehicle on four posts under each wheel, with each post acting as an actuator that simulates the road input. Each actuator is controlled ultimately by a computer that dictates the generated road input and records the measurements to later correlate the two sets of data. These tests normally are conducted by generating movements within a range of frequencies, so it becomes relatively easy to determine the response based on a frequency.

The four-post rig can provide information that avoids some limitations of circuit-logged data from sensors. This poses several advantages:

- It provides sensors for tire contact patch load (not possible to record on track).
- It provides sensors for tire deflection (input road known and wheel position measurable).
- There is no lap/circuit dependency.
- It is cost-effective.

Nevertheless, the four-post rig also has limitations:

- Static tire behavior is quite different from a rolling tire.
- Car balance assessment is unreliable.
- Aerodynamic load simulation requires additional actuators and cannot generate the interaction of aerodynamic forces with the suspension.

In general, the information recorded during a rig test is useful for comparing the performance of different suspension configurations quantitatively and for detecting anomalies or major deviations from the expected behavior. This information is used normally to provide the best damper settings for the spring configurations chosen after tests on the circuit track.

The test rig data analysis relies heavily on frequency analysis. The road input is generated by sweeping a known range of relatively pure frequencies. This makes the analysis simpler and more reliable.

Test rig engineers use rig test data in the form of the following:

- main frequency transfer functions
  - body movement amplitude versus road input amplitude
  - contact patch load fluctuations versus road input amplitude
- Scale parameters (extracted from frequency analysis)
  - suspension elasticity rates versus frequency
  - suspension damping rates versus frequency
- modal components

Measured parameters of a suspension, taken on the circuit or at the four-post rig, generally are used to find the optimal configuration and make decisions about changes to be made.

To set up a suspension, the engineer ideally seeks to do the following:

- minimize the energy absorbed by the vehicle,
- minimize the energy absorbed by the suspension components (dissipated in the dampers),
- maintain the body movement response within acceptable limits for race driving parameters (constraints related to camber and castor angle changes),
- maintain the vehicle height within acceptable limits (under aerodynamic conditions), and
- avoid tire load fluctuation to prevent tire contact breakaway.

Such decisions are easier to make with a proper understanding of the vehicle dynamics' dependency upon the suspension.

Frequency Analysis Versus Time-Space Analysis

An analysis of the actual movement as a function of time is often difficult to use because it is linked too often to the event under analysis. Time-space analysis is used for simulations that attempt to predict the exact response of the system to a very specific event. This can only be useful in the context of an extremely powerful simulation effort, which otherwise can be very expensive and not useful for understanding the basics of the problem to generalize conclusions.

Although not the most intuitive method, frequency analysis is used effectively to understand the behavior and identity of a suspension configuration.

A study of discrete inputs offers a large amount of information about a suspension system. The two graphs in ***Figure 10.11*** illustrate the one-dimensional movement of suspended body (mxy) as a time response to the input movement (xy). This could be the oversimplification of the vehicle movement induced by road input, a step in height (first graph), or a bump (the second graph).

These two graphs reveal much about the suspension behavior but are too specific to the selected input. The graphs may change significantly if the amplitude or the steepness of the input ramps is changed. Therefore, they are not very useful for understanding if the suspension response will be adequate under other input conditions.

Nevertheless, these tests make it easy to obtain an intuitive sense of damper effectiveness. ***Figures 10.12*** and ***10.13*** show an overdamped and an underdamped configuration, respectively, that are obvious from the graph, but a good measure of damper effectiveness is complex to achieve. A much more refined method is required that provides the measurable effectiveness of the suspension and characterizes the response to any input.

The frequency analysis generates the spectrum of frequencies that a signal contains. Road input is basically a noise signal and is understood as a combination of many different pure frequencies. The transformation from the input signal to the frequency spectrum is known as the Fourier transform.

Named after Joseph Fourier, the Fourier transform is one of many mathematical ways used to understand the world better through complex tricks. Fourier transforms have many scientific applications in signal processing, acoustics, optics, physics, and many other areas. In signal processing, it is used systematically to deconstruct a signal into its component frequencies and amplitudes.

The Fourier transform acts as a filter. It combines the input signal with each harmonic function for every frequency to determine how that signal matches with that frequency. Given the input signal [x(t)], the Fourier transformation is calculated with the square integral of ***Equation 10.7***.

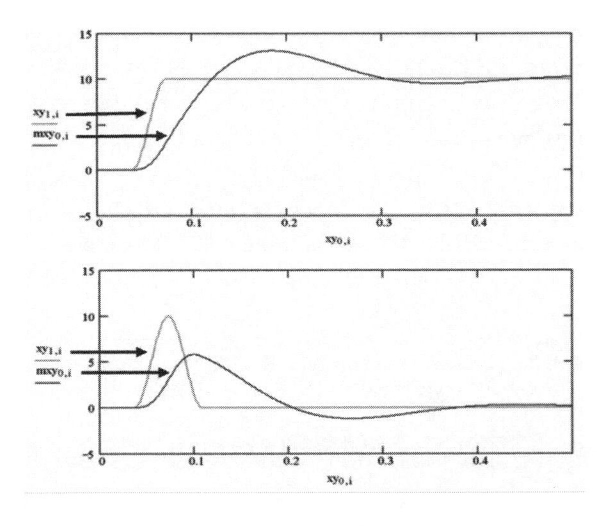

Figure 10.11
Response of suspended mass to a discrete input (stepped input and bump)

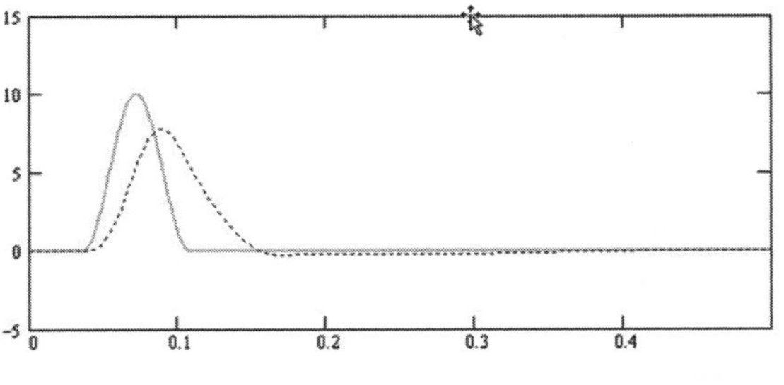

Figure 10.12
Overdamped configuration

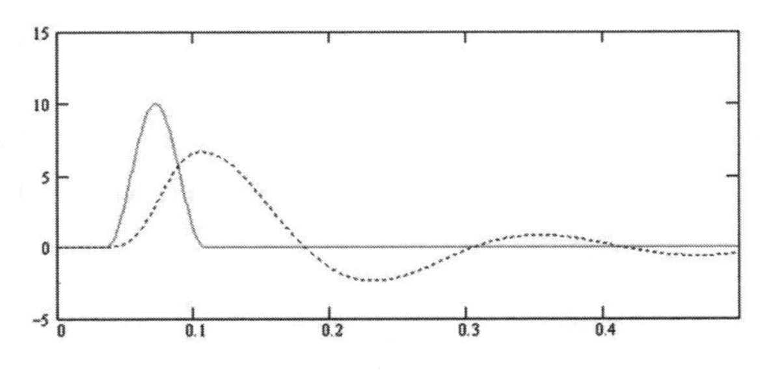

Figure 10.13
Underdamped configuration

$$X(\omega) = \frac{1}{\sqrt{2 \cdot \pi}} \int_{-\infty}^{\infty} x(t) \cdot e^{-i\omega t} dt \qquad \textit{(Eq. 10.7)}$$

The result is another function [X(ω)] that contains the amplitude and phase of each frequency component. In other words, this transformation shows the content of each pure frequency in the input signal [x(t)].

As an example, ***Figure 10.14*** shows the four Fourier transforms of the suspension movement channels, measured by suspension potentiometers. These sensors actually measure the difference between the wheel and body movement. It can be argued that minimizing the peaks in these graphs optimize the suspension settings.

Theoretical Analysis

In any theoretical analysis, one begins with a model of what needs to be analyzed. Four-wheel vehicles and racecars are quite complex systems, so choose an adequate model that characterizes the system with the acceptable degree of detail.

Most suspension analyses begin with a simple model of a suspended mass ***(Figure 10.15)*** that is held by an elastic and damped link. This model is adequate for the first approach but soon is found to be oversimplified.

Figure 10.14 The Fourier transforms of suspension potentiometer signals provide detailed information about vehicle suspension characteristics.

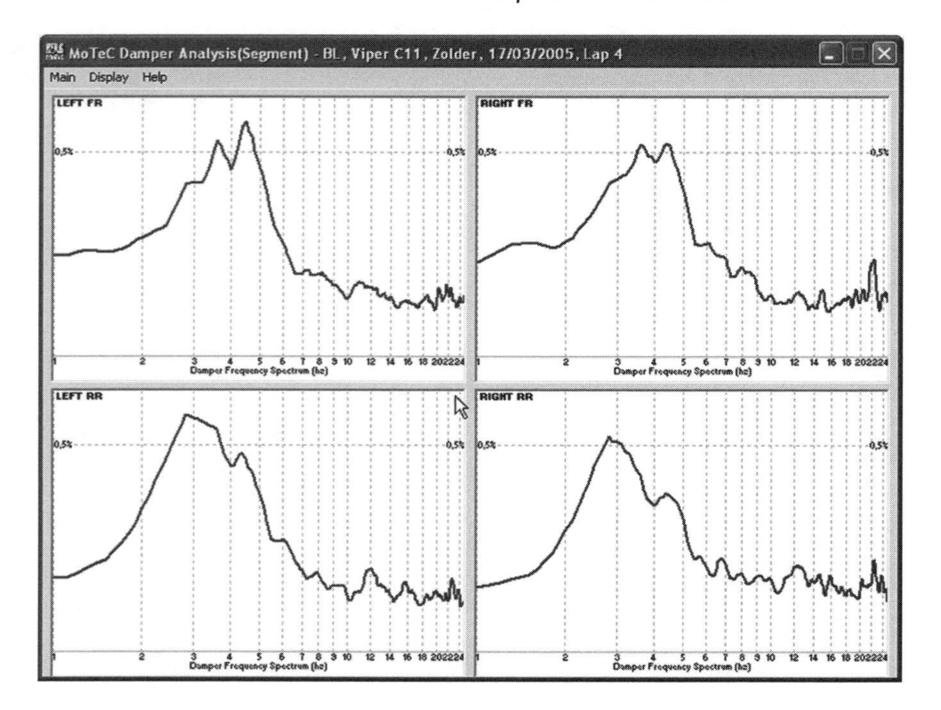

In a vehicle, two important issues that require extra elements:

- The vehicle actually is suspended over four wheels (four suspension links). These links provide three degrees of freedom that may require separate analysis. In addition, the hyperstatic configuration (one more link than degrees of freedom) introduces a fourth component related to weight distribution.
- Because suspension links are far more complex, they require consideration of additional masses (e.g., the wheel hub) and an additional spring and damping to account for tire deflection.

The first issue requires a modal analysis that separates each wheel movement combination:

- vertical movement (all wheels moving in the same direction),
- pitch movement (front and rear wheels moving in opposite directions),
- roll movement (right and left wheels moving in opposite directions), and
- warp (unrelated to body movements but responsible for weight distribution changes when the vehicle is over a nonplanar surface).

The modal analysis is discussed later in this chapter. Discussion continues here about the frequency analysis of a single suspended mass that subsequently can be translated to each modal movement.

The second issue concerning the complexity is covered in most quarter-vehicle models and at a minimum must address the fact that the wheel mass cannot be neglected and the tire deflects under load. This model ***(Figure 10.16)*** takes into account that the wheel mass is a secondary suspended mass with a link to the ground, the tire, and the suspension link to the vehicle body.

Normally, the wheel mass is much smaller than the body mass. Therefore, it is associated with higher frequencies and the actual contact patch load fluctuations.

The single suspended mass system is easy to characterize mathematically. In this system, consider the position, velocity, and acceleration of the suspended mass and the road surface (the input signal). Position ***(Equation 10.8),*** velocity ***(Equation 10.9),*** and acceleration ***(Equation 10.10)*** are derived as time functions.

$$x = x(t) \quad \textit{(Eq. 10.8)}$$

$$V = \frac{d}{dt}x(t) \quad \textit{(Eq. 10.9)}$$

$$a = \frac{d}{dt}v(t) = \frac{d^2}{dt^2}x(t) \quad \textit{(Eq. 10.10)}$$

To solve this system, find how the suspended mass reacts when the road input is excited with a pure harmonic signal.

In this system, the physics of the spring and damper verify ***Equations 10.11*** and ***10.12***.

$$F_m = K \cdot (x_r - x_m) + C \cdot (V_r - V_m) \quad \textit{(Eq. 10.11)}$$

$$F_m = m \cdot a_m \quad \textit{(Eq. 10.12)}$$

F_m is the force applied to the suspended mass resulting from the spring's compression and the damper's reaction. The constant weight of the mass is ignored for the dynamic analysis. If this system is linear (not true in reality but a useful approximation), a pure harmonic signal in the road induces a pure harmonic movement in the mass. ***Equations 10.13*** and ***10.14*** describe the pure harmonic movements.

$$x_r = Xa_r \cdot \sin(\omega \cdot t) + Xb_r \cdot \cos(\omega \cdot t) \quad \textit{(Eq. 10.13)}$$

$$x_m = Xa_m \cdot \sin(\omega \cdot t) + Xb_m \cdot \cos(\omega \cdot t) \quad \textit{(Eq. 10.14)}$$

If the previous differential equations are solved assuming the time functions, one can find a solution that calculates the mass movement parameters (Xa_m and Xb_m) from the road input movement parameters (Xa_r and Xb_r) as ***Equation 10.15***.

$$\begin{pmatrix} Xa_m \\ Xb_m \end{pmatrix} = \frac{M_{KC}}{M_{KC} + M_M} \cdot \begin{pmatrix} Xa_r \\ Xb_r \end{pmatrix} \quad \textit{(Eq. 10.15)}$$

with $M_M = \begin{pmatrix} -m \cdot \omega^2 & 0 \\ 0 & -m \cdot \omega^2 \end{pmatrix}$

$M_{KC} = \begin{pmatrix} K & -C \cdot \omega \\ -C \cdot \omega & K \end{pmatrix}$

This solution is very interesting because it produces a quick way to calculate the reaction of the system to a particular input frequency. Later in this chapter, how this translates into a system frequency analysis is covered.

This method applies to more complex systems to obtain not only the suspended mass induced movement but also the movements of other components of the suspension (wheel) and the fluctuations of the forces involved.

Once the solution to the previous equation is determined, one can represent the suspended mass response to different frequencies. ***Figure 10.17*** illustrates the amplitude ratio between input and induced movement, while ***Figure 10.18*** gives the phase angle between input and induced movement.

These graphs are logarithmic-scaled representations of the suspended mass oscillation amplitude and phase angle given a unitary input signal amplitude. The graph in Figure 10.17 is called the

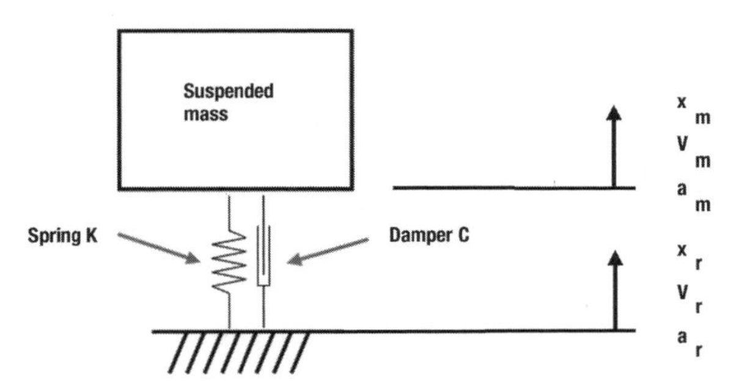

Figure 10.15
Simple suspended mass system with one spring and one damper

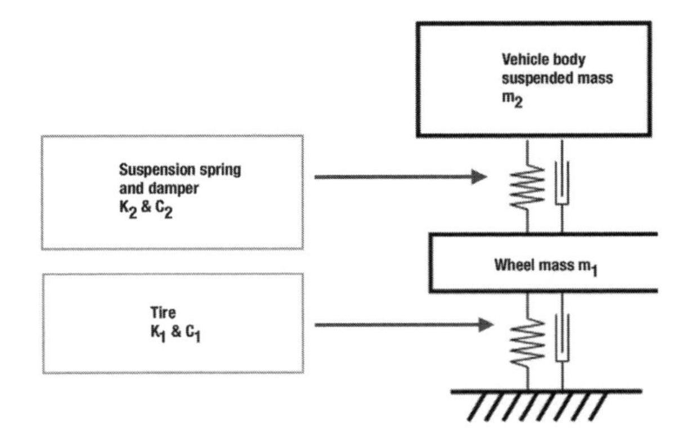

Figure 10.16
Mass-spring system taking into account the nonsuspended wheel mass and tire spring and damping rate

transfer function. This graph is particularly comprehensive as it covers an important range of frequencies. For very low frequencies, the suspended mass copies the input signal (transfer function value near 1). For very high frequencies, the suspended mass remains isolated from the input signal; the transfer function value is near 0. There is, however, a particular frequency in which the transfer function value is at a maximum. This is the resonant frequency, and if there is no damping, the function tends toward the infinite.

This graph calculates the ratio between the input signal magnitude and the suspended mass movement amplitude. The transfer function ***(Equation 10.16)*** is calculated from Equation 10.15.

$$\left|\frac{M_{KC}}{M_{KC}+M_M}\right| \qquad (Eq.\ 10.16)$$

This function in the frequency domain permits identifying the critical points of the vehicle suspension relative to human perceptions:

- resonant frequency,
- resonant amplitude, and
- noise level.

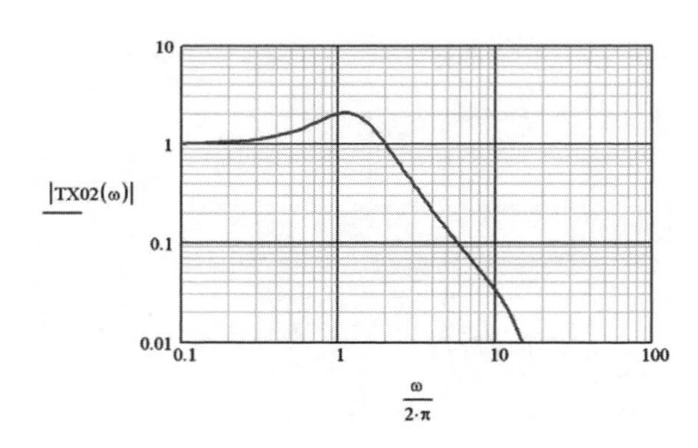

Figure 10.17 *Suspended mass response to different frequencies*

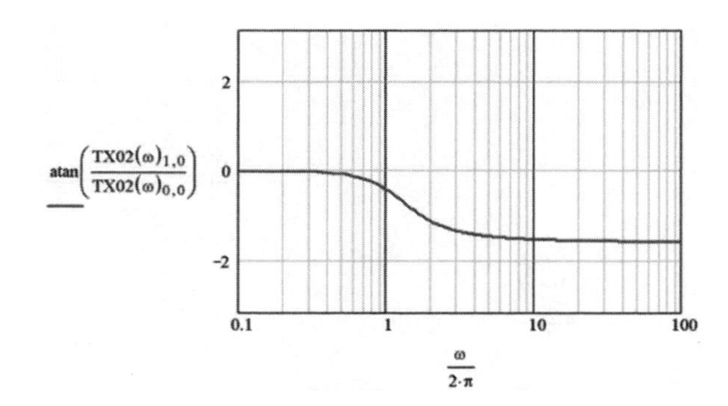

Figure 10.18 *Phase difference between input signal and induced suspended mass movement for different frequencies*

Normally, the simplified model is a first approximation that is only valid when the wheel mass is small (relative to the body mass) and the tire stiff (relative to the suspension stiffness). The main problem is the tire has too little damping, and most of the suspension damping relays only in the suspension damper itself. This, in addition to other considerations, requires that the model include the wheel and tire.

The effect of the wheel is multiple. Since the tire has little damping, it must be supplied with slightly stiffer damping in the suspension. The wheel mass helps isolate the vehicle body from the noise of the road input at the expense of more contact patch load fluctuations. All in all, beyond comfort issues, reduce the wheel weight primarily to avoid the undesired effects of the lack of a damping component for the tire.

In the transfer function, the presence of the wheel creates some modifications in the range of frequencies associated with the wheel mass and the added spring rates that link it to the system.

In this model, the three useful transfer functions are as indicated in ***Figure 10.19***.

- The road-to-body transfer function shows the effect of road input on the vehicle body movement.
- The road-to-wheel transfer function helps to show how the wheel movement follows the road.
- The road-to-wheel deflection transfer function indicates the suspension's capability to maintain the contact patch load and therefore guarantee grip on irregular surfaces.

As an example, a quarter-vehicle with the following configuration is considered:

suspended mass = 300 kg
spring rate = 20 N/mm
tire spring rate = 300 kN/mm
damping rate = 3800 N/ms^{-1}
tire damping rate = 300 N/ ms^{-1}

The effect of the wheel mass is observed in ***Figure 10.20***. The plots illustrate two clear effects of the wheel mass increase:

- The transfer function increases near the wheel natural resonant frequency to approximately 10 Hz, although it quickly decreases at higher frequencies (noise).
- Road-tire deflection increases significantly near the wheel natural resonant frequency of approximately 10 Hz.

These plots clearly show the negative effects of wheel mass, therefore justifying the pursuit of light wheel hubs, resulting in tighter regulations in many racecar competitions.

When the wheel and tire effects are taken into account, identifying the problems associated with inadequate damping rates is easier. The plots in ***Figure 10.21*** show the effect not only on body control but also on tire contact patch load fluctuations when the damper rate deviates from normal.

The effects of underdamped configurations are obvious, and the body frequency response function (FRF) peak is considerably higher. Nevertheless, overdamped configurations produce larger FRF values in the range of frequencies between 1 and 10 Hz, which have a negative impact on grip and comfort. Similar considerations should be taken into account when increasing the suspension stiffness or using softer tire spring rates.

In reality, the suspension link between the road and the vehicle body incorporates many more elements that contribute to the response to the road input. In general, parasitical springs such as nonrigid links or underdamped bushes generate problems at certain frequencies. These problems can be identified with rigorous study of all suspension components. A complete suspension model can include all these elements plus the aerodynamic damping in the scheme in ***Figure 10.22,*** which produces transfer functions illustrated in ***Figure 10.23***.

Suspension Optimization Using Frequency Analysis

To optimize suspension performance, the parameters that best reduce the body movements and tire load fluctuations must be found. These two goals may not respond to the same suspension parameters, so finding the ideal solution requires a compromise.

The transfer function provides a useful tool for characterizing and eventually evaluating the suspension performance. The transfer functions provide two ratios important for evaluating the movement isolation (stability) of the vehicle body and the tire load fluctuations (grip) separately. These ratios are the following:

- transfer function between the road input and body movements, and
- transfer function between the road input and tire deflection.

Figure 10.19 *Suspension system transfer functions*

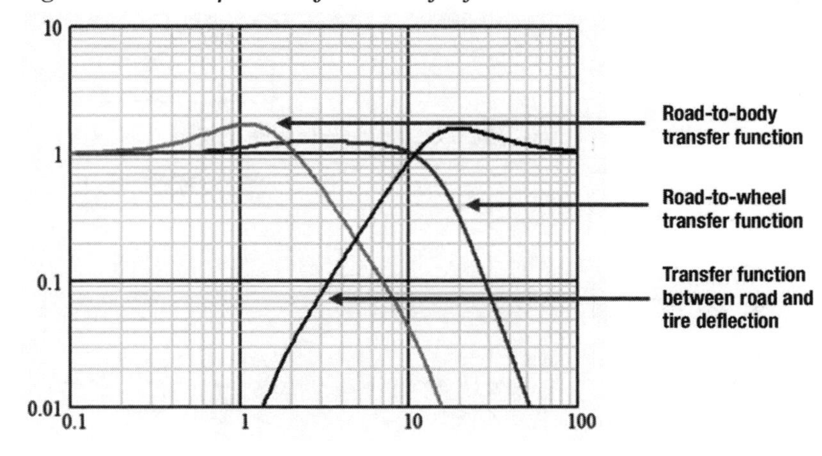

Figure 10.20 *The effect on the transfer functions of nonsuspended mass*

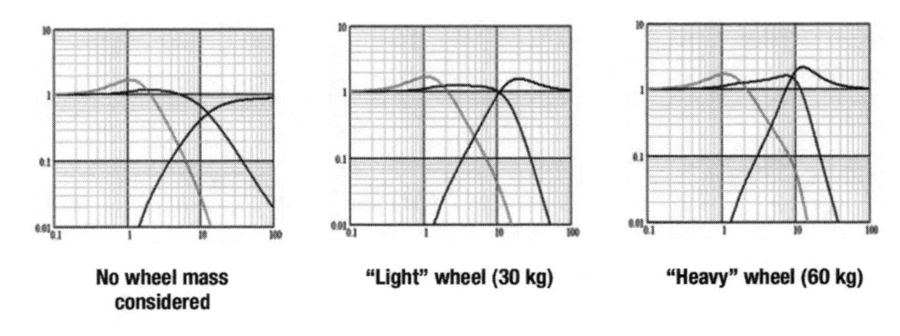

Figure 10.21 *Effect of different damping rates on the transfer functions*

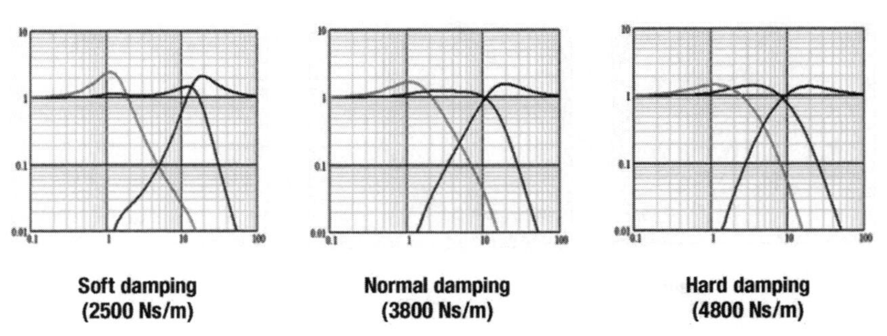

Figure 10.22
Detailed suspension model

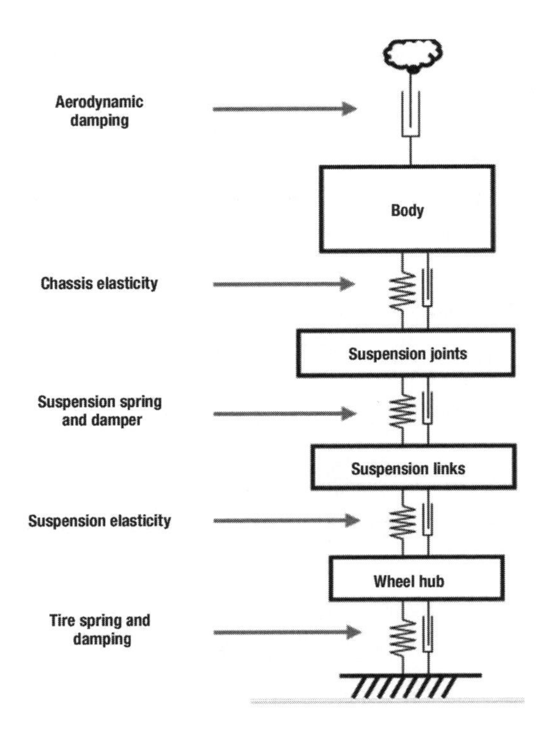

Figure 10.23
Transfer functions of a detailed suspension model as pictured in Figure 10.22

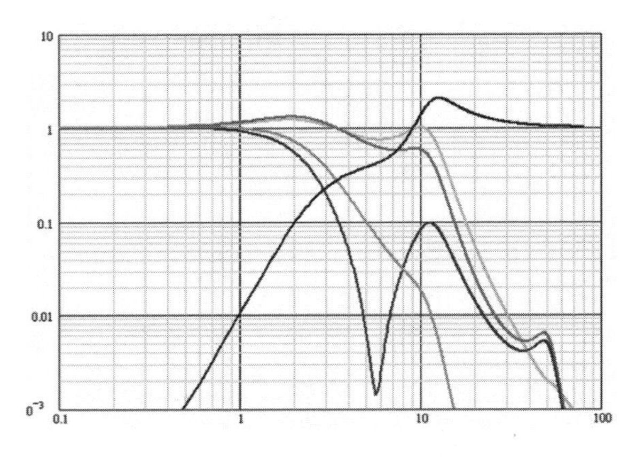

Figure 10.24 *Grip and stability quantification using the transfer functions*

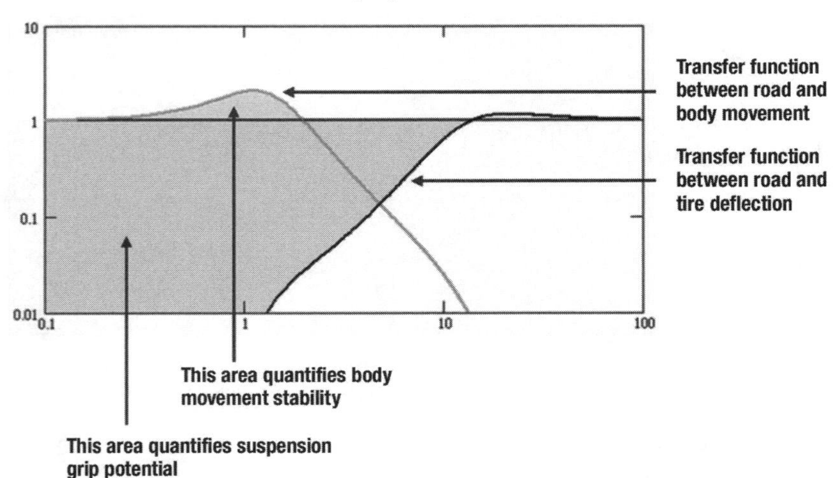

Figure 10.24 illustrates these two transfer functions. Stability is associated inversely to either the maximum value body movement transfer function or the area between the curve and the unitary value (1). Grip can be associated inversely to either the maximum value of the tire deflection transfer function or the area between the curve and the unitary value (1).

The criteria based on the above-mentioned areas can be more precise. Nevertheless, this can be quite dependent on the input signal (road) spectrum. Therefore, the simpler criteria based on maximum values of the curves is used normally. Using the maximum value of each curve, the optimal values of dampers that minimize these values can be calculated. ***Figure 10.25*** shows the optimization curves for both criteria in the same plot.

This graph illustrates that the optimal damper values employed to optimize stability and grip are different. In fact, this situation coincides with the popular idea that a soft suspension provides better grip, although the optimal values are for relatively stiff settings.

Modal Analysis

The quarter-vehicle model utilized thus far is a very useful model for analyzing the vertical movement of the car and, to a certain extent, the pitch movement. These two movements have more to do with the inertia of the vehicle and involve the greatest dynamic loads on the suspension. However, roll movement involves initially a much smaller inertia and then a much stiffer elasticity when the antiroll bars are added to the suspension springs. With this in mind, a different analysis must be performed for each body movement.

Modal analysis[11] combines the four independent wheel movements into something that corresponds more precisely to the body movements known as heave, roll, and pitch. Since these three movements imply combinations of wheel movements, it makes sense that these combinations are used.

The four wheel movements define a system with four degrees of freedom. When transforming the movements into modal combinations, the four modal movements are obtained, three that correspond to the body movements and a fourth associated with additional movement among the wheels ***(Figure 10.26)***.

Considering each wheel movement ***(Equations 10.17–10.20),*** modal movements can be calculated as the average of the appropriate wheel movement combinations ***(Figure 10.27)***.

$$x_H = \frac{1}{4} \cdot (x_{LF} + x_{RF} + x_{LR} + x_{RR}) \quad \text{(Eq. 10.17)}$$

$$x_P = \frac{1}{4} \cdot (x_{LF} + x_{RF} - x_{LR} - x_{RR}) \quad \text{(Eq. 10.18)}$$

$$x_R = \frac{1}{4} \cdot (x_{LF} - x_{RF} + x_{LR} - x_{RR}) \quad \text{(Eq. 10.19)}$$

$$x_X = \frac{1}{4} \cdot (x_{LF} - x_{RF} - x_{LR} + x_{RR}) \quad \text{(Eq. 10.20)}$$

The movement combinations are expressed easily with a linear transformation represented by a matrix shown in ***Equation 10.21***.

$$\begin{bmatrix} x_H \\ x_P \\ x_R \\ x_X \end{bmatrix} = \begin{bmatrix} 1 & 1 & 1 & 1 \\ 1 & 1 & -1 & -1 \\ 1 & -1 & 1 & -1 \\ 1 & -1 & -1 & 1 \end{bmatrix} \cdot \begin{bmatrix} x_{LF} \\ x_{RF} \\ x_{LR} \\ x_{RR} \end{bmatrix} \quad \text{(Eq. 10.21)}$$

With this notation, the four modal movements are defined in the simplest way ***(Equation 10.22)***.

$$M = \begin{bmatrix} 1 & 1 & 1 & 1 \\ 1 & 1 & -1 & -1 \\ 1 & -1 & 1 & -1 \\ 1 & -1 & -1 & 1 \end{bmatrix} \quad \text{(Eq. 10.22)}$$

It is possible to use different coefficients that define off-centered pivot points for each movement, but this adds unnecessary complexity at this point.

This modal analysis is extended easily not only to movements but also to velocities, accelerations, and forces that are combined in a modal fashion as ***Equations 10.23–10.25*** clearly show.

$$X_M = M \cdot X_W \quad \text{(Eq. 10.23)}$$

$$V_M = M \cdot V_W \quad \text{(Eq. 10.24)}$$

$$F_M = M \cdot F_W \quad \text{(Eq. 10.25)}$$

When forces and movements are combined, defining modal elasticities and modal damping ratios becomes possible. For heave movement, the modal elasticity can be defined as ***Equation 10.26***.

$$F_{LF} + F_{RF} + F_{LR} + F_{RR} = K_H \cdot (x_{LF} + x_{RF} + x_{LR} + x_{RR}) \quad \text{(Eq. 10.26)}$$

Figure 10.25 *Optimization curves for body movement and tire deflection transfer functions*

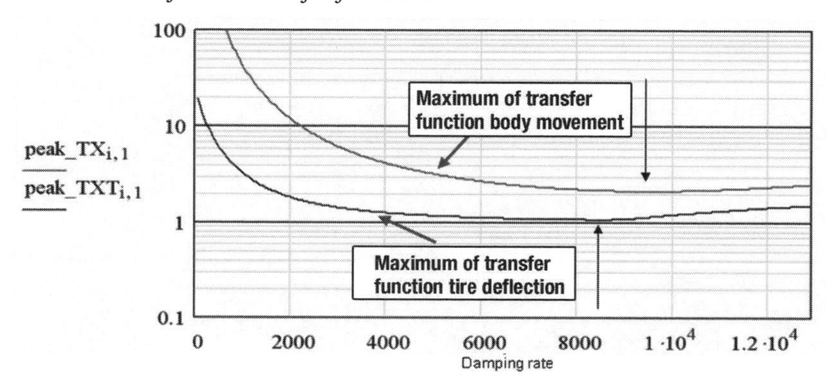

Figure 10.26 *The four suspension modes (Courtesy of Creuat S.L.)*

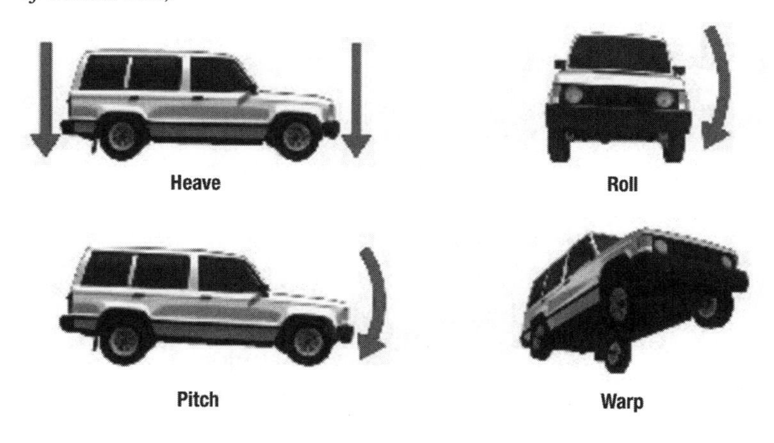

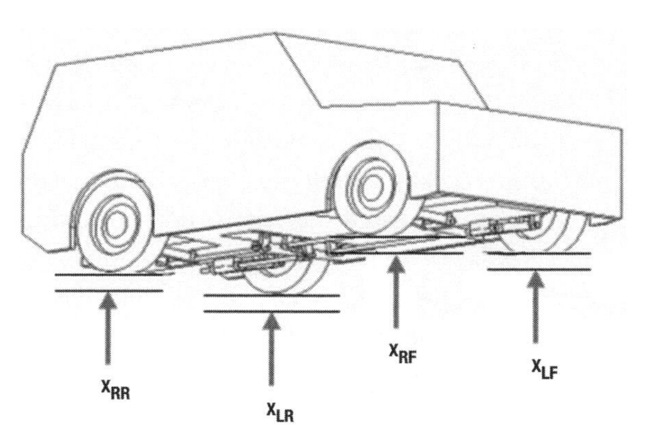

Figure 10.27 *Individual wheel movements (Courtesy of Creuat S.L.)*

The vertical spring rate (K_V) relates the modal vertical movement to the modal vertical force. This is generalized to all movements as ***Equation 10.27***.

$$F = K \cdot X \qquad \text{(Eq. 10.27)}$$

where

$$F = \begin{bmatrix} F_H \\ F_P \\ F_R \\ F_x \end{bmatrix} = \begin{bmatrix} 1 & 1 & 1 & 1 \\ 1 & 1 & -1 & -1 \\ 1 & -1 & 1 & -1 \\ 1 & -1 & -1 & 1 \end{bmatrix} \cdot \begin{bmatrix} F_{LF} \\ F_{RF} \\ F_{LR} \\ F_{RR} \end{bmatrix}$$

and

$$X = \begin{bmatrix} x_H \\ x_P \\ x_R \\ x_x \end{bmatrix} = \begin{bmatrix} 1 & 1 & 1 & 1 \\ 1 & 1 & -1 & -1 \\ 1 & -1 & 1 & -1 \\ 1 & -1 & -1 & 1 \end{bmatrix} \cdot \begin{bmatrix} x_{LF} \\ x_{RF} \\ x_{LR} \\ x_{RR} \end{bmatrix}$$

The elasticities matrix (K) in the modal space and the elasticities on the wheel movement space are related with ***Equations 10.28*** and ***10.29***.

$$K = M^{-1} \cdot R \cdot M \qquad \text{(Eq. 10.28)}$$

$$K = \begin{bmatrix} K_H & a & 0 & 0 \\ b & K_P & 0 & 0 \\ 0 & 0 & K_R & c \\ 0 & 0 & c & K_X \end{bmatrix} \qquad \text{(Eq. 10.29)}$$

The matrix K offers the advantage that it shows more intuitive values when addressing the vehicle movement analysis. The main modal elasticities are in the matrix diagonal:

K_H	**vertical elasticity**
K_P	**pitch elasticity**
K_R	**roll elasticity**
K_X	**warp elasticity**

In addition, the off-diagonal coefficients *a, b, c,* and *d* represent the differences in elasticities between the front and rear wheel springs. In a conventional suspension, *a* and *b* relate to the front/rear differences in spring rates and *c* and *d* to the differences of *a* and *b* plus the front/rear antiroll bar stiffness. These parameters are quite important because they define the car's balance. Specifically, *d* is related to the oversteer or understeer tendency of the car.

The same transformation can be applied to damper rates ***(Equation 10.30)***.

$$C = \begin{bmatrix} C_H & a & 0 & 0 \\ b & C_P & 0 & 0 \\ 0 & 0 & C_R & c \\ 0 & 0 & c & C_X \end{bmatrix} \qquad \text{(Eq. 10.30)}$$

This matrix tool is useful for nonconventional suspension systems as well as for easy integration of the different chassis components that interact with the suspension (e.g., tires or chassis rigidity).

In a conventional suspension, the elasticities are defined primarily by the springs, antiroll bars, tires, and chassis. These components are combined either in series or parallel. The following text shows how to portray them together easily with the matrix notation.

When two elements are in parallel, the net result is the addition of elasticities. In other words, the springs and antiroll bars are added to provide the antiroll elasticity ***(Equation 10.31)***.

$$K_{roll} = K_{rollSPRINGS} + K_{rollARB} \qquad \text{(Eq. 10.31)}$$

As an example, consider the following configuration:

spring rate front = 24 N/mm
spring rate rear = 18 N/mm
antiroll bar rate front = 22 N/mm
antiroll bar rate rear = 10 N/mm
tire spring rate front = 350 N/mm
tire spring rate rear = 300 N/mm

With these spring and roll bar rates, the elasticities matrix shown in ***Equation 10.32*** are obtained.

When two elements are in series, such as the suspension spring and the tire spring, the total spring rate can be calculated as ***Equation 10.33***. In fact, if this is done with matrices, the expression is slightly different ***(Equation 10.34)***. With the previous example, one has ***Equation 10.35***. This calculation method is very useful for understanding the influence of tires and chassis rigidity on the car balance.

Modal Frequency Issues

When following the quarter-vehicle model, some assumptions are made that are partially valid. However, as stated previously, the dynamics of vertical movement cannot be applied always to the other body movements such as pitch and roll.

The differences of inertias and spring rates (the stiffer roll spring rate resulting from the anti roll bars) must be taken into account to analyze the dynamics of such movements.

Additionally, the warp movement requires a different analysis because it does not involve any vehicle body movements. In this case, the dynamic analysis takes into account the only moving elements, which are the wheels.

The vehicle body normally is considered to be a rigid solid and is analyzed as such. This is a good assumption for all dynamic considerations except for car balance, which is analyzed later. The vehicle body, as with any solid, has six degrees of freedom, but the suspension has only three: heave, pitch, and roll. For this analysis, the other degrees of freedom are considered solid.

The vertical inertia is the vehicle mass and its distribution. There can be different front/rear weights, and this has other implications. However, for roll and pitch, the mass distribution is important. These movements do not direct the same mass on each wheel because normally the mass is concentrated in the center of the vehicle, thereby reducing the inertia to roll and pitch.

When calculating the effect of the mass concentration, it is useful to identify the distance where the central mass should be separated to obtain the vehicle inertia, as indicated in ***Figure 10.28***.

Normally, this distance (W_r) is smaller than the wheel track (W) distance. In this situation, when considering the roll movement, the movement of the two analogous masses is less than if they are acting on the vertical of the wheels. Therefore, they effectively behave as if the wheel equivalent mass is a fraction of the quarter-vehicle mass used for the vertical movement calculation.

The pitch movement indicates a similar problem. Due to the small overhanging masses in a car, distance L_p is less than the wheel base (L) distance (***Figure 10.29***). The results from the previous paragraph also apply here.

Given the previous considerations, it is necessary to calculate the wheel equivalent mass for roll and pitch movements. Considering that these two movements have an instantaneous center of rotation above the ground and not factoring in other considerations, a first approximation of the mass reduction is ***Equations 10.36*** and ***10.37***.

$$M_R = \frac{M \cdot \left(W_R^2 + 4 \cdot (CMH - CMR)\right)}{W^2} \qquad \text{(Eq. 10.36)}$$

$$M_P = \frac{M \cdot \left(L_R^2 + 4 \cdot (CMH - CMP)\right)}{L^2} \qquad \text{(Eq. 10.37)}$$

with M_R = wheel equivalent mass for roll
M_P = wheel equivalent mass for pitch
M = total vehicle mass

On many street cars, the mass reduction for roll is less than 70% of the vehicle mass, with much lower values for racecars (e.g., between 40% and 20%). The mass reduction for pitch can be occasionally more than 100% of the vehicle mass but is

$$\begin{bmatrix} 21 & 3 & 0 & 0 \\ 3 & 21 & 0 & 0 \\ 0 & 0 & 21 & 3 \\ 0 & 0 & 3 & 21 \end{bmatrix} + \begin{bmatrix} 0 & 0 & 0 & 0 \\ 0 & 0 & 0 & 0 \\ 0 & 0 & 16 & 6 \\ 0 & 0 & 6 & 16 \end{bmatrix} = \begin{bmatrix} 21 & 3 & 0 & 0 \\ 3 & 21 & 0 & 0 \\ 0 & 0 & 37 & 9 \\ 0 & 0 & 9 & 37 \end{bmatrix} \qquad \text{(Eq. 10.32)}$$

$$K_{Total} = K_{SPRING} \cdot K_{TIRE} \cdot \left(K_{SPRING} + K_{TIRE}\right)^{-1} \qquad \text{(Eq. 10.33)}$$

$$K_{Total} = K_{SPRING} \cdot (K_{SPRING} + K_{TIRE})^{-1} \cdot K_{TIRE} \qquad \text{(Eq. 10.34)}$$

$$\begin{bmatrix} 21 & 3 & 0 & 0 \\ 3 & 21 & 0 & 0 \\ 0 & 0 & 37 & 9 \\ 0 & 0 & 9 & 37 \end{bmatrix} \cdot \left(\begin{bmatrix} 21 & 3 & 0 & 0 \\ 3 & 21 & 0 & 0 \\ 0 & 0 & 37 & 9 \\ 0 & 0 & 9 & 37 \end{bmatrix} + \begin{bmatrix} 325 & 25 & 0 & 0 \\ 25 & 325 & 0 & 0 \\ 0 & 0 & 325 & 25 \\ 0 & 0 & 25 & 325 \end{bmatrix} \right)^{-1} \cdot \begin{bmatrix} 325 & 25 & 0 & 0 \\ 25 & 325 & 0 & 0 \\ 0 & 0 & 325 & 25 \\ 0 & 0 & 25 & 325 \end{bmatrix} = \begin{bmatrix} 19.7 & 2.7 & 0 & 0 \\ 2.7 & 19.7 & 0 & 0 \\ 0 & 0 & 33.1 & 7.5 \\ 0 & 0 & 7.5 & 33.1 \end{bmatrix} \qquad \text{(Eq. 10.35)}$$

normally close to 90%; with lower values for racecars, it can be between 60% and 80%.

The fact that the wheel equivalent mass is less than the quarter-vehicle mass has important implications for calculating the dynamics of every specific movement. Taking into account the effect of mass reduction and different spring rates ***(Figure 10.30)***, the analysis of modal movements yields quite interesting results.

The graphs in ***Figure 10.31*** are from a street car. The transfer functions for roll indicate a relatively important increase in body movements because the damper settings cannot cope with the antiroll bar stiffness added to that movement.

Racecars behave in a much more radical manner to roll. Roll inertia is reduced by design and roll stiffness increased because of race-specific requirements. In these circumstances, the wheel mass relative to the wheel-equivalent body mass is much more important, the tire spring compared to the roll stiffness is comparable, and the system gets trapped with the low damping rate from the tire ***(Figure 10.32)***.

The fourth movement associated with wheels usually is referred to as warp or axle crossing. This is not associated to any body movement and is only responsible for the distribution of load among the wheels.

This movement is usually ignored. Conventional suspensions cannot modify the warp resistance, which involves all suspension components. Four-post rig tests use the warp movement to perform measurements for the roll stiffness, as the two are relatively similar if the body rigidity is great enough.

Wheel load distribution is important for two reasons:

- It influences grip as the fluctuations of loads reduce grip.
- Lateral load transfers define the car balance and are therefore fundamental in the car setup. The difference of weight transfers front/rear is the direct result of converting the roll movement into warp forces.

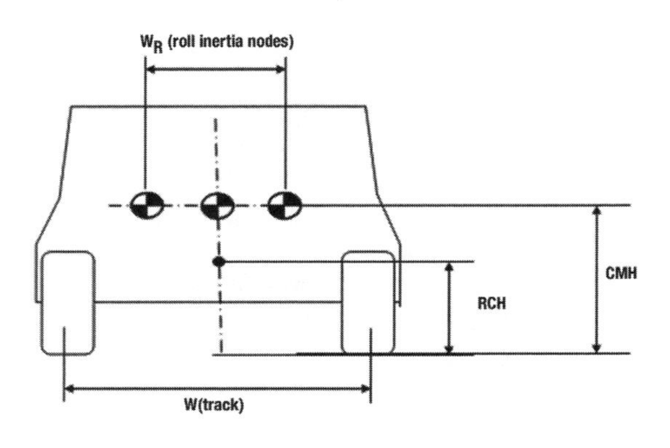

Figure 10.28
Roll inertia calculation (Courtesy of Creuat S.L.)

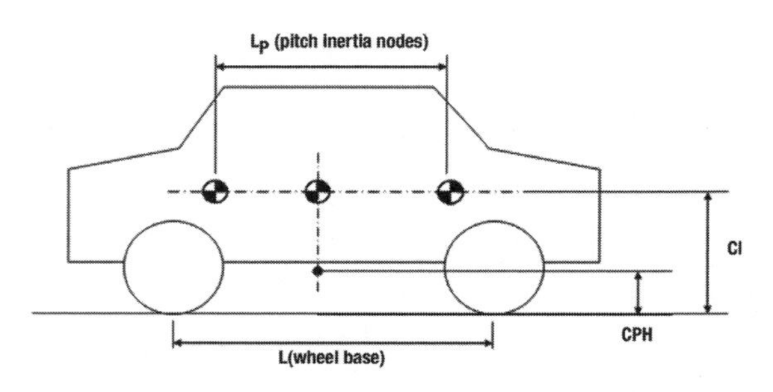

Figure 10.29
Pitch inertia calculation (Courtesy of Creuat S.L.)

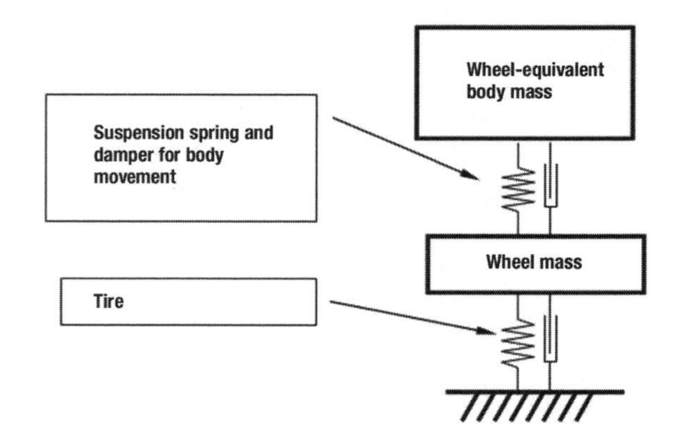

Figure 10.30
Mass-spring system with wheel-equivalent body mass

The car balance can be calculated with the modal matrix in the following way. Given the suspension modal matrix K (Equation 10.33), the car balance (weight transfer distribution) is directly proportional to ***Equation 10.38***.

$$\frac{K_R + d}{K_R - d} - 1 \qquad \textit{(Eq. 10.38)}$$

If the suspension modal matrix for the suspension, tires, and chassis is known, the system matrix can be calculated and the car balance found.

As an example, with the following suspension elasticities matrix ***(Equation 10.39)***, the apparent balance ratio is shown in ***Equation 10.40***. Then factor in the tires and chassis stiffness ***(Equations 10.41*** and ***10.42)***. Then the total suspension matrix is calculated as ***Equation 10.43***.

Therefore, one can assume ***Equation 10.44***. The matrix in ***Equation 10.45*** yields a much lower weight transfer factor. The matrix demonstrates that the finite chassis stiffness softens the warp mode.

The dynamic analysis of the warp movement must consider that a pure warp movement involves no body movement, so the transfer function is only between the road input and wheel movement. Tire deflection is calculated easily from the difference between the wheel movement and road input itself.

The first graph in ***Figure 10.33*** illustrates the transfer functions of the wheel movement and tire deflection of a street car. The comfort considerations in a street car explain the insufficient damping rates that cause some loss of grip at certain wheel-tire resonant frequencies. The peak in the black curve indicates a high load fluctuation linked to the tire deflections.

In a racecar (bottom graph in Figure 10.33), the lower wheel relative mass and the much higher damping rates reduce the grip problems at the wheel resonant frequency. Nevertheless, the higher warp stiffness indicates a fixed transfer function value for lower frequencies that can become important if the road input contains a low-frequency spectrum such as undulations of the road surface.

Nonlinear Considerations

Bearing in mind that linear models are useful as a simple approximation of reality, suspension nonlinearities that limit suspension travel affect any analysis. Dampers normally are neither linear nor symmetric. Nonlinearity implies that the system

Figure 10.31 *Transfer functions for heave, pitch, and roll for a street car*

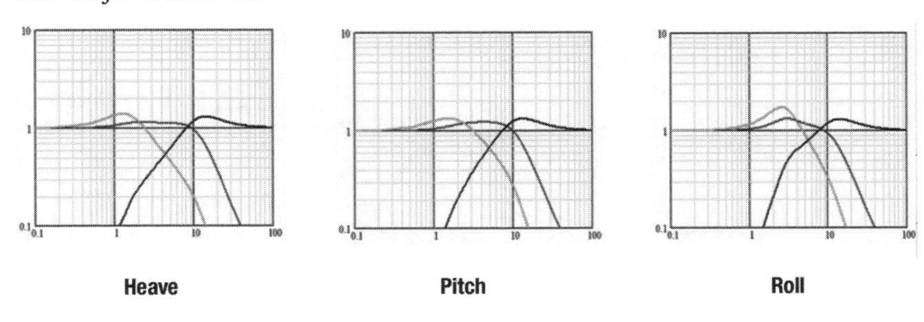

Figure 10.32 *Transfer functions for heave, pitch, and roll for a racecar*

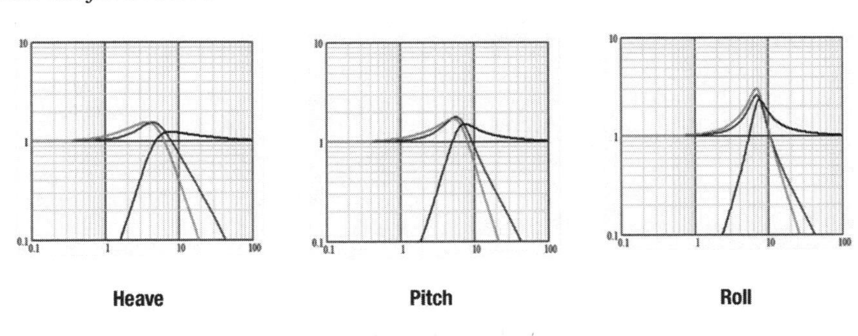

$$M_{Suspension} = \begin{bmatrix} 75 & 15 & 0 & 0 \\ 15 & 75 & 0 & 0 \\ 0 & 0 & 300 & 135 \\ 0 & 0 & 135 & 300 \end{bmatrix} \quad \text{(Eq. 10.39)}$$

$$\frac{300+135}{300-135} - 1 = 163.6\% \quad \text{(Eq. 10.40)}$$

$$M_{Tire} = \begin{bmatrix} 360 & 20 & 0 & 0 \\ 20 & 360 & 0 & 0 \\ 0 & 0 & 360 & 20 \\ 0 & 0 & 20 & 360 \end{bmatrix} \quad \text{(Eq. 10.41)}$$

$$M_{Chassis} = \begin{bmatrix} inf & 0 & 0 & 0 \\ 0 & inf & 0 & 0 \\ 0 & 0 & inf & 0 \\ 0 & 0 & 0 & 3000 \end{bmatrix} \quad \text{(Eq. 10.42)}$$

$$M_{Total} = M_{Chassis} \cdot \left[M_{Chassis} + M_{Tire} \cdot (M_{Tire} + M_{Suspension})^{-1} \cdot M_{Suspension} \right]^{-1} \cdot \left[M_{Tire} \cdot \left(M_{Tire} + M_{Suspension} \right)^{-1} \cdot M_{Suspension} \right] \quad \text{(Eq. 10.43)}$$

$$M_{Total} = \begin{bmatrix} 62 & 11 & 0 & 0 \\ 11 & 62 & 0 & 0 \\ 0 & 0 & 156 & 44 \\ 0 & 0 & 44 & 149 \end{bmatrix} \quad \text{(Eq. 10.44)}$$

$$\frac{156+44}{156-44} - 1 = 78.6\% \quad \text{(Eq. 10.45)}$$

reacts differently depending on the amplitude of the signal. Different high- and low-speed damping also affects the way the vehicle jacks in response to road inputs.

In general, more linear damper characteristics imply that a vehicle responds in a more consistent way to road inputs. Linear dampers can make the vehicle faster, but the driver typically has a more difficult time taking advantage of the increased speed. Spring progressivity affects vehicle roll and pitch angles during maneuvers and can change the lateral balance of the vehicle along the corner.

Figure 10.33 Transfer functions for warp mode. The upper graph is taken from a street car, the lower one from a racecar.

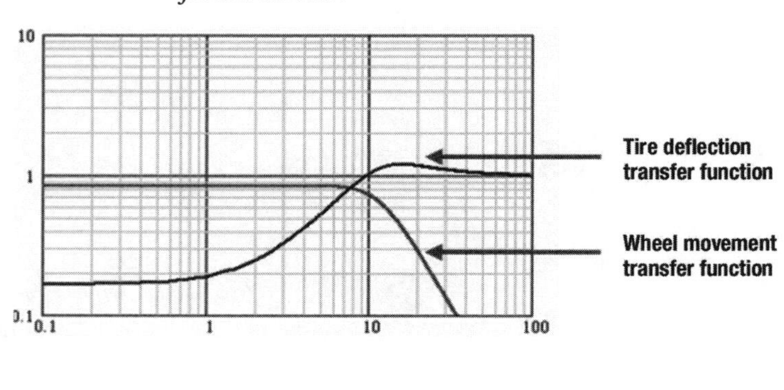

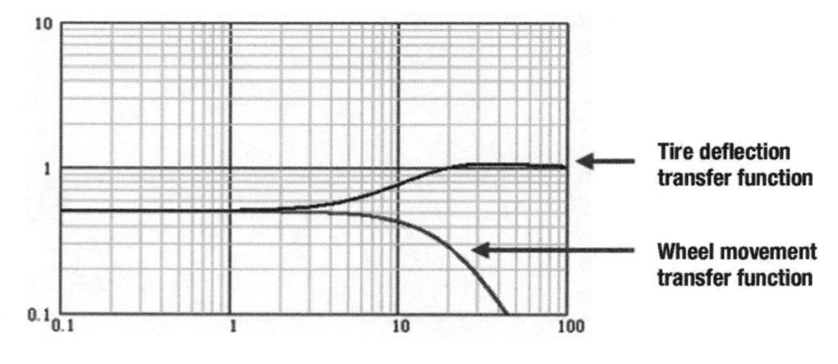

Figure 10.34 Test data illustrated in the time domain

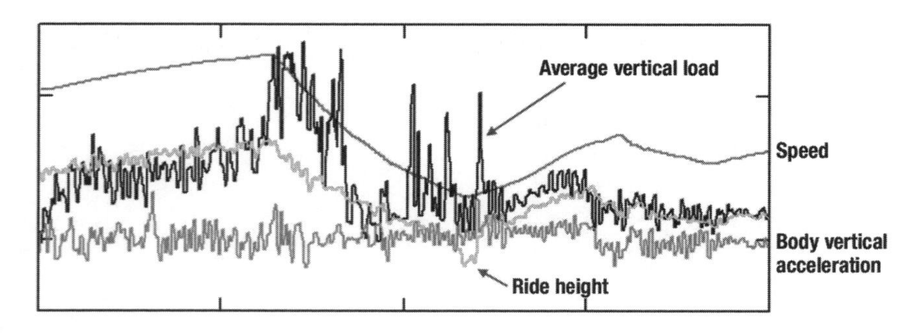

Assuming linear components generally is acceptable as long as the measured circumstances are valid within a predetermined window. This can still be useful for analyzing the entire system in operation.

Frequency Analysis from Sensor Data

Currently, various post-rig layouts perform a frequency analysis by inducing movements through the suspension and analyzing the outcome. These tests are relatively simple and are routine for many teams.

Because post-rig tests do not test the actual car on the track, there are a few issues not easily resolved such as aerodynamics or wheel behavior. In addition, chances are that the car on the track is not necessarily identical to the car tested months ago on the rig.

The alternative is using the sensors on the car to perform the rig test on the track. Many cars have load and position sensors for each wheel. Unfortunately, chassis accelerometers primarily are limited to one 3-axial accelerometer, which for this purpose results in useful data for only the vertical modal movements.

Using on-vehicle sensors, measurements of the wheel position and wheel load (not the tire load) are obtained easily. For vertical movement, the suspension input is the average of the wheel position and load. In addition, there is the body vertical acceleration measured by the accelerometer. ***Figure 10.34*** shows the following signals as a function of time:

- average load,
- average ride height,
- body vertical acceleration (G_{vert}), and
- reference speed.

The sample in Figure 10.34 was obtained from an actual racecar at a 100-Hz sampling rate. The graph indicates 20 sec from a lap. This data is included to demonstrate that raw data does not provide much meaningful information regarding actual car behavior.

The frequency analysis works with the second derivatives of the load and the position of the wheels. To perform this analysis, the following values are taken into account:

the ¼ body mass ($M_{1/4}$)	**220 kg**
the wheel mass (m_{wheel})	**35 kg**
the tire spring rate (K_{tire})	**480 N/mm**

These values are necessary to relate the measurements of wheel movement and force with the body movement to infer the actual tire load. These are considered with the equations of dynamics ***(Equations 10.46–10.48)***.

$$F_{CP} - F_{Suspension} = m_{wheel} \cdot a_{hub} \qquad \text{(Eq. 10.46)}$$

$$F_{Suspension} = M_{1/4} \cdot G_{vert} \qquad \text{(Eq. 10.47)}$$

$$a_{hub} = G_{vert} + a_{suspension} \qquad \text{(Eq. 10.48)}$$

with F_{CP} = tire contact patch force
$F_{Suspension}$ = suspension force
a_{hub} = hub vertical acceleration
G_{vert} = body vertical acceleration
$a_{suspension}$ = suspension acceleration

These dynamic equations relate forces and accelerations with the relevant masses of the car body and the wheel hub.

Hub acceleration (a_{hub}) also can be obtained directly with a specific sensor mounted on the wheel upright. In this example, it is calculated from the sum of body acceleration and the suspension movement acceleration derived from the position sensors. This process impairs high-frequency data because of the low sampling rates in time and space (sensor resolution) of the position sensors. ***Figure 10.35*** shows a_{hub} and G_{vert} in the time domain, a graph that does not provide indication of vehicle behavior. To draw any significant conclusions, these measurements must be illustrated in the frequency domain.

The contact patch load can be calculated from the wheel hub dynamics. Once the tire load is known, assuming the tire behaves as a spring, it is possible to determine the contact patch position ***(Equations 10.49*** and ***10.50)***.

$$F_{CP} = F_{Suspension} + m_{wheel} \cdot \left(G_{vert} + a_{suspension}\right) \qquad \text{(Eq. 10.49)}$$

$$F_{CP} = K_{tire} \cdot \left(x_{CP} - x_{hub}\right) \qquad \text{(Eq. 10.50)}$$

with x_{CP} = contact patch movement
x_{hub} = hub movement

From the last equation follows ***Equation 10.51***.

$$\frac{d^2}{dt^2} F_{CP} = K_{tire} \cdot \left(a_{CP} - a_{hub}\right) \qquad \text{(Eq. 10.51)}$$

That provides the tire's contact patch acceleration ***(Equation 10.52)***.

$$a_{CP} = \frac{\frac{d^2}{dt^2} F_{CP}}{K_{tire}} + a_{hub} \qquad \text{(Eq. 10.52)}$$

Armed with the contact patch acceleration (a_{CP}), the hub acceleration (a_{hub}), and the body acceleration (a_{body}), the FRF now can be calculated.

It is interesting to see the frequency domain of these three accelerations, as indicated in ***Figure 10.36***.

a_{hub} and G_{vert} are the data obtained from sensors. a_{CP} is the calculated acceleration of the contact patch, assuming that the tire behaves as a pure spring.

The actual shape of these graphs is track- and speed-specific and depends on the track surface profile. It is the relationship between these graphs that characterizes the vehicle's response through its suspension.

Figure 10.35 Hub and body vertical acceleration in the time domain

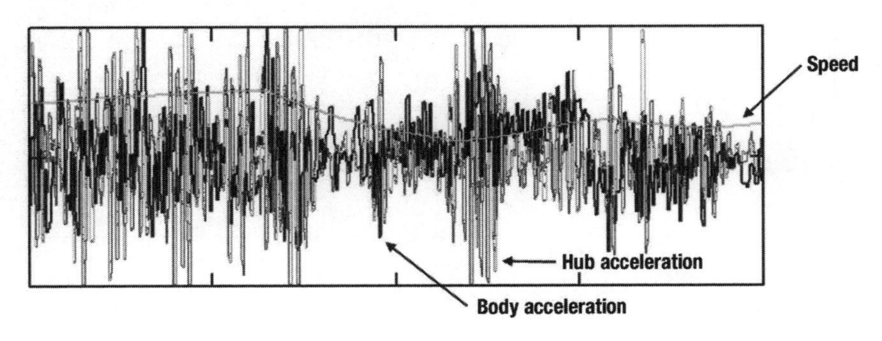

The graphs are coherent; the low frequencies indicate the suspension is not absorbing the movement, allowing the body to follow the track profile. At higher frequencies, the body has much less response, and the suspension absorbs most of the movements induced by road irregularities.

With these frequency response functions, the FRF diagrams can be calculated as in the earlier theoretical cases. From the frequency domain data, one can calculate ***Equations 10.53*** and ***10.54***.

Figure 10.36 Frequency domain graph of body, hub, and tire contact patch acceleration

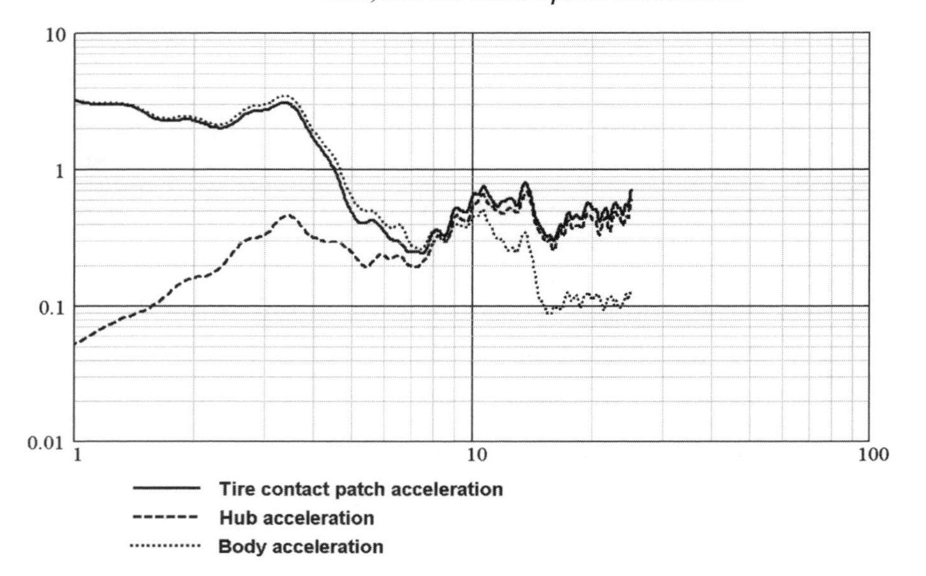

Figure 10.37 Body transfer and tire frequency response functions

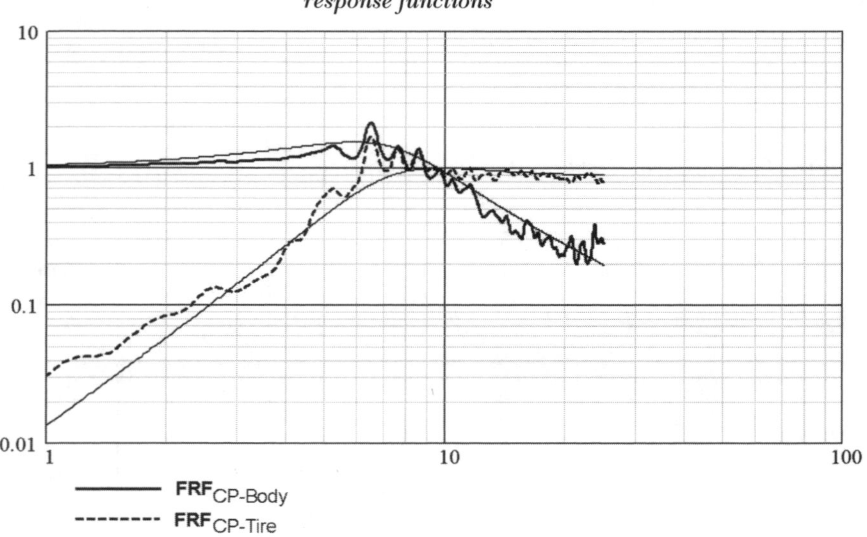

$$FRF_{CP-body} = \frac{Ft(G_{vert})}{Ft(a_{CP})} \quad (Eq.\ 10.53)$$

$$FRF_{CP-hub} = \frac{Ft(a_{hub})}{Ft(a_{CP})} \quad (Eq.\ 10.54)$$

with $FRF_{CP\text{-}body}$ = frequency response function tire contact patch (body)

$FRF_{CP\text{-}hub}$ = frequency response function tire contact patch (hub)

Figure 10.37 shows the two FRFs for the body transfer ($FRF_{CP\text{-}body}$) and the FRF for the tire ($FRF_{CP\text{-}tire}$.) The thin graphs are the theoretically calculated FRFs based on the known values of masses and spring rates of the vehicle suspension.

What is interesting about this is that relatively accurate measurements can be obtained resulting in a suspension FRF that is reasonably consistent with those of a vehicle on the track. These calculations include some data filtering, windowing techniques, and data segmentation based on vehicle speed. Further refinements can be done, taking into account suspension geometry details such as the antidive that in this vehicle accounts for some suspension-added loads during braking.

This analysis was performed only for vertical movement. It is feasible to produce the same analysis for other body movements such as pitch and roll, provided one has accelerometers in the vehicle that can measure the accelerations of these movements. A sensible configuration requires one vertical acceleration sensor on each corner of the car body so that it is possible to calculate the acceleration of the extension of the body over the vertical of the wheel contact patch.

CHAPTER 11

AERODYNAMICS

Aerodynamics are a key factor in the overall performance of a racecar. Aerodynamic downforce increases the tires' capability to develop cornering force, while drag reduces the engine power available for accelerating the vehicle. This chapter offers techniques for measuring aerodynamic forces with the data acquisition system.

Aerodynamic Measurements

Every object moving through the atmosphere experiences an aerodynamic force proportional to its shape, size, and speed as well as the density of the air surrounding it. The direction of the aerodynamic force generally varies from the direction of travel. It has a vertical and horizontal directional component. This also applies to a racecar moving through the air, which experiences an aerodynamic force that can be divided into a horizontal (drag) and vertical component (downforce). These are the two most common aerodynamic measurements in racecar engineering.

Drag is the resisting force acting on the vehicle, which is parallel and in the opposite direction to the direction of travel, primarily influencing the vehicle's top speed on the straights. Downforce is the vertical component of the aerodynamic force experienced by the vehicle. A rise in downforce increases the cornering potential of the tires and potential cornering speeds ***(Figure 11.1)***.

Analogous to the center of gravity in mechanical calculations, a point is defined as the area where the aerodynamic forces act on the body. This point is the center of pressure, and at its location there is no aerodynamic moment. The longitudinal location of the center of pressure represents the downforce distribution between the front and rear axles (i.e., the aerodynamic balance). This balance can be very sensitive to changes in ride height.

The two aerodynamic force components can be calculated using ***Equations 11.1*** and ***11.2***.

$$D = \frac{1}{2} \cdot \rho \cdot V^2 \cdot C_D \cdot A \qquad \text{(Eq. 11.1)}$$

$$L = \frac{1}{2} \cdot \rho \cdot V^2 \cdot C_L \cdot A \qquad \text{(Eq. 11.2)}$$

with
- D = aerodynamic drag force
- L = aerodynamic downforce (L = lift)
- V = vehicle speed
- C_D = drag coefficient
- C_L = downforce coefficient (or lift coefficient)
- A = vehicle frontal area

In these equations the term $\frac{1}{2} \cdot \rho \cdot V^2$ is the dynamic pressure, which is proportional to the difference between the static pressure away from the car and local air pressure at the point where a measurement is taken. The measurement and consequences of dynamic pressure are covered later in this chapter. Frontal area can be estimated as illustrated in ***Figure 11.2***.

If the dimensionless aerodynamic coefficients (C_D and C_L) are known (e.g., from a wind

__Figure 11.1__ Aerodynamic forces acting on a moving vehicle (Courtesy of www.bertlongin.com)

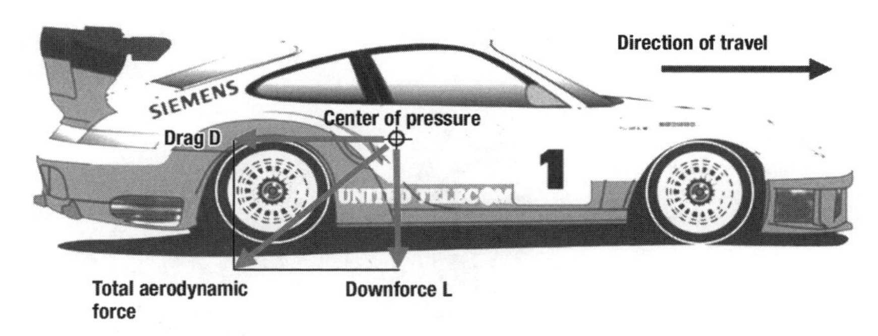

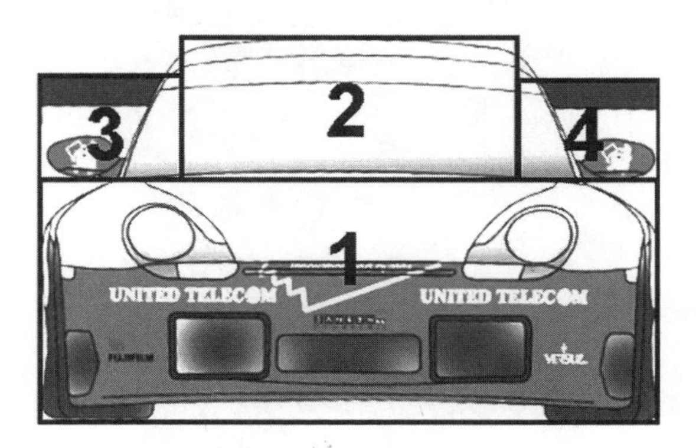

__Figure 11.2__ Estimation of a car's frontal area (Courtesy of www.bertlongin.com)

tunnel test), the aerodynamic forces can be calculated by measuring the dynamic pressure. Conversely, if these coefficients are unknown or the wind tunnel tests require verification, they can be determined by directly measuring the aerodynamic forces. It gets complicated because C_D and C_L are dependent on the vehicle's front and rear ride heights. This is why measuring ride height accurately is an inherent part of aerodynamic analysis.

The effects of a change to the vehicle's aerodynamic configuration are analyzed easily by looking at cornering speeds and segment times. A change resulting in more downforce should show an increase in cornering speed in the faster corners, while the induced drag due to this change is manifested in a slower straight-line segment time and a lower top speed. The effect on lap time should be easily assessable.

Air Density

One determining factor in Equations 11.1 and 11.2 is the density of air (ρ), which refers to the weight of a cubic meter. A variation in air density due to changing atmospheric conditions also alters the aerodynamic forces acting on the car. A thorough understanding of these variations is necessary.

Figure 11.3
Digital weather station used to measure ambient temperature, pressure, and humidity

Atmospheric conditions can vary significantly between test sessions. As a result, racecar engineers place a high priority on understanding the impact of changing weather conditions on the performance of racecars.

Air is composed of 75.54% nitrogen, 1.3% argon, and 23.1% oxygen. The amount of air per unit of volume (air density) depends on air temperature, pressure, and humidity. This affects various vehicle performance parameters:

- Cold, dense air means a greater mass of oxygen. If the air-fuel mixture is calibrated properly for the conditions, the engine's power output is greater.
- A higher air density increases the aerodynamic forces acting on the car. Higher density means more downforce and more drag.

By carefully measuring atmospheric conditions, the density of the ambient air can be tracked to gain an understanding of the effects these changing conditions have on racecar performance.

With a simple digital weather station similar to that shown in ***Figure 11.3,*** atmospheric conditions can be monitored accurately during race weekends or test sessions.

For racing, it is important to measure the following parameters:

- ambient air temperature,
- ambient absolute air pressure, and
- relative air humidity.

The ideal gas law is considered using ***Equation 11.3***.

$$p \cdot V = n \cdot R \cdot T \qquad \text{(Eq. 11.3)}$$

with p = pressure
V = volume
n = number of moles
R = gas constant
T = temperature

Density is the number of molecules of an ideal gas given as a specific volume, which is mathematically expressed in ***Equation 11.4***.

$$\rho = \frac{n}{V} \qquad \text{(Eq. 11.4)}$$

Then, by combining Equations 11.3 and 11.4, density is expressed in ***Equation 11.5***.

$$\rho = \frac{P}{R \cdot T} \quad \text{(Eq. 11.5)}$$

As an example, use standard conditions P = 101325 Pa and T = 20 °C (= 20 + 273.15 °K). The gas constant for dry air is 287.05 J/kg°K. Density can be calculated in ***Equation 11.6***.

$$\rho = \frac{101325}{287.05 \cdot (20 + 273.15)} = 1.187 \text{ kg/m}^3 \quad \text{(Eq. 11.6)}$$

This example assumes the air to be dry. However, in real-world situations, it is necessary to understand how air moisture affects the density.

Equation 11.7 expresses the density of a mixture of dry air and water molecules.

$$\rho = \left(\frac{P_a}{R_a \cdot T}\right) + \left(\frac{P_w}{R_w \cdot T}\right) \quad \text{(Eq. 11.7)}$$

where P_a = pressure of dry air (Pa)
R_a = gas constant for dry air (287.05 J/kg°K)
P_w = pressure of water vapor (Pa)
R_w = gas constant for water vapor (461.495 J/kg°K)
T = temperature (°K = °C + 273.15)

The actual water vapor pressure can be calculated from the relative humidity indicated by the digital weather station. Relative humidity is the ratio of the actual vapor pressure to the saturation vapor pressure at a given temperature. Saturation vapor pressure can be determined from ***Figure 11.4***.

Therefore, continuing with the previous example, a weather station assumedly provides a relative humidity reading of 45%. From the graph, the saturation vapor pressure at 20°C is 2350 Pa. Actual vapor pressure then becomes ***Equation 11.8***.

$$2350 \text{ Pa} \cdot 0.45 = 1057.5 \text{ Pa} \quad \text{(Eq. 11.8)}$$

All the variables required for Equation 11.7 are now known ***(Equation 11.9)***.

$$\rho = 1.187 + \left(\frac{1057.5}{461.5 \cdot (20 + 273.15)}\right) = 1.195 \text{ kg/m}^3 \quad \text{(Eq. 11.9)}$$

Enter this quantity into the equations for calculating aerodynamic forces and engine output power. This may explain why a racecar is not reaching the top speed it did the last time it was on a specific track or why downforce numbers are lower than expected. Being aware of the weather situation can prevent a lot of tail-chasing in situations like these.

Finally, air temperature, pressure, and humidity are not the only weather parameters affecting the racecar's performance. Asphalt temperature influences the rolling resistance of the tires, and wind speed and direction alter the drag and downforce of the racecar. Air temperature affects the cooling of the engine and transmission fluids, which modifies the friction in the driveline.

Dynamic Pressure

The scientific definition of dynamic pressure is the pressure of a fluid due to its motion. It is the difference between total pressure and static pressure. In racecar aerodynamics, dynamic pressure is the pressure acting on the car as it travels through the air. It typically is measured with a pitot tube, as on the Ferrari Formula One car shown in ***Figure 11.5***. The operating principle of a pitot tube is covered in more detail in ***Chapter 15.***

As indicated earlier in this chapter, dynamic pressure ***(Equation 11.10)*** is half of the air density multiplied by the vehicle speed squared.

$$q = \frac{1}{2} \cdot \rho \cdot V^2 \quad \text{(Eq. 11.10)}$$

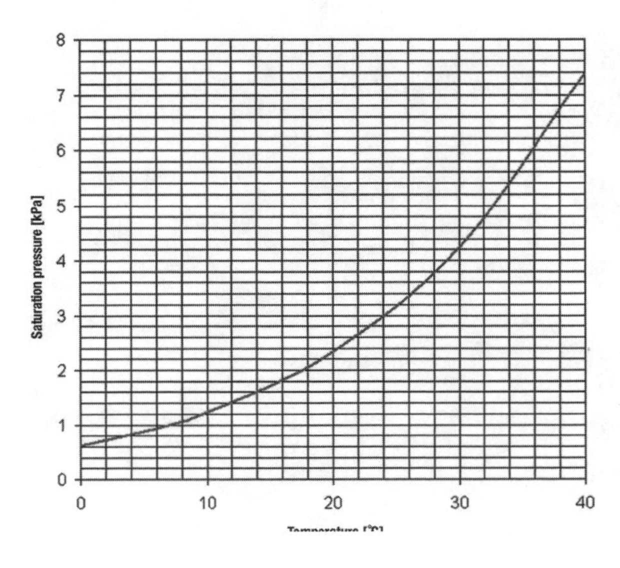

Figure 11.4
Water vapor saturation pressure versus temperature

Dynamic pressure can be estimated from the wheel-speed channel or a GPS-based speed channel. Air density remaining equal, the shape of the resulting math channel is exactly the same as the speed channel.

The dynamic pressure trace measured with a pitot tube resembles the calculated dynamic pressure. However, the traces deviate from one another when headwinds or tailwinds come into play or when the car is slipstreaming another car. In the case of a headwind, the actual dynamic pressure is higher than estimated. This increases drag, something that is experienced when riding a bicycle against the wind. Downforce levels are higher as well for the same reason. The opposite is true for a tailwind, which increases the vehicle's top speed but results in less downforce.

The effect of wind is determined not only by measuring dynamic pressure. By entering this value into Equations 11.1 and 11.2, the aerodynamic forces can be estimated. To do this, the drag and downforce coefficients must be known. If not available, they can be determined by directly measuring drag and downforce. The procedures are covered in the following section.

Figure 11.5 Pitot tube mounted on a Ferrari F1 car to measure dynamic pressure (Courtesy of Jaqueline Perreira do Nascimento Gramke)

The Coastdown Test

The procedure for calculating external resistances on the car was explained in Chapter 4. The most significant component of these forces is aerodynamic drag. In the Dodge Viper example, an aerodynamic drag coefficient (C_D) was assumed. The ideal way to determine the C_D of a vehicle is to test it in a wind tunnel. Because this is not a financially viable option for most motor racing teams, the coastdown test presents an affordable alternative. This test involves driving the car on a straight asphalt road, preferably as smooth as possible. The car is accelerated to a certain speed and then shifted into neutral and simply allowed to coast down to a lower predetermined speed with minimum steering fluctuations. The data logging system provides the data necessary for calculating the total external forces (and also the driveline friction and inertia). Only a speed signal is required for the calculation, but a measurement of dynamic pressure is preferable.

Instead of determining the absolute drag force (or downforce), aerodynamic performance is primarily expressed by C_D (or C_L in the case of downforce). Dimensionless coefficients are used because they are independent of speed or ambient conditions. For instance, if the atmospheric pressure changes between two runs, it becomes difficult to correlate setup modifications with changes in downforce or drag because the ambient conditions influence the result.

When the car is coasting, there is no transfer of power to the wheels by the engine. The forces slowing the car are aerodynamic Also contributing to the loss of speed are the rolling resistance of the wheels, the inertia and friction in the driveline, and the slope of the track on which the car is traveling.

As discussed in the previous section, aerodynamic forces relate to the vehicle speed squared. To measure the C_D of the car and have a minimal influence on the result from the rolling resistance and friction, begin coasting at a speed that is as high as possible. Influences resulting from track slope can be determined and subtracted from the result by measuring vertical acceleration.

Figure 11.6 is a graph obtained from a 2001 coastdown test in Zandvoort using a Dodge Viper

GTS-R. On the start/finish straight, the driver accelerates until the next reference point on the track is reached (in this case, a bridge crossing the racetrack). Here, the driver lifts the throttle, shifts into neutral, and lets the resisting vehicle forces slow the vehicle down until it is necessary to brake for the next corner. The x-axis represents the elapsed time in seconds, and the vehicle speed is indicated on the y-axis.

After 2 sec, the car decelerates to a speed of 205.5 km/h. At this point, Newton's Law ensures that the resisting force acting on the car may be calculated using ***Equation 11.11*** where G_{long} is obtained from the data acquisition's longitudinal G graph if it is present or calculated from the speed signal.

$$F_{res} = m \cdot G_{long} \qquad \text{(Eq. 11.11)}$$

The average deceleration between 1 and 3 sec is defined in ***Equation 11.12***.

$$\frac{213 \text{ km/h} - 198.5 \text{ km/h}}{3.6 \cdot (3 \text{ s} - 1 \text{ s})} = 2.01 \text{ m/s}^2 \qquad \text{(Eq. 11.12)}$$

At the time this test was conducted, the total weight of the car (including the driver) was 1,260 kg. Plug this information in for ***Equation 11.13***.

$$F_{res} = 1260 \cdot 2.01 = 2532 \text{ N} \qquad \text{(Eq. 11.13)}$$

For simplicity, the rolling resistance and inertias are ignored for the moment. This means that $F_{res} = F_{aero}$, which gives ***Equation 11.14*** for C_D with $A = 2.3$ m² and $\rho = 1.187$ kg/m³. C_D becomes 0.571.

$$C_D = \frac{m \cdot a_{long}}{\frac{1}{2} \cdot \rho \cdot V^2 \cdot A} \qquad \text{(Eq. 11.14)}$$

For the Dodge Viper in Chapter 4, a rolling resistance of 325 N and a 6% driveline loss totaling approximately 123 N were calculated. When these are subtracted from the 2,532 N measured in the coastdown test and C_D is calculated again, the result is 0.470.

This test was conducted with the rear wing at its minimum angle. To illustrate the feasibility of this method, C_D is calculated again with the rear wing at its maximum angle. The speed signal is shown in ***Figure 11.7***. The signal for the minimum wing also is pictured.

Look again at the speed after 2 sec, which is 202.2 km/h. A second before and after this moment, the speeds are 210.4 and 194.3 km/h, respectively, meaning longitudinal deceleration is 2.24 m/s². F_{res} becomes 2,822 N (or 2,374 N when the rolling resistance and inertia forces are subtracted).

Finally, a C_D of 0.656 (or 0.553) is obtained. This means that between the minimum and maximum rear wing angle, the total vehicle drag increases by 11%.

Drag measurement using coastdown testing is valuable for determining this racecar parameter. However, some factors should be considered:

- Drag (and downforce) varies with air density. Preferably, comparisons should be

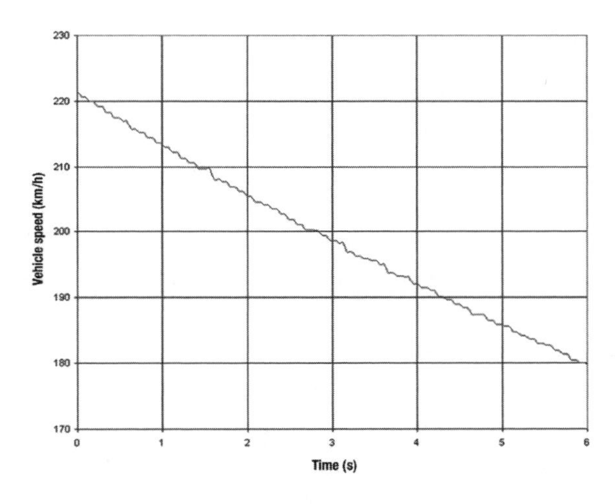

Figure 11.6 *Speed signal versus time. Coastdown test performed at Circuit Zandvoort in 2001 with a Dodge Viper GTS-R*

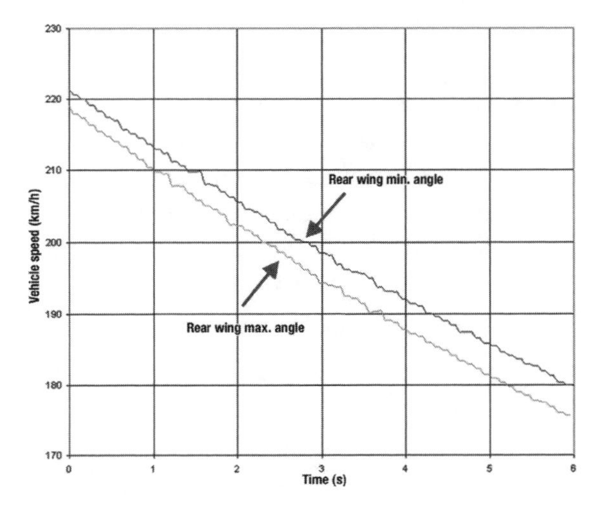

Figure 11.7 *Speed signal versus time. Coastdown test was performed at Zandvoort in 2001 with a Dodge Viper GTS-R*

made for coastdown tests performed on the same day or differences in air density should be corrected for.

- Between runs, ensure that the car configuration (specifically tire pressures and temperatures as well as engine and driveline temperatures) remains as constant as possible so that it does not influence the measurements.
- Always perform the coastdown test in two directions to take into account the influence of the wind and track gradient.

Coastdown tests are standardized SAE J1263.[12] This standard states the following initial conditions and prerequisites that must be met:

- Tests should be conducted at temperatures between –1°C and 32°C. Data obtained at temperatures outside this range cannot be reliably adjusted to standard conditions.
- Tests shall not be run during foggy conditions.
- Tests shall not be run when wind speeds average more than 16 km/h or when peak wind speeds are more than 20 km/h. The average of the component of the wind velocity perpendicular to the test road may not exceed 8 km/h.
- Roads shall be dry, clean, smooth and must not exceed 0.5% grade. In addition the grade should be constant and the road should be straight since variations in grade or straightness can significantly affect results.
- Tires shall have accumulated a minimum of 160 km prior to coastdown testing.
- The tires should have at least 75% of the original tread depth remaining.
- Vehicle tires should be inflated to the manufacturer's recommended cold inflation pressure corrected for the difference between ambient temperature and tire temperature.

A final note on tire pressures: Increasing the tire pressure during a coastdown test decreases the vehicle's rolling resistance. However, it also increases the tires' spring rate. A greater aerodynamic downforce at higher speeds results in more tire compression, and this also affects the aerodynamic properties of the vehicle. Maintaining a higher-than-normal tire pressure therefore introduces an inconsistent variable into the test.

Measuring the Aerodynamic Downforce

When wind tunnel downforce data is unavailable or when evaluating wind tunnel data on the track is desired, downforce numbers can be extracted from wheel load measurements. As with drag measurements, the dimensionless force coefficient (C_L) is the main focus. Also important is how the total downforce is distributed between the front and rear axles (i.e., the location of the center of pressure).

Aerodynamic forces depend on the vehicle's ride height. To evaluate the downforce a vehicle develops, the wheel loads on the front and rear axles through the car's ride height range must be measured.

The wheel loads can be measured with strain gauges in the suspension or, alternatively, by calculating the suspension loads from suspension deflection measurements. Downforce measurements are conducted preferably as steady-state tests at a constant speed in order to minimize the influence of inertial forces, as outlined in Reference 13.

Rearranging Equation 11.2 results in the definition for C_L ***(Equation 11.15)***.

$$C_L = \frac{L}{\frac{1}{2} \cdot \rho \cdot V^2 \cdot A} \qquad \text{(Eq. 11.15)}$$

To solve for C_L at a determined front and rear ride height, the total vertical load and dynamic pressure at those ride heights should be measured at a constant vehicle speed. The run pictured in ***Figure 11.8*** was obtained from a test performed on an airstrip. It shows the front and rear vertical load and vehicle speed for two runs at constant speeds of 50 and 195 km/h, respectively. The run at 50 km/h established the wheel load when aerodynamic downforce is not present, leaving only the static weight of the car. This figure can be compared to the weights measured when the car is placed on the corner weight scales to ensure there is no downforce present at the test speed. Taking the average over the time interval in which a speed of 50 km/h is maintained results in the following vertical axle loads.

front static vertical load	8,250 N
rear static vertical load	8,200 N
total static vertical load	16,450 N

Doing the same for the time spent at a speed of 195 km/h gives the following total axle loads.

front vertical load	8,775 N
rear vertical load	9,375 N
total vertical load	18,150 N

Subtracting the static loads from these figures results in the amount of downforce developed at a speed of 195 km/h at the tested front and rear ride heights.

front downforce	525 N
rear downforce	1,175 N
total downforce	1,700 N

With a dynamic pressure 1,740 Pa and a frontal area of 2.3 m², at these ride heights the downforce coefficient C_L becomes ***Equation 11.16***.

$$C_L = \frac{1700\ \text{N}}{1740\ \text{Pa} \cdot 2.3\ \text{m}^2} = 0.425 \qquad \textit{(Eq. 11.16)}$$

Finally, the aerodynamic center of pressure is located at (1,175/1,700) x 100% = 69.1 % of the vehicle's wheelbase from the front wheel centerline.

These tests should be repeated until the desired ride height interval is covered. When all the necessary data is collected, the following aeromaps then can be created:

- downforce coefficient as a function of front and rear ride heights, and
- center of pressure location as a function of front and rear ride heights.

The more sophisticated analysis software packages can define lookup tables. Each aeromap is entered as a lookup table. Then a math channel is created that assumes the table value with front and rear ride heights as indices. As an example, ***Figure 11.9*** graphs the aeromap of center-of-pressure location versus front and rear ride heights. Every value is stored in a spreadsheet, which then is imported into the analysis software as a lookup table. From this table, a math channel can be created that takes the front and rear ride height values and assumes the corresponding center-of-pressure location ***(Figure 11.10)***. This channel offers a good indication of the variation in the location of the center of pressure and allows the investigation of any abnormality in the aerodynamic balance of the vehicle.

The same can be done with the downforce coefficient table. With a measurement of dynamic pressure, the total aerodynamic downforce can be calculated at each point on the track. Combining this channel with the center-of-pressure channel indicates the absolute downforce on the front and rear axles. Alternatively, when the lookup table feature is not available in the analysis software, a lookup table can be generated using a spreadsheet and the data export function in the analysis soft-

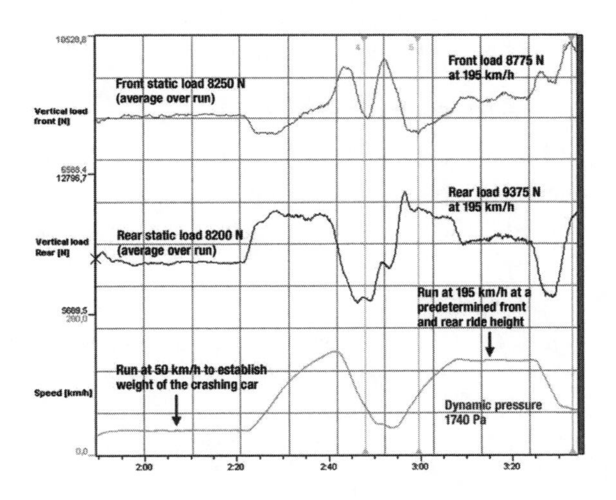

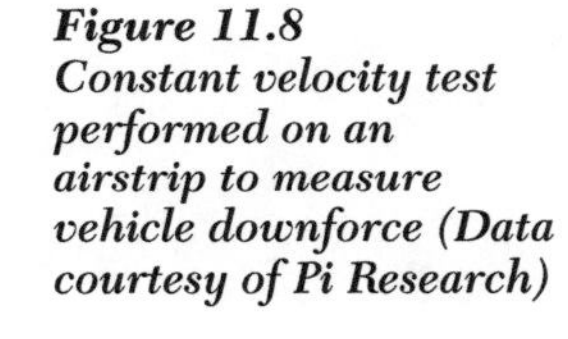
Figure 11.8
Constant velocity test performed on an airstrip to measure vehicle downforce (Data courtesy of Pi Research)

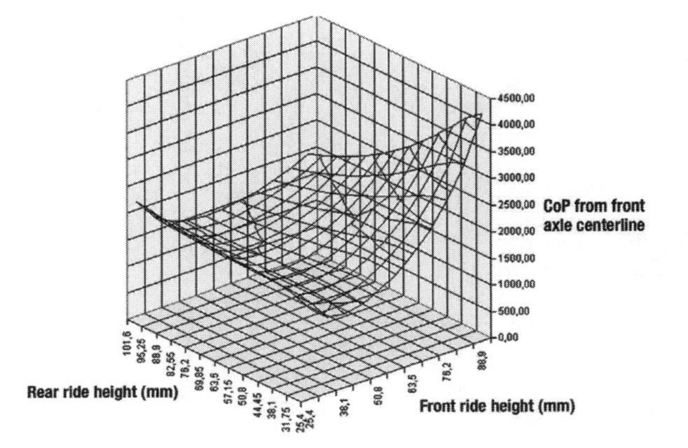

Figure 11.9
A graphical representation of a center-of-pressure aeromap. The center-of-pressure location is expressed here as the distance in millimeters from the front axle centerline.

ware. Front and rear ride height channels can be imported into the spreadsheet and associated with the corresponding location of the center of pressure and downforce coefficient.

As with coastdown testing, a constant velocity test requires a specific vehicle setup to maximize accuracy. The downforce produced by the vehicle depends on its ride height. Therefore, this is a very important parameter to control.

- Springs and damping should be as soft as possible to minimize vertical load variation resulting from bumps in the track.
- Any influence of packers or bump rubbers should be removed.
- Antiroll bars preferably are disconnected.
- Static ride height should be adjusted so that the target ride height is reached at the target speed. This is necessary to prepare aeromaps of different ride heights at different speeds (Figure 11.9).
- Tests should be performed in two opposite directions to minimize or remove the influence of wind direction and track gradient.

Airbox Efficiency

What determines the power potential of an engine? All power comes from the amount of fuel burned in the engine. The more fuel the engine can burn per time interval, the greater the power output is. However, to burn fuel, oxygen is required, and this comes straight from the atmosphere. The amount of air the engine can process depends primarily on the design of the cylinder head and the inlet manifold.

Figure 11.10 The center-of-pressure location math channel as a function of front and rear ride heights

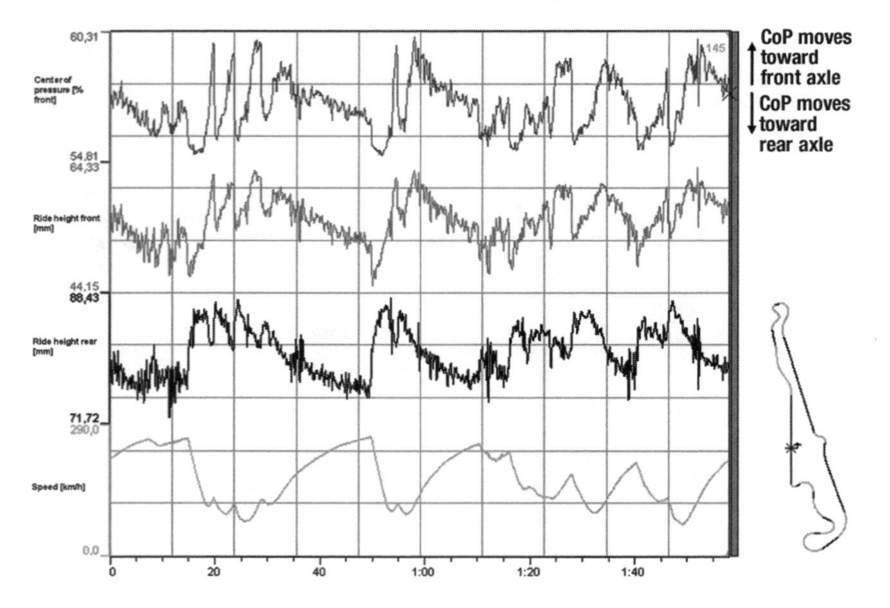

Atmospheric engines use the ambient atmospheric pressure to push air into the engine's cylinders. When regulations provide inlet restrictors, the air usually is guided through these restrictors into the inlet manifold and then into the engine. When the ambient air pressure is known and the absolute pressure in the inlet manifold is determined, this serves as a measure for the inlet system efficiency.

Figure 11.11 illustrates the manifold air pressure (MAP) signal of a GT car with the throttle position, engine RPM, and vehicle speed.

An approximate atmospheric pressure can be obtained from the graph by taking a point where the throttle is closed completely. In this case, it is 100 kPa. As soon as the throttle opens, the MAP drops proportional to the engine RPM. Shift points are indicated as upward spikes, when the throttle is closed momentarily. Note that these upward spikes are much more subtle when the car is equipped with a power-shift system that enables the driver to keep his foot down on the accelerator during upshifts.

An idea of how well the inlet system is performing is indicated when the minimum air pressure in the graph (which is approximately 87 kPa) is subtracted from the atmospheric pressure. As the throttle is opened, air flows into the engine cylinders and the rise in RPM increases the flow velocity so air pressure drops. However, with increasing RPM, the engine literally asks for more air to be fed into the manifold. As vehicle speed increases, the dynamic pressure before the manifold increases quadratic to the vehicle velocity, which in this case is a good thing. The lower the pressure drop in the manifold, the better the air supply to the engine is. This is why engine builders try to design air inlets in such a way that the air pressure before the throttle is as close to the atmospheric pressure as possible.

Figure 11.11 shows a pressure drop of 100 – 87 = 13 kPa at an engine speed of 6,100 RPM. To illustrate the effect of manifold pressure, a test was performed in which the air inlet restrictors were removed from the engine air inlet. The results are given in ***Figure 11.12***.

Tests on the engine dyno resulted in a power increase of approximately 80 brake horsepower (bhp) for this car by removing the restrictors. Figure 11.12 illustrates why this is the case. Atmospheric pressure is 100 kPa and the MAP reading at 6,100 RPM is 91 kPa, which means a drop of 9 kPa.

Figure 11.11 *MAP signal for a GT car around Zolder*

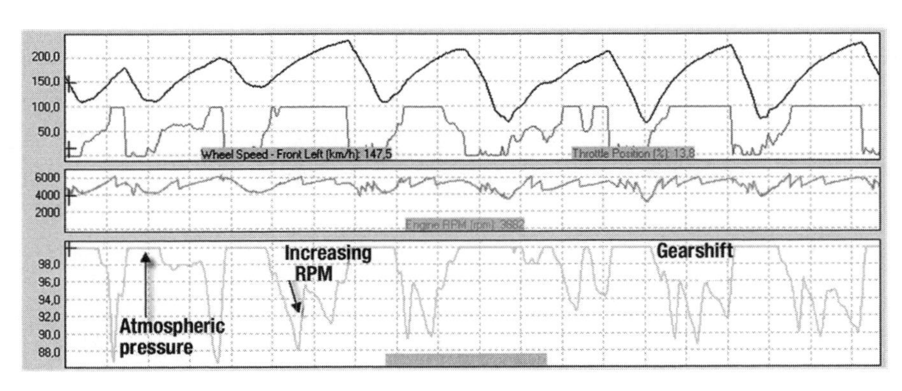

Figure 11.12 *MAP signal for a GT car around Zolder, with no air inlet restrictors*

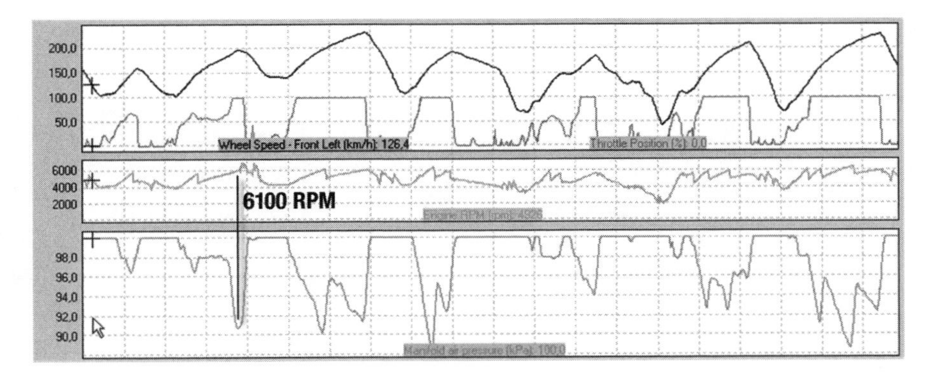

CHAPTER 12

ANALYZING THE DRIVER

Racecar engineering is not limited to tuning a vehicle to its maximum dynamic performance. Racecars do not drive themselves; the driver is a very important part of the performance equation. Logging driver activities provides a detailed record of what is happening in the cockpit and furnishes the driver with tools that help improve performance.

Improving Driver Performance

When evaluating a driver, the data acquisition engineer must be aware of two significant pitfalls. The first is that driver activity and chassis balance are interrelated closely. Something diagnosed as driver deficiency may be the result of an unbalanced chassis. The opposite can be true as well, so the driver's comments should be interpreted in combination with detailed data analysis.

Realize that a racecar driver is a human being with a finite tolerance to criticism. Even founded, well-intended, and tactfully offered suggestions are not accepted always in a graceful manner. In the worst-case scenario, pointing out to drivers what they are doing wrong can potentially undermine their self-confidence, something that should be avoided at all costs.

A solution that often works is letting drivers ascertain (to the best of their ability and in conjunction with subtle guidance from the engineer) what they are doing wrong. In addition, encourage them to utilize data analysis as a developmental tool. Drivers have everything they did on the the track stored in their brain. Driver activity recordings made by the data logger should make sense to them. To ensure that this occurs, there are some conditions. First, drivers should know how to operate the analysis software. Second, they should be educated in the analysis techniques necessary to evaluate their performance. And third, they should possess some basic knowledge about the dynamics of the vehicle to correlate their performance with that of the vehicle. All of this depends on the willingness and motivation of drivers to learn and develop themselves. At the end of the day, this is nothing more than part of their job.

In practice, drivers should be able to analyze data independently on their own computer and should have access to the data quickly while everything is still fresh in their memory. Modern technology, such as USB storage devices and wireless pitbox networks, can facilitate this accessibility.

In the cockpit, the driver has five main controls to help get the car around the track as quickly as possible: throttle pedal, brake pedal, steering wheel, clutch pedal, and gear lever. A driver's activities consist of acceleration, braking, cornering, and shifting gears. ***Figure 12.1*** illustrates a display template that includes the important channels to begin the driver evaluation: (front) brake pressure, gear, engine RPM, vehicle speed, and throttle position. This screen template contains the basic information for evaluating driving style. Drivers usually want to compare their laps with previous outings or with the data of other drivers. The techniques for overlaying and comparing data from different laps were covered in Chapter 3.

Figure 12.1 *A typical user-friendly display of driver activity channels*

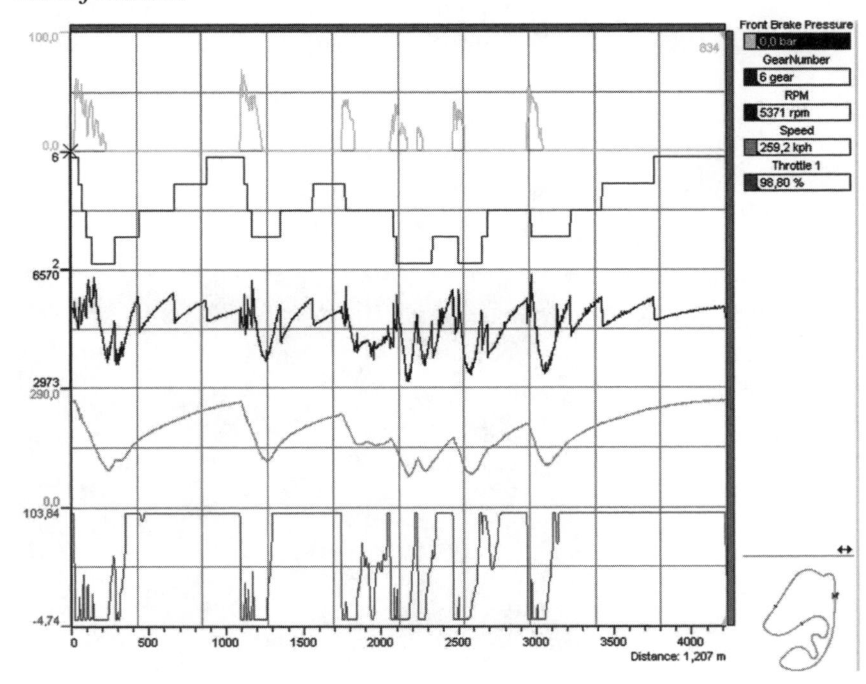

Modern dashboard systems often allow the display of messages to the driver or output signals to be sent (e.g., to a lamp) once a condition is met. Different applications exist to assist drivers during their time on the track. Data logged on previous runs provide the information to program the dashboard. Here are some examples:

- Programmable shift lights, which are a series of LEDs placed on the dashboard to indicate when the driver should shift to a higher gear. For each gear, an independent shift RPM can be programmed in the dashboard.
- Brake balance indication on the dashboard
- Visual indication on the dashboard when a wheel is blocking, underbraking, or slipping under acceleration
- Sector times from virtual beacons, which are placed at fixed distances from the infrared beacon indicating the start and finish of a lap. When the car reaches the respective distance of a virtual beacon, the sector time is displayed on the dashboard display. GPS integration has improved greatly the accuracy of this feature. Often, a reference lap can be programmed into the dashboard, enabling drivers to observe during their lap if they are faster or slower than the reference lap ***(Figure 12.2)***.
- Predictive lap time calculation in which a previously logged reference lap can be loaded into the dashboard. The time/distance data of the current lap is compared constantly to that of the reference lap; from the difference between the two, the lap time of the current lap can be predicted.

Figure 12.2 *Virtual beacons compare the driver's current lap to a reference lap. (Courtesy of Jean-Michel le Meur)*

Driving Style Evaluation

For driver evaluation, the following characteristics should be investigated:

- Performance: This is a measure of achieved results. Performance improvement is the concept of measuring the output of a particular process and then modifying this process to increase its output, efficiency, or effectiveness. Examples of performance measurements are cornering speed, maximum lateral G during cornering, maximum longitudinal G during braking, and gearshift times. Lap time is also a performance measurement.
- Smoothness: *The American Heritage® Dictionary of the English Language* defines *smooth* as "having a texture that lacks friction; not rough."[14] A driver should maintain a certain degree of smoothness to avoid upsetting the car in transient phases. For instance, accelerating out of a corner with a fluctuating throttle pedal causes abrupt changes in longitudinal weight transfer resulting in load fluctuations at the tire contact patches.
- Response: This is an output resulting from an input. Opposite steering lock can be a response to the rear end of the car stepping out during cornering. Response is also the time delay between two actions (e.g., the delay between coming off the throttle and applying the brakes).
- Consistency: This refers to the repeatability of performance factors (e.g., consistency in gearshift times, throttle blips, braking effort, and lap time consistency).

Driver improvements in smoothness, response, and consistency have a beneficial effect on overall performance. Some techniques for measuring and quantifying these driver characteristics are discussed in the following sections.

Throttle Application

The accelerator pedal is the driver's primary interface with the vehicle. It has a simple function—to accelerate the vehicle. However, the driver can apply too little or too much throttle or apply it too slow or too fast. The more available engine power, the thinner this too much/too little line becomes.

The first performance indicator is the throttle histogram (Chapter 7). It shows how much time the driver spends at full throttle and part throttle. More time spent at full throttle improves lap times.

When the driver exits a low- to medium-speed corner, the point where full throttle is reached should follow shortly after the G-peak. The value of the lateral acceleration channel at the full throttle point is a measure of the assertiveness of the driver. ***Figures 12.3*** and ***12.4*** illustrate a late and early throttle application respectively.

In Figure 12.3, the driver exits a 60-km/h corner in which a lateral acceleration maximum of 1.81 G was reached. At the point where the throttle is open fully, lateral acceleration has already decreased to 0.70 G. To compare different corners, these figures can be expressed as a percentage in ***Equation 12.1***.

$$\frac{0.70\text{ G}}{1.81\text{ G}} \cdot 100\ \% = 38.7\% \qquad \textit{(Eq. 12.1)}$$

Different cars produce different figures, and driver experience factors largely into this as well. However, it is safe to say that 38.7% can be classified as being a bit careful on the gas. Figure 12.4 shows the results from same driver in the same car at another location on the track. Here, the driver is much more assertive on the throttle pedal. The lateral Gs peak at 1.74 G, and at 1.60 G the driver has the throttle fully open again. ***Equation 12.2*** expresses this as a percentage.

$$\frac{1.60\text{ G}}{1.74\text{ G}} \cdot 100\ \% = 91.9\% \qquad \textit{(Eq. 12.2)}$$

Author Buddy Fey offers a target value for the percentage of lateral acceleration when the driver should be at full throttle.[6] These target values depend on how much power is available at the driven wheels ***(Table 12.1)***.

A late throttle application, as illustrated in Figure 12.3, is observed when the car is oversteering at a corner exit. Investigate other channels and talk to the driver to diagnose this. When there is no balance or traction problem, the driver is probably being too careful on the accelerator.

The driver also can apply full throttle too early. The result of this is probably the rear breaking out. The driver corrects with an opposite steering lock and backs off the throttle. Look for oversteer and traction problems here when the dif-

***Table 12.1** Target values for percentage of lateral acceleration where the driver should be at full throttle*

Power output	% lateral G at 100% throttle
<150 hp	95%
150–250 hp	90%
250–400 hp	85%
>400 hp	80%

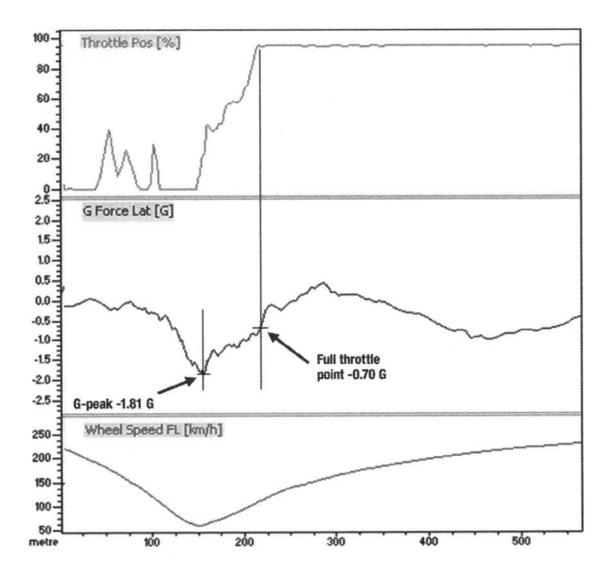

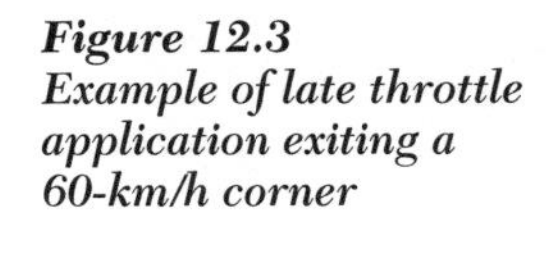

***Figure 12.3** Example of late throttle application exiting a 60-km/h corner*

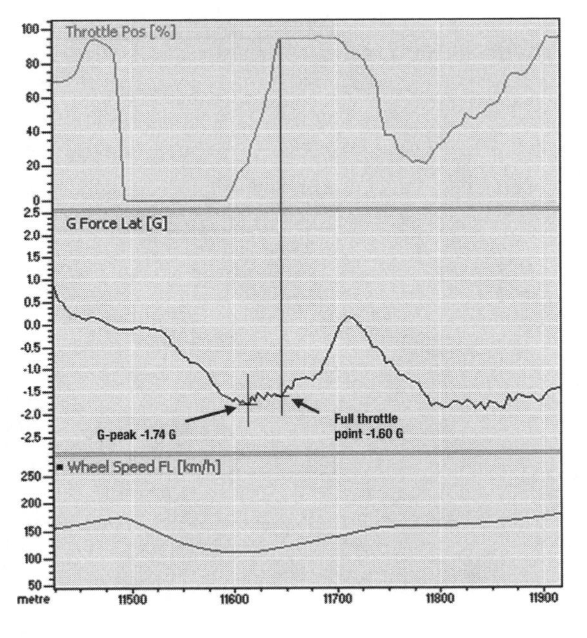

***Figure 12.4** Assertive throttle application exiting a 120 km/h corner*

ference between the G-peak and lateral acceleration at the full throttle point becomes greater.

The smoothness with which the driver treats the throttle pedal is also a parameter affecting lap times. When driving at the traction limit, it is vital to not upset the car with a fluctuating throttle. A good example of this is provided in ***Figure 12.5,*** where the throttle position signals of two drivers driving the same car are overlaid. The corner sequence pictured is indicated in the little track map. A tight righthander is followed by a sweeping fast left-hand corner. The dark trace produced the quickest lap time.

The driver who produced the lighter color signal trace is clearly dancing on the throttle pedal, inducing an upsetting longitudinal weight transfer while searching for traction. The dark trace shows that the driver uses one progressive throttle lift to negotiate the fast lefthander and then returns to full throttle.

The gray trace is an exaggerated example of a car upsetting throttle behavior. In more subtle cases, the speed of the pedal movement can be observed. Throttle speed can be calculated by taking the first derivative of the throttle position signal. The throttle position often is expressed as a percentage. When this channel is derived, its unit is typically expressed as percent per second. A throttle speed of 50% per second means the throttle pedal is being reduced to half of its maximum travel in 1 sec. The lower portion of Figure 12.5 shows the throttle speed for both drivers. In the section where the slower diver is fighting to keep the car under control, the throttle speed trace illustrates much higher peaks than that of the faster driver.

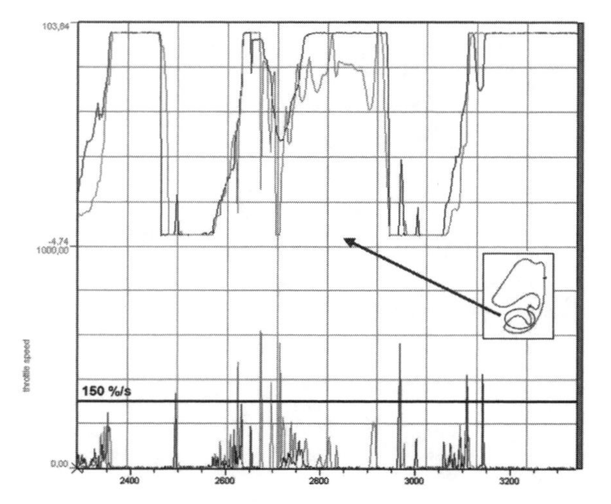

Figure 12.5
The throttle position and throttle speed for two different drivers in the same car

Experience has shown that throttle applications in a corner that occur at a speed of more than 150% per second do not have a beneficial effect on lap times. Note that this is true during cornering, but throttle blips and lifts while shifting gears can produce throttle speeds of 500% per second and more.

Braking

A driver's braking analysis should include the following:

- braking point location and consistency,
- total braking distance and braking distance consistency,
- reaction time between the moment the driver comes off the throttle and steps on the brake,
- quickness in building up maximum deceleration,
- how hard the driver is braking,
- brake pressure modulation to compensate for changes in friction between the tires and track surface, and
- brake pressure variation during throttle blips for downshifting (driver footwork).

Techniques for investigating how hard, how quickly, and where the driver starts braking are covered in Chapter 5. Braking, as with throttle application, requires a certain degree of smoothness. Too much brake pressure modulation upsets the chassis, making it even more difficult to stop the car in time. On the other hand, brake pressure modulation is necessary to compensate for changes in the traction between the tires and track surface, constantly at the limit of wheel lockup. This is another thin line to tread.

In the following example, three drivers are compared braking their way into Senna's S, the first (downhill) corner after start/finish on Brazil's Interlagos track (known also as the Autódromo José Carlos Pace). All three drivers were driving the same car model. Using the method for determining throttle speed from the previous paragraph, the first derivative of the (front) brake pressure signal is calculated to determine the brake application speed. In this case, the braking speed is expressed in bar per second. Another way to do this is to differentiate the longitudinal G channel (Chapter 5, braking quickness). To analyze pedal work, however, use the channel that directly relates to the pedals, which in this case is the brake pressure channel.

The first driver's activities are indicated in ***Figure 12.6***. The pictured channels are braking speed, (front) brake pressure, speed, and throttle position. The driver must reduce the speed of the car from 260 km/h to a minimum cornering speed of 88 km/h; this is achieved in a braking distance of 240 m (a performance measurement!). During this braking maneuver, the driver shifted from sixth to third gear, which does not help the situation.

The braking maneuver begins where the driver's right foot is removed from the throttle pedal and transferred to the brake pedal. This action takes the driver 0.1 sec. The brake pressure channel and its derivative shows that the brake pressure increases relatively slowly. Maximum brake pressure is 45 bar and is achieved only after one-third of the total braking distance is covered. The driver's footwork is rather sloppy, something that is indicated by a seriously fluctuating brake speed trace. During the driver's downshift from fifth to fourth gear, nearly all brake pressure was lost as the driver blips the throttle. The braking sequence ends abruptly as the driver's foot is removed from the pedal. This is followed by a short coasting period when the driver is not on the brake or the throttle.

Figure 12.7 provides an example of a much better braking maneuver performance, proven by a heroic braking distance of 195 m and a braking point that is located 25 m further down the track. The second driver takes Senna's S in second gear, so there is one downshift more than in the previous example. There is absolutely no delay between the moment the driver's foot is removed from the throttle and when the brake is engaged. A maximum brake pressure of 69 bar (at a speed of 301 bar/s) is increased instantaneously. From that moment, the driver gradually decreases pedal pressure to avoid wheel lock due to decreasing aerodynamic drag. This is achieved by neatly modulating the brake pressure below 90 bar/s. Peak speeds in this area are caused mainly by footwork during downshifting. This is how it should be done!

The third driver is a left-foot braker, illustrated by the short transition period between coming off the throttle and engaging the brake in ***Figure 12.8***. For a very short period, the driver is simultaneously on the throttle and building up brake pressure. Pressure is being increased to a maximum of 50 bar at a rate of 250 bar/s. The driver brakes 34 m earlier than the driver in Figure 12.7 and achieves a braking distance of 209 m. This driver also shifts down to second gear. After the first peak, the driver modulates the brakes at a rate below 70 bar/s, which is considerably lower than the drivers in the previous two examples. The third driver achieves this because his left foot is on the brakes, which keeps the right foot free to blip the throttle during downshifts. Here, the possible

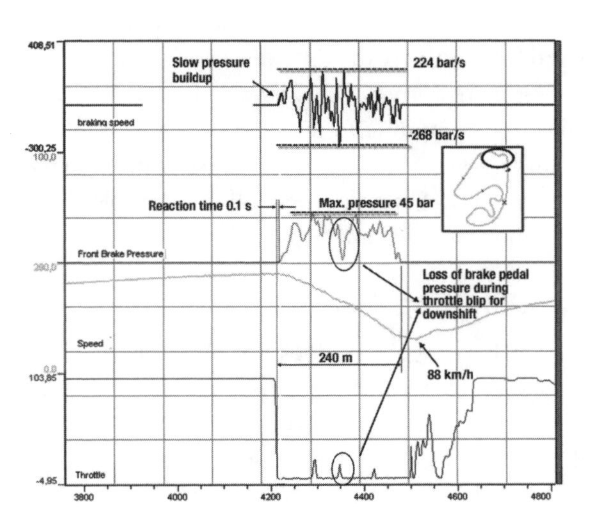

Figure 12.6
Driver 1 braking into Senna's S

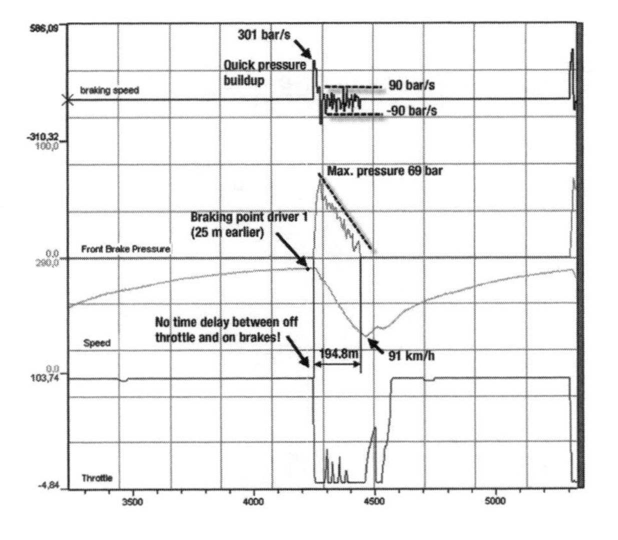

Figure 12.7
Driver 2 braking into Senna's S

advantage of left-foot braking is illustrated. Although the driver's brake pressure modulation is better, this driver misses an opportunity in this braking zone for two reasons: selecting an early braking point and correcting pedal pressure halfway through the braking zone (see the indicated area in the brake pressure trace). The driver soon realizes here the pressure is decreasing at too high a rate and therefore increases pedal pressure.

The previous examples showed different techniques used in a very hard braking zone, where the primary goal was to slow the car down sufficiently in as short a distance as possible. However, the brake pedal is also a way for the driver to influence the vehicle's transient handling. The location of braking points, maximum braking effort, and the way brake pressure is modulated determines at what rate longitudinal weight transfer takes place and the attitude of the car in the corner. A sudden reduction in brake pressure creates an instant forward longitudinal weight shift, momentarily providing more cornering grip to the front wheels. In addition, beginning to brake with less than maximum braking effort may allow a more predictable corner entry. Always observe all the variables and, more importantly, talk to the driver!

Shifting Gears

Changing gears does not make the driver's life easier. Downshifting can be frustrating at times because it comes when the car must be slowed down into a corner over a very short distance.

Figure 12.8 Driver 3 braking into Senna's S

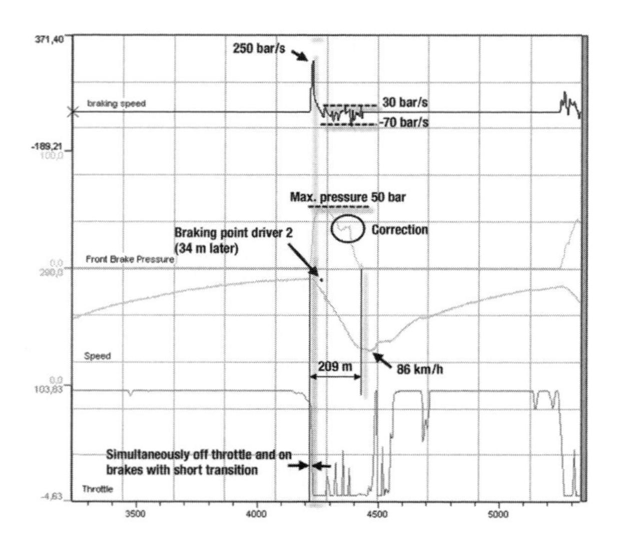

Concentrating on two different activities performed simultaneously requires skill.

Gear-change analysis focuses on the following issues:

1. **Upshift**
   - At which engine speed does the driver change to a higher gear?
   - Duration of the upshift
2. **Downshift**
   - At which engine speed does the driver change to a lower gear?
   - Throttle blipping
   - Brake modulation during downshifting (i.e., the driver's footwork; see previous section)

The techniques used to analyze these issues are covered extensively in Chapter 6.

Cornering

The physics behind getting a racecar around a corner are discussed in Chapter 7. The traction circle is a tool used to determine the cornering potential of a car. The traction circle can also determine if the driver effectively uses the available cornering power.

Just like the throttle and brake pedals, the steering wheel is a driver control that influences the dynamic attitude of the vehicle. Turning the wheel, therefore, requires the driver to exercise a degree of smoothness to not upset the chassis with unnecessary movements.

A rough steering angle signal results from steering feedback due to road irregularities and driver response to changes in the vehicle balance. The steering wheel gives a driver an indication of what is happening at the front tire contact patches. Any changes in the tires' self-aligning torque are felt by the driver's hands on the steering wheel and provide an idea of how much grip is available at the front axle. For the rear axle, the driver senses changes in the lateral acceleration to estimate the grip level.

These brain inputs result in a driver response at the steering wheel. This response should be of such magnitude that it keeps the car on its designated trajectory. Too much response upsets the chassis.

Steering smoothness is evaluated in exactly the same way as throttle and brake inputs. The steering angle signal is differentiated to obtain steering speed.

An example is provided in ***Figure 12.9***. The steering wheel angle and steering speed traces are given for two different drivers going through Senna's S at Interlagos. The driver producing the black steering speed trace has a lower average amplitude, which also is proven by a smoother steering angle trace.

A car with handling problems requires more steering corrections from the driver and creates higher steering speed amplitudes. When a corner is taken at a lower speed, steering speed amplitudes decreases. First find out from the driver if the corner is being approached at a maximum allowable speed and if the balance of the car is acceptable before thinking about steering smoothness.

Driving Line

The line selected by the driver when negotiating the corner is determined by the maximum speed of a car being driven at the limit during cornering. To achieve the greatest possible cornering speed, the path through the corner should be an arc with the greatest possible radius. The speed maintained through the corner also determines the speed on the following straightaway. (However, it does not determine necessarily the top speed; see Chapter 4.) Because cornering speed and straightaway speed are the performance factors determining lap time, it is important to find out if the driver is locating the ideal path around the racetrack.

The analysis software can calculate the corner radius by defining the mathematical channel ***(Equation 12.3)***.

$$R = \frac{V^2}{G_{lat}} \qquad \text{(Eq. 12.3)}$$

The result looks as illustrated in ***Figure 12.10***.

According to Equation 12.3, the corner radius is the smallest where lateral acceleration reaches its maximum. With this in mind, the location of the apex can be determined from the corner radius graph. The apex is the point in the corner where the car is closest to the inside edge and is determined by the driving line chosen by the driver. Possible apex locations are in ***Figure 12.11***.

Assuming that the track width at the corner entry and exit are the same, a constant radius line translates to an apex located in the middle of the corner. A mid-corner apex allows the driver to maintain a straight-line speed as long as possible without sacrificing corner exit speed.

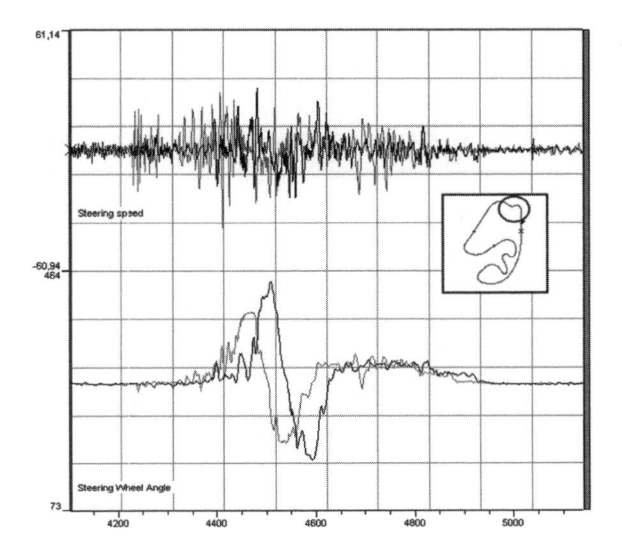

Figure 12.9
Steering wheel angle and steering speed for two drivers through Senna's S

Figure 12.10 *Speed and corner radius overlay of two lap sections of the Paul Ricard High Tech Test Track*

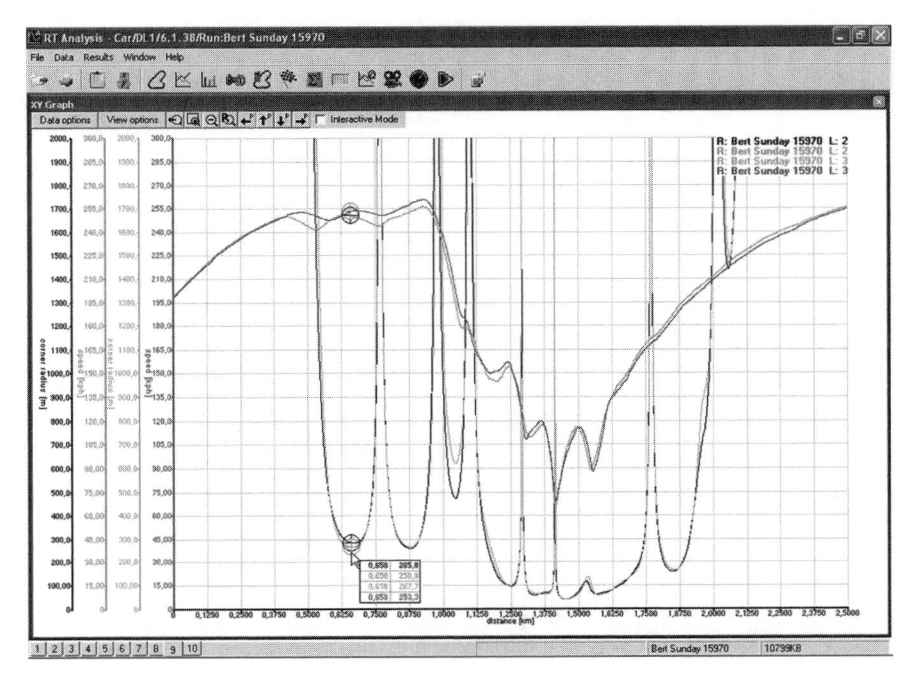

Because the geometric location of the apex is the middle of the corner, the corner radius is at its minimum at this point. This results in a symmetrical corner radius trace, which is illustrated in the upper left drawing in Figure 12.11. Mid-corner apexes often show a flat, lateral G plateau in this section of the corner.

Late corner entry and late apex go hand in hand as indicated in the right upper drawing in Figure 12.11. This results in an excess amount of track left at the corner exit, which is not utilized when accelerating the car, and compromises speed on the following straight. In reality, the driver unwinds the steering wheel at this point to use all available grip for forward acceleration. Also, by performing the largest amount of turning at the corner entry, the car is in a better position for the exit, which makes it easier to accelerate out of the corner. However, the drawback of a late apex is sacrificing corner entry speed. Achieving a minimum corner radius in the early part of the corner gives away a late apex in the corner radius graph. Lateral Gs increase quickly, peaking early following corner entry, after which they decrease slowly until the driver starts to unwind the steering wheel.

Finally, an early corner entry results in an early apex. The driver turns the steering wheel more (to decrease the radius of the path) at the corner exit, which inevitably sacrifices the corner exit speed. In this case, the minimum corner radius is reached in the later stage of the corner, which creates a trace resembling the lower drawing in Figure 12.11. As the steering angle increases, lateral acceleration increases to a late peak ***(Figure 12.12)***. Early apexing also results in a later throttle application point, as much of the cornering is done so late in the corner.

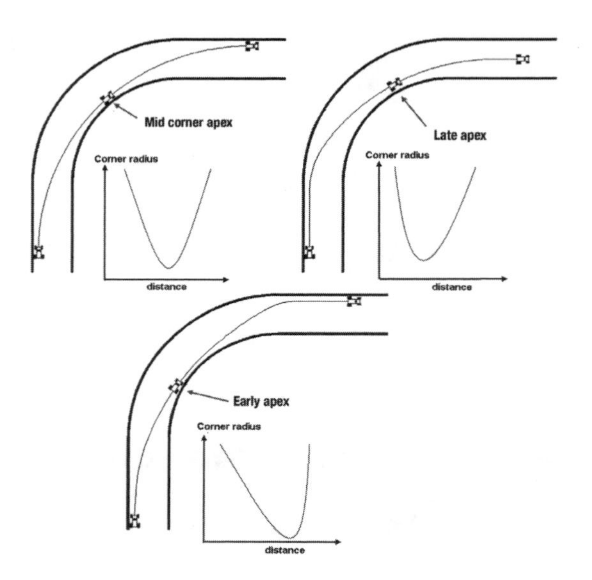

Figure 12.11
Mid-corner, late, and early apex

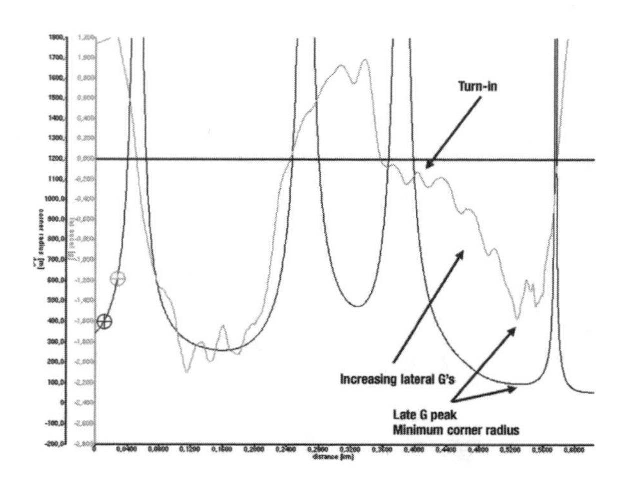

Figure 12.12
An early apex indicated by increasing lateral Gs to a late peak and minimum corner radius

To determine where the driver places the apex, know what type of corner one is dealing with. A constant radius corner, one defined by one single radius, favors an apex placed close to the middle of the corner. A decreasing radius corner, where the latter part of the turn is tighter, normally experiences a late apex, while in an increasing radius corner the apex is placed earlier.

Hairpins are corners exceeding 120 deg and are special cases. These usually are tackled with a single, late apex to maximize corner exit speed, but an approach with a double apex is also possible. Much depends on the car and how it handles the mid-corner change of direction. To complicate the matter even more, grip level changes, banked corners, and bumps also influence apex placement.

The data logging system reveals which line was taken to handle a corner, but the driver and engineer must decide if it was the correct one. Look at the minimum cornering speed and corner exit speed to evaluate different lines. Always remember that a higher corner exit speed minimizes the time spent on the following straight. Therefore, driving line analysis should concentrate primarily on the corners that matter, the ones followed by a significant acceleration zone. Sector time analysis can be very helpful here as well.

Driving Line Analysis Using GPS

Track mapping with GPS is becoming more useful in racecar data logging systems. The accuracy of these track maps make them very suitable for driving line analysis. When overlaying track maps from different laps ***(Figure 12.13)***, a difference in driving line is visualized clearly without calculating and analyzing a corner radius channel. In combination with an analysis of track segment times, it is possible to determine which driving line was the fastest.

This type of visualization gives the person behind the laptop screen more of a feeling for where the car actually drives on the track. The track borders also can be shown on the track map by first doing a lap around the circuit on the outside track border followed by a lap on the inside border. This way, it is easy to visualize how much of the track the driver is using. An example is given in ***Figure 12.14***, which shows a GPS-created track map of the Circuit Duivelsberg Maasmechelen rallycross circuit. The black lines representing the track perimeters were logged by strapping the datalogger together with a GPS antenna and an external battery to the back of a bicycle. These lines are not as fluent as the actual driving lines logged in the vehicle. This is primarily because of the absence of high lateral and longitudinal acceleration levels on the bicycle. The position signals normally are calculated from the raw GPS signal and the acceleration channels, but in the case of the bicycle the data logger relies exclusively on the GPS data. The indicated area in Figure 12.14 illustrates a part of the track where the GPS signal was temporarily blocked by a group of trees, which is detrimental for positional accuracy in this case. However, the black lines provide a fine representation of the track surface. Downhill skiing has similar use for this type of analysis. A GPS receiver with inertial sensors is attached to the athlete to analyze the downhill trajectory. Data logging is not limited exclusively to motor racing.

The example in ***Figure 12.15*** goes even further. Here the GPS trajectory logged by a Race Technology DL1 data logger is exported from the analysis software and projected on top of a satellite image of the track obtained from the Internet application Google Earth™.

Driving Line Analysis Using Video Feed

The next best thing to putting the driving line on a two-dimensional map is recording it on video. Modern data logging systems often can synchronize the logged data with video recordings. Watch the images with the speed trace to find out which line through a corner is the fastest ***(Figure 12.16)***.

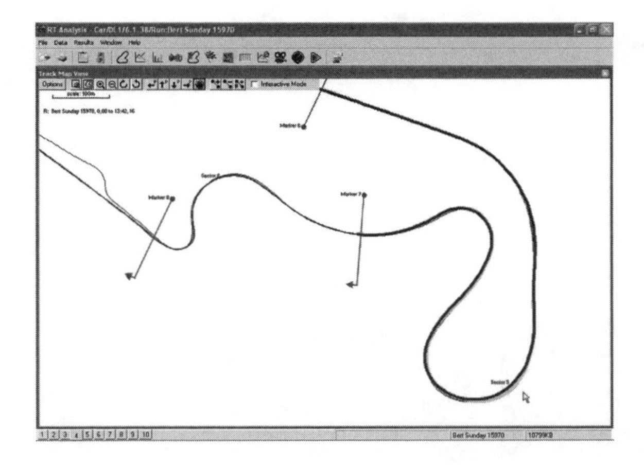

Figure 12.13
A comparison of driving lines on a track map overlay created from a GPS signal

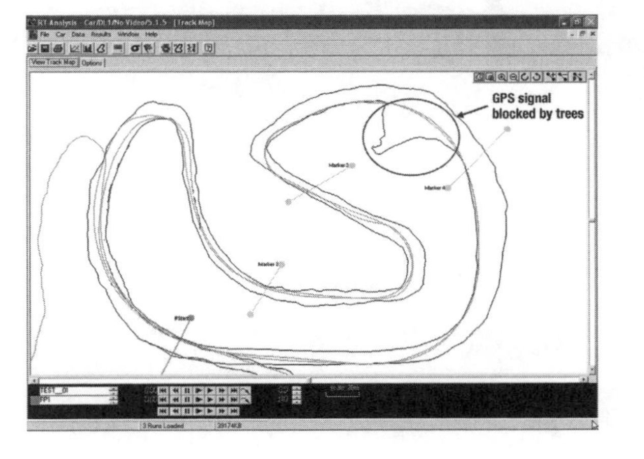

Figure 12.14
Track map of the Maasmechelen rallycross circuit. The black lines represent the track perimeters that were logged at low speed with a Race Technology DL1 GPS receiver strapped to the back of a bicycle.

Figure 12.15
Driving lines in Google Earth™

Driver Consistency over Multiple Laps

Driver consistency is easily assessable by observing lap times. Statistical calculations (e.g., average and standard deviation) can be performed to quantify consistency and use as measures for tactical decisions.

***Figure 12.16** Synchronizing video recordings with the logged data can be very revealing when analyzing driving lines*

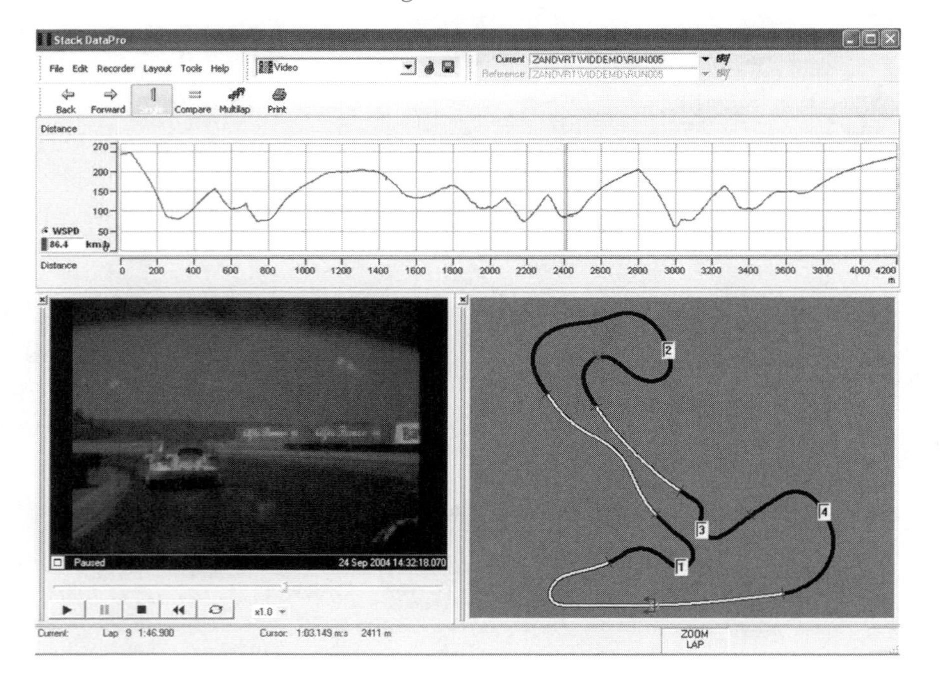

***Figure 12.17** Sector time report of 12 laps during a GT race on the Zolder racetrack*

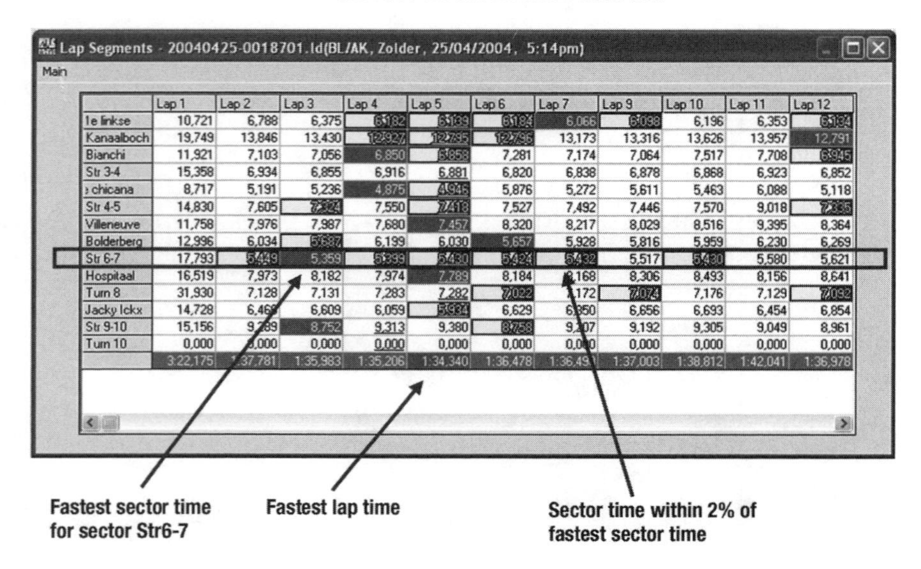

Lap Segments - 20040425-0018701.ld(BL/AK, Zolder, 25/04/2004, 5:14pm)

	Lap 1	Lap 2	Lap 3	Lap 4	Lap 5	Lap 6	Lap 7	Lap 9	Lap 10	Lap 11	Lap 12
1e linkse	10,721	6,788	6,375	6,182	6,139	6,184	6,066	6,098	6,196	6,353	6,134
Kanaalboch	19,749	13,846	13,430	12,927	12,795	12,796	13,173	13,316	13,626	13,957	12,791
Bianchi	11,921	7,103	7,056	6,850	6,858	7,281	7,174	7,064	7,517	7,708	6,945
Str 3-4	15,358	6,934	6,855	6,916	6,881	6,820	6,838	6,878	6,868	6,923	6,852
chicana	8,717	5,191	5,236	4,875	4,946	5,876	5,272	5,611	5,463	6,088	5,118
Str 4-5	14,830	7,605	7,324	7,550	7,418	7,527	7,492	7,446	7,570	9,018	7,335
Villeneuve	11,758	7,976	7,987	7,680	7,457	8,320	8,217	8,029	8,516	9,395	8,364
Bolderberg	12,996	6,034	5,687	6,199	6,030	5,657	5,928	5,816	5,959	6,230	6,269
Str 6-7	17,793	5,449	5,359	5,399	5,430	5,424	5,432	5,517	5,430	5,580	5,621
Hospitaal	16,519	7,973	8,182	7,974	7,789	8,184	[illegible]	8,306	8,493	8,156	8,641
Turn 8	31,930	7,128	7,131	7,283	7,282	7,022	[illegible]	7,074	7,176	7,129	7,092
Jacky Ickx	14,728	[illegible]	6,609	6,059	5,934	6,629	[illegible]	6,656	6,693	6,454	6,854
Str 9-10	15,156	[illegible]	8,752	9,313	9,380	8,758	[illegible]	9,192	9,305	9,049	8,961
Turn 10	0,000	[illegible]	0,000	0,000	0,000	0,000	[illegible]	0,000	0,000	0,000	0,000
	3:22,175	[illegible]	1:35,983	1:35,206	1:34,340	1:36,478	[illegible]	1:37,003	1:38,812	1:42,041	1:36,978

Sometimes, on short courses with many competitors on the track, it is difficult to determine how consistent the driver was from only the timesheets. In this case, rely on the sector time report.

Figure 12.17 shows a sector time report from the 12 first laps of a GT race on the Circuit Zolder racetrack. In this race, forty-five competitors shared a 4-km circuit. The track was divided into fourteen segments in which each corner and straight was defined as a separate segment.

In this report, the fastest segment time for each sector is indicated automatically with a black background, and every segment time coming within 2% of this time is highlighted in gray. The more gray area appearing in this report, the more consistent the driver is.

In sector Str6-7 (Figure 12.17) over the 12 laps, a fastest sector time of 5.359 sec was reached. This means that every segment time within 5.359 sec x 1.02 = 5.466 sec is highlighted in gray. To be consistent, a driver needs to be within one-tenth of his fastest sector time every time.

Of course, this all depends on what one considers consistent. In this example, the driver is supposed to be within 2% of the best performance. The driver's fastest lap during the race was 94.344 sec. This means that a consistent lap time is anything within 94.344 sec x 1.02 = 96.227 sec, which is already a big difference in lap time.

CHAPTER 13

SIMULATION TOOLS

In all motor racing disciplines, track time is limited because race organizers must divide practice and race time among various classes to offer a suitable program to spectators. The testing time available between races is primarily a budget question. Because of this, racecar behavior simulation is becoming more popular, and simulation software is now available to the wider public. This chapter offers a short introduction to simulation and how it can interact with the data acquisition system.

Introduction

Simulation gives an engineer the ability to predict racecar behavior without putting the car on a racetrack. Various techniques are available for predicting the performance of a vehicle in different areas. These include the following:

1. Computational Fluid Dynamics (CFD)

This comprises the numerical solution obtained from computational methods of fluid flow properties. CFD allows simulation of the aerodynamic behavior of racecar components or the entire vehicle. It is much like having a virtual wind tunnel at one's disposal. Other applications of CFD include heat transfer simulation, engine combustion process simulation, and fluid driven systems analysis.

2. Engine Simulation

Software packages are available to simulate the behavior of various engine configurations. Engine power and torque are predictable to a high degree of accuracy, and different solutions can be tested without putting an engine on the dyno.

3. Suspension Kinematics Simulation

Suspension geometry simulation packages allow the user to input all relevant suspension pickup points, the three-dimensional coordinates, and other necessary vehicle dimensions. The software's output typically includes dynamic roll center location, camber change, wheel rates, and bump steer. This type of simulation is particularly useful for creating mathematical channels relating to suspension behavior for the data acquisition software.

4. Vehicle Dynamics Simulation

Simulating the dynamic behavior of a vehicle allows the user to verify the performance of different vehicle configurations. In motorsports, these packages can be used to predict the lap time achievable on any given racetrack.

The remainder of this chapter concentrates on kinematics and lap time simulation. One can have different objectives when simulation software is integrated into the racing team structure. The most obvious reason for using these tools is to virtually test changes to the racecar without having to physically test them on the track. When the vehicle model is accurate enough to perform these virtual tests, money and time is saved ***(Figure 13.1)***.

Another (possibly not so evident) reason for using these tools is determining certain vehicle parameters. This can be achieved by attempting to match data measured by the data acquisition system to the output of the simulation by tweaking the simulation's input parameters, a bit like solving an equation with more than one unknown variable. Much of the recent lap time simulation program's output is very similar to the graphs created by the data acquisition analysis software and offers the

Figure 13.1 Racecar behavior simulation

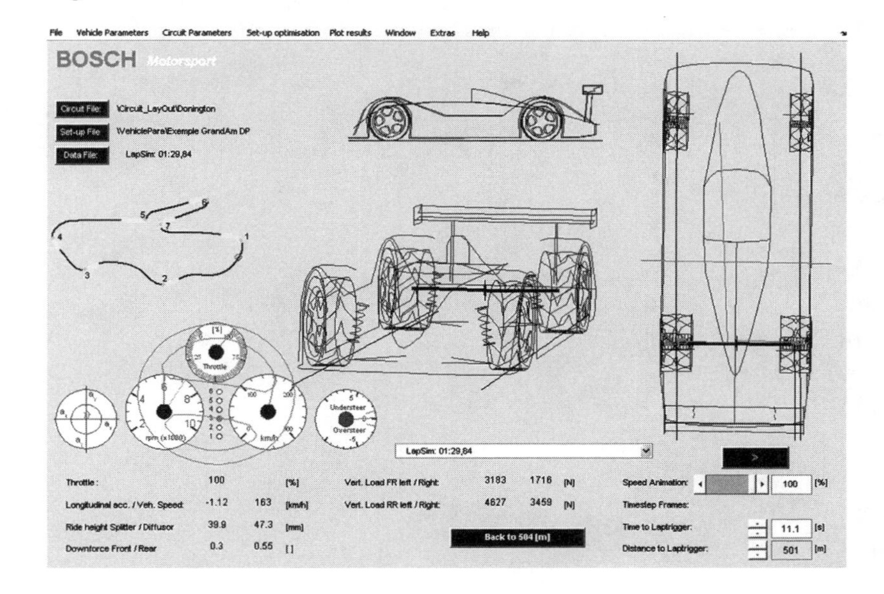

capability to import data and directly overlay them onto the simulated data.

In addition, vehicle parameters not directly measured can be simulated. A simple example is to use kinematics simulation to create the wheel's bump steer curve at different wheel travel values. Subsequently, a mathematical channel can be created in the data analysis software that relates this calculated curve to the wheel travel data measured during the car's track time.

The interaction between the data acquisition system and the applied simulations works in two directions. The simulation software requires input that in some cases must be measured on the track; the recorded data validates the model afterward. Once a reliable model is created, its output can be used not only to predict the effect of changes to the vehicle but also to calculate channels not directly measured on the car.

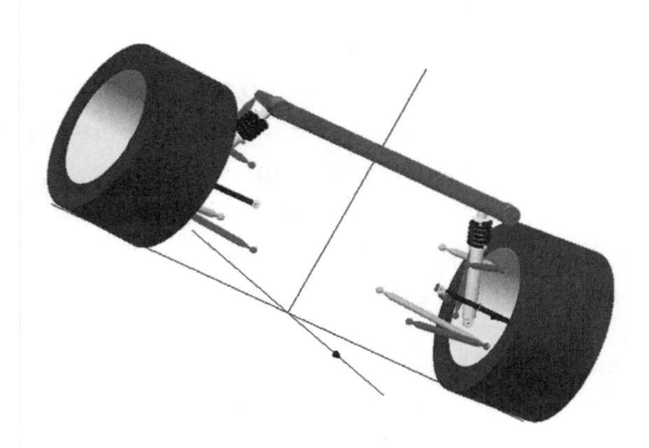

***Figure 13.2** Dodge Viper GTS-R front suspension geometry*

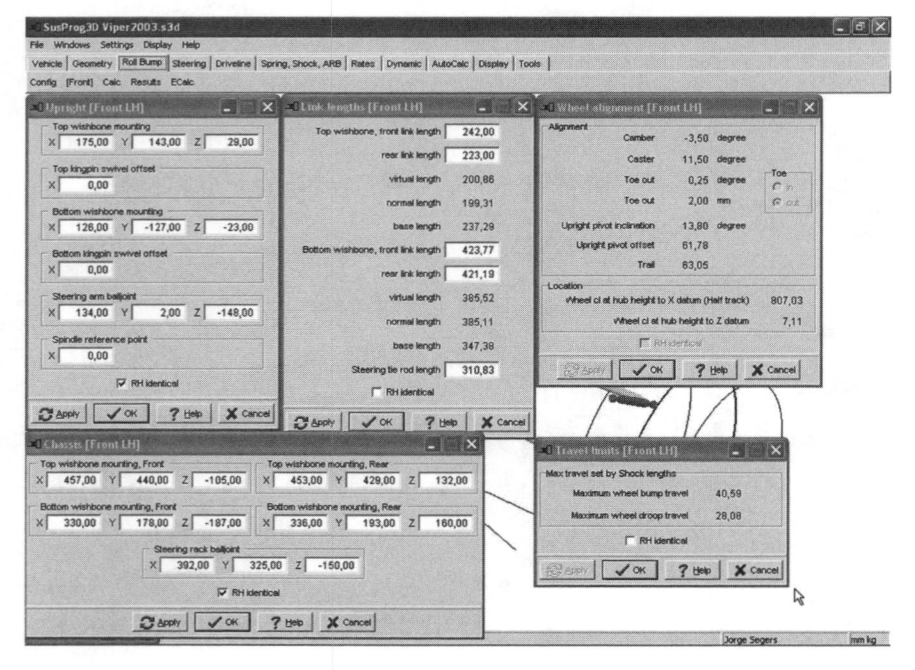

***Figure 13.3** A kinematics model is created by entering all relevant suspension pickup point coordinates and wheel alignment settings.*

Suspension Kinematics Simulation

Suspension kinematics describes the way that the sprung and unsprung mass of the vehicle relate to one another and to the relative motion of the various suspension and chassis components. A typical kinematics software package requires the coordinates of each suspension pickup point to be entered ***(Figure 13.2)***. The typical output parameters include the following:

- roll-center height from the ground and offset from the axle centerline,
- swing axle length and angle,
- dynamic camber gain,
- dynamic caster,
- dynamic kingpin inclination,
- tire scrub,
- dynamic toe (bumpsteer),
- roll steer,
- wheel movement versus shock absorber movement (motion ratio),
- antilift/antisquat,
- dynamic track width,
- Ackerman steering, and
- driveshaft angle and plunge.

These parameters can be calculated as a function of ride height, wheel travel, roll angle, and steering angle. When the software allows vehicle dynamics parameters such as spring rates, corner weights, and center of gravity heights to be input, dynamic weight transfer and wheel loads can be calculated also.

When the location of the suspension pickup points are not supplied by the manufacturer, care must be taken when measuring them on the car. The output obtained from the software calculations is only as accurate as that of the coordinate measurements. Manufacturer drawings of the various suspension parts (e.g., hubs, uprights, wishbone, chassis) are useful. Occasionally, parts must be disassembled to obtain a proper measurement.

Typically, most of the time required to develop a good kinematics model goes toward measuring the suspension pickup points (if one is not designing the suspension from scratch). It is also necessary to ensure that the relative positions of the pickups are measured from a known ride height.

When an appropriate kinematics model is selected, the effect of modifying various suspension parameters can be investigated. In addition, the data acquisition engineer has access to more information to better understand the behavior of the vehicle. The output of the kinematics model can be input into the data analysis software through the use of math channels ***(Figure 13.3)***.

Figure 13.4 shows (among other parameters) the dynamic camber values for the front and rear suspension of a Dodge Viper GTS-R that were calculated with a kinematics model created in Susprog3D. If the analysis software allows this function, these values can be entered directly as a lookup table with wheel travel as an input variable (see the example in Chapter 11). Alternatively, the camber values can be exported to a spreadsheet where they are graphed against wheel travel and an equation (i.e., a trendline) is calculated from this curve ***(Figure 13.5)***. The equations calculated by the spreadsheet software are indicated in the graph. The last argument in both equations equals the static camber angle (3.5 deg on the front wheels, 1.8 deg on the rear). In other words, the front and rear dynamic camber angles in this example are defined by ***Equations 13.1*** and ***13.2***.

These two equations can be entered into the analysis software as mathematical channels and treated like any channel logged by the data acquisition system. ***Figure 13.6*** portrays the values of the above channels for a lap around Circuit Zolder in a Dodge Viper GTS-R, for which this kinematics model was created. This technique can be used to relate all parameters calculated by the kinematics software to measured track data.

Lap Time Simulation

The primary performance indicator of a racecar is the lap time that it can achieve on any given racetrack. Most changes made to the car aim to decrease lap time. Lap time simulation packages are available for simulating racing laps. Changes to the car can be evaluated beforehand. It is doubtful that simulation will replace completely circuit testing, but it reduces the amount of setup variables and gives the team more track time so team members can focus on other issues. In addition, the more accurate the model, the more setup parameters can be optimized without actually testing them on the racetrack.

There are different software packages available. The following are the more popular ones:

- PiSim (Pi Research),
- LTS (Lap Time Simulation, Milliken Research Associates Inc.),

Figure 13.4 Kinematics model output

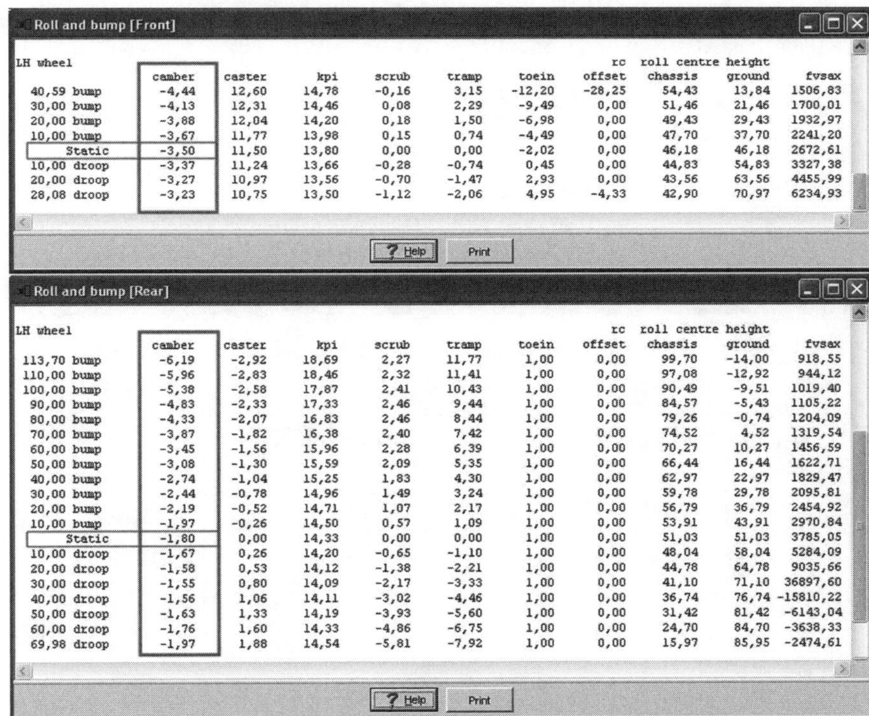

Roll and bump [Front]

LH wheel	camber	caster	kpi	scrub	tramp	toein	rc offset	roll centre height chassis	roll centre height ground	fvsax
40,59 bump	-4,44	12,60	14,78	-0,16	3,15	-12,20	-28,25	54,43	13,84	1506,83
30,00 bump	-4,13	12,31	14,46	0,08	2,29	-9,49	0,00	51,46	21,46	1700,01
20,00 bump	-3,88	12,04	14,20	0,18	1,50	-6,98	0,00	49,43	29,43	1932,97
10,00 bump	-3,67	11,77	13,98	0,15	0,74	-4,49	0,00	47,70	37,70	2241,20
Static	-3,50	11,50	13,80	0,00	0,00	-2,02	0,00	46,18	46,18	2672,61
10,00 droop	-3,37	11,24	13,66	-0,28	-0,74	0,45	0,00	44,83	54,83	3327,38
20,00 droop	-3,27	10,97	13,56	-0,70	-1,47	2,93	0,00	43,56	63,56	4455,99
28,08 droop	-3,23	10,75	13,50	-1,12	-2,06	4,95	-4,33	42,90	70,97	6234,93

Help Print

Roll and bump [Rear]

LH wheel	camber	caster	kpi	scrub	tramp	toein	rc offset	roll centre height chassis	roll centre height ground	fvsax
113,70 bump	-6,19	-2,92	18,69	2,27	11,77	1,00	0,00	99,70	-14,00	918,55
110,00 bump	-5,96	-2,83	18,46	2,32	11,41	1,00	0,00	97,08	-12,92	944,12
100,00 bump	-5,38	-2,58	17,87	2,41	10,43	1,00	0,00	90,49	-9,51	1019,40
90,00 bump	-4,83	-2,33	17,33	2,46	9,44	1,00	0,00	84,57	-5,43	1105,22
80,00 bump	-4,33	-2,07	16,83	2,46	8,44	1,00	0,00	79,26	-0,74	1204,09
70,00 bump	-3,87	-1,82	16,38	2,40	7,42	1,00	0,00	74,52	4,52	1319,54
60,00 bump	-3,45	-1,56	15,96	2,28	6,39	1,00	0,00	70,27	10,27	1456,59
50,00 bump	-3,08	-1,30	15,59	2,09	5,35	1,00	0,00	66,44	16,44	1622,71
40,00 bump	-2,74	-1,04	15,25	1,83	4,30	1,00	0,00	62,97	22,97	1829,47
30,00 bump	-2,44	-0,78	14,96	1,49	3,24	1,00	0,00	59,78	29,78	2095,81
20,00 bump	-2,19	-0,52	14,71	1,07	2,17	1,00	0,00	56,79	36,79	2454,92
10,00 bump	-1,97	-0,26	14,50	0,57	1,09	1,00	0,00	53,91	43,91	2970,84
Static	-1,80	0,00	14,33	0,00	0,00	1,00	0,00	51,03	51,03	3785,05
10,00 droop	-1,67	0,26	14,20	-0,65	-1,10	1,00	0,00	48,04	58,04	5284,09
20,00 droop	-1,58	0,53	14,12	-1,38	-2,21	1,00	0,00	44,78	64,78	9035,66
30,00 droop	-1,55	0,80	14,09	-2,17	-3,33	1,00	0,00	41,10	71,10	36897,60
40,00 droop	-1,56	1,06	14,11	-3,02	-4,46	1,00	0,00	36,74	76,74	-15810,22
50,00 droop	-1,63	1,33	14,19	-3,93	-5,60	1,00	0,00	31,42	81,42	-6143,04
60,00 droop	-1,76	1,60	14,33	-4,86	-6,75	1,00	0,00	24,70	84,70	-3638,33
69,98 droop	-1,97	1,88	14,54	-5,81	-7,92	1,00	0,00	15,97	85,95	-2474,61

Help Print

$$\text{Camber}_{\text{Front}} = -5 \cdot 10^{-8} \cdot x^3_{\text{WheelF}} - 2 \cdot 10^{-3} \cdot x^2_{\text{WheelF}} - 0.0151 \cdot x_{\text{WheelF}} - \text{Camber}_{\text{StaticF}} \qquad \textit{(Eq. 13.1)}$$

$$\text{Camber}_{\text{Rear}} = 2 \cdot 10^{-7} \cdot x^3_{\text{WheelR}} - 2 \cdot 10^{-3} \cdot x^2_{\text{WheelR}} - 0.0147 \cdot x_{\text{WheelR}} - \text{Camber}_{\text{StaticR}} \qquad \textit{(Eq. 13.2)}$$

- MSC.ADAMS/Motorsports (MSC Software),
- CarSim (Mechanical Simulation Corporation),
- CALLAS Motorsports (Sera-CD),
- RaceSim (D.A.T.A.S. Ltd.),
- FastLapSim (ProRacingSim),
- LapSim (Bosch Motorsport GmbH), and
- Chassissim (Chassissim Technologies).

***Figure 13.5** Dodge Viper GTS-R front and rear dynamic camber curves*

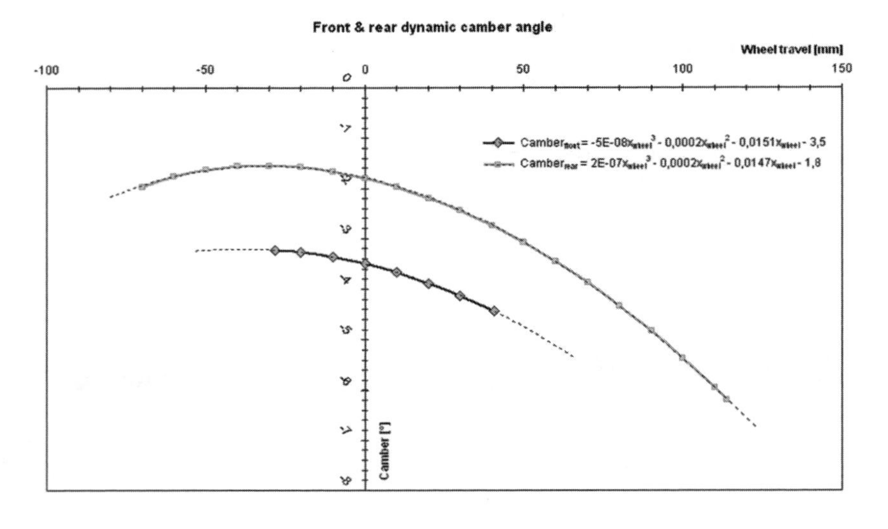

***Figure 13.6** Dynamic camber channels created from a kinematics model during a lap around Zolder*

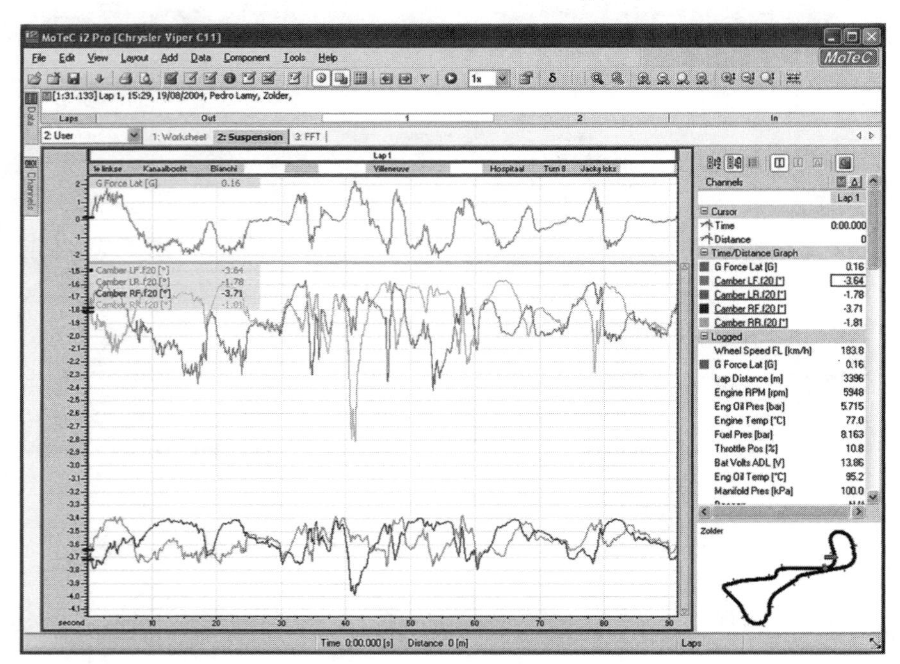

Lap time simulation is a technique formerly used only in the upper echelons of motorsports. Many of these high-end users actually develop their own programming code to set up a model of their cars. This is often quicker than using commercially available packages. Also, when the software is written in-house, the programmer also know its limitations.

Not everyone has access to these programming facilities. However, as with data acquisition technology, simulation software has become more accessible to all racing disciplines. In fact, some illustrations in this chapter were created with LapSim, which can be downloaded free of charge from www.bosch-motorsport.com.

The approach used to create a simulation model differs among various software packages, but the fundamental principles remain the same.[15] A racetrack is divided into many individual sectors, each defined by length and (corner) radius. The maximum speed in this sector is determined by the friction between the tires and road surface as well as the resistance against forward acceleration experienced by the vehicle. Once the sector speed is known, the segment time can be calculated; the sum of all these segment times yields the theoretical lap time.

The maximum force that can be transmitted through the tires' contact patches is defined by ***Equation 13.3***.

$$F_{max} = \mu \cdot F_N \qquad \text{(Eq. 13.3)}$$

μ is the coefficient of friction between the tire and the road surface and F_N the normal force acting on the tire.

These two parameters present a challenge when creating a mathematical racecar model. The coefficient of friction partially depends on tire characteristics and alignment settings, pressure, temperature, and the state of the road surface, parameters that may not remain constant over the duration of a lap. Normal load depends on vehicle weight, the location of the center of gravity, roll stiffness distribution, aerodynamic configuration, and suspension kinematics. Therefore, the quality of the vehicle model and the characteristics of the various software packages primarily depends on the number of parameters and the accuracy of the tire model.

A racecar normally is driven near the boundary of the tires' traction circle, where the tire behavior is strongly nonlinear and load sensitivity is significant. It is very difficult to acquire such data from tire manufacturers, and if the data is available it often requires extensive manipulation to be useful. In many of today's simulation software packages, the tire is modeled using Pacejka's magic formula, a technique developed by H.B. Pacejka from Delft University of Technology. This model can represent the primary tire properties, such as side force, longitudinal force, or self-aligning torque.

Once a suitable model is created, it can be used to investigate, among others, the following typical applications:

- determining suitable gear ratios for a given racetrack,
- analyzing the effect of shift RPM and duration on lap time,
- compromising between aerodynamic downforce and drag,
- determining engine output characteristics,
- assessing the effect of vehicle weight and weight distribution on lap time (e.g., fuel load and ballast),
- determining the effect of a series regulations on maximum vehicle performance (e.g., restrictor size versus vehicle weight, and effect of penalty weight),
- comparing the performance envelope of different cars,
- optimizing brake balance,
- optimizing suspension setup,
- assessing the effect of environmental circumstances on lap time (e.g., wind, and ambient temperature and pressure),
- measuring the effect of tire deterioration on lap time, and
- determining nonmeasured vehicle parameters (e.g., drag and downforce figures).

The data acquisition system in this context is a validation tool. By comparing the measured data with the simulation output, the model's accuracy can be evaluated and improved where necessary. Some parameters required for the model must be deduced from the logged data. Most simulation packages have an export function for transferring the simulation data into the data acquisition software, and vice versa. By doing this, a real lap can be overlaid onto a simulated lap, and the analysis techniques of the overlaid data can be used for comparison.

The vehicle model is defined by a number of parameters, that depend on the quality of the simulation and the effect under investigation. For instance, to determine the optimum gear ratios for a given track, not as many parameters are required as for investigating the transient cornering behavior of a vehicle.

Some vehicle parameters, such as vehicle weight and distribution as well as spring rates, can be obtained directly from the car's setup sheet. Others must be measured statically (e.g., track width, wheelbase, and center of gravity height), or estimated from the logged data (e.g., aerodynamic configuration and tire characteristics).

A Simulated Example

In this paragraph, a simple vehicle model is created for a GT car using Bosch's LapSim. The simulated output is linked to actual recorded data from the car while on the Circuit de Spa-Francorchamps. The vehicle model is validated with the recorded data.

The LapSim software uses a quasi steady-state model in which the car consists of a body with six degrees of freedom plus four wheels, each with four degrees of freedom.[16] The model is described with twenty-six vehicle parameters. The tire model is a variation of the Pacejka formula, in which each tire's characteristics are defined by thirteen coefficients. In the standard version of LapSim, these coefficients cannot be modified. The simulation model first calculates the minimum speed at which the car can negotiate each corner. Subsequent to that, the model simulates acceleration out of the first corner, while braking backward from the second corner. Where these two calculations converge is the top speed between the first and second corner. This sequence repeats itself until the complete lap has been simulated.

The first step in modeling the vehicle is entering all parameters defining the racecar. The driveline parameters include a power/torque curve of the engine and the ratios installed in the gearbox and differential. In addition, the dynamic tire radius and the upshift duration are required to create an accurate acceleration model ***(Figure 13.7)***.

Total weight and distribution, wheelbase, track width, and height of the center of gravity are used to determine the static weight on the four tires, the longitudinal load transfer during acceleration and braking, and the lateral load transfer during cornering. The stiffness of the springs and antiroll bars, with the brake force distribution, determine how this load transfer is distributed over the four wheels. The brake balance should be calculated from the dimensions of the brake system components (i.e., the master cylinder size and brake caliper dimensions).

Figure 13.7 Baseline model parameters

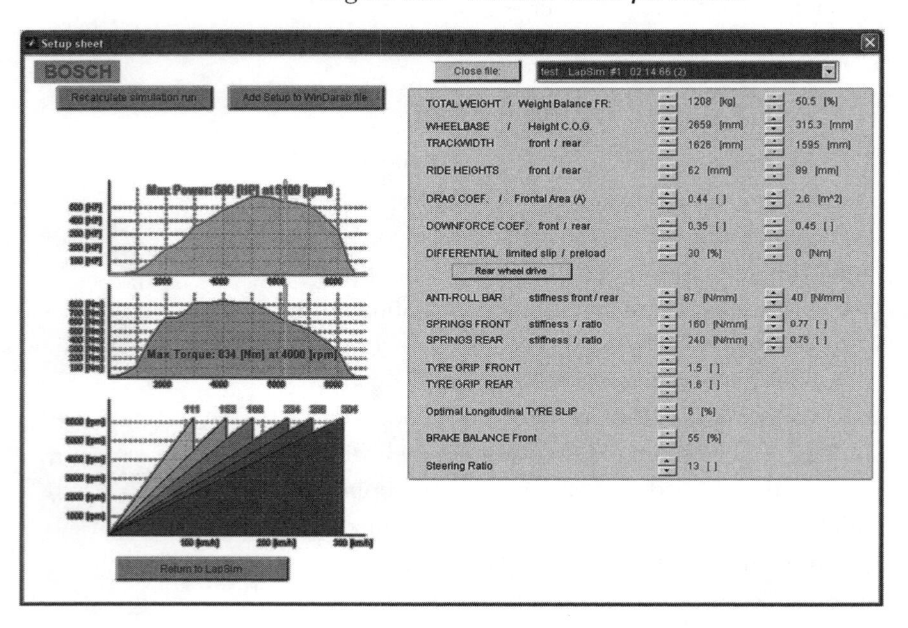

Figure 13.8 A file with onboard recorded data can be imported into the simulation software

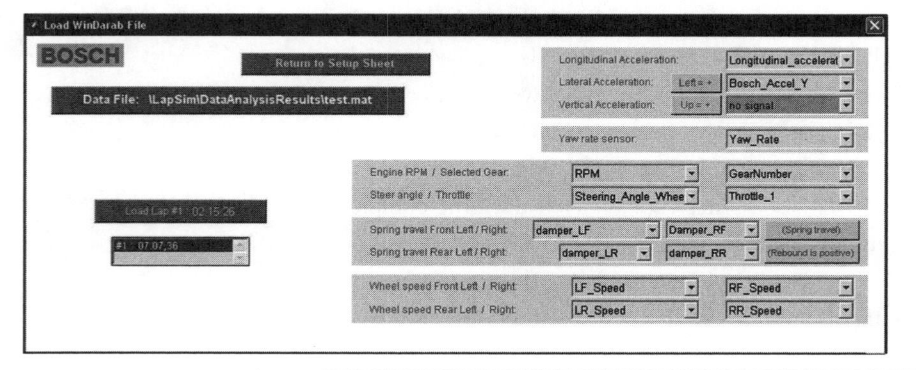

The suspension motion ratio is entered to calculate the relationship between the wheel and suspension travel. This enables the model to calculate the car's roll and pitch movement.

The aerodynamic configuration of the car is defined by a drag coefficient and two lift coefficients, one for the front axle and the other for the rear axle. These coefficients are assumed for a static ride height and are independent of dynamic ride height or speed. They are multiplied in the calculations by the frontal surface of the vehicle. Static ride height is used in the model as a reference for calculating an output channel of dynamic ride height around the track.

The LapSim model uses a limited slip differential with an equal percentage of limited slip for braking and acceleration. In addition, a preload value can be entered.

The light version of LapSim uses a Pacejka tire model that cannot be modified by the user. The user input is limited to two grip factors, one for the front wheels and one for the rear wheels.

The tire grip factors in the software are used to tune the model to real-life tire characteristics. Selecting a grip factor similar to the maximum lateral acceleration of the vehicle in a slow corner with little aerodynamic significance results in a reasonable estimation. If the front tires are smaller than the rear tires, the front grip factor entered into the software should be a bit lower than that of the rear tires. These two coefficients can be tweaked by overlaying the simulated steering angle channel on the trace of a reference lap. By varying the grip level of the front tires compared to that of the rear tires, the steering angle of the simulation can be tuned to that of the real vehicle. The longitudinal tire characteristics are defined by a longitudinal slip value at which the maximum longitudinal tire force occurs.

Next, the simulation model requires a racetrack. In this example, the car behavior is simulated around the Spa track. A data file with the channels is pictured in ***Figure 13.8***. This file is exported from the data analysis software and imported into LapSim. To calculate a track map, the software requires at a minimum the vehicle speed and lateral acceleration. The calculated track is two-dimensional, but the user can define the height of each point on the track manually as well as a banking angle for each corner.

The imported lap was obtained from a qualification session with the test car, and the lap time achieved was 2'15"36. The additional channels are imported to correlate the simulation model with the measured data.

The model is now ready to calculate the duration of a lap around the Spa track. The predicted lap time is 2'14"66, or 0.7 sec faster than the actual achieved lap time. The speed, RPM, and used gear traces of the simulated lap (black trace) and the real lap (white trace) are illustrated in ***Figure 13.9,*** while the lower graph depicts the time difference between the two laps.

By overlaying different channels, the model now can be tuned so that the time difference is minimal. In this way, the value of some parameters that were originally estimated can be more realistically determined. The following are some examples:

- The drag coefficient should be chosen in such a way that the speed signals match between the simulated and real lap at the end of high-speed straights. The two traces in Figure 13.9 indicate a reasonable correlation between the real and simulated car.
- To estimate the downforce coefficients, the front and rear ride heights should be compared ***(Figure 13.10)***. Of course, the spring rates of the model and the real car should be the same. At this point, observe the ride heights at higher speeds, where aerodynamic influences are pronounced more. When the simulated ride height on one axle is lower than the recorded height, the selected downforce coefficient on that axle is probably too high. Once the drag and downforce coefficients have been satisfactorily determined, the aerodynamic forces acting on the vehicle can be calculated from the simulation results ***(Figure 13.11)***.
- The amount of limited slip of the differential can be estimated by comparing the difference in wheel speed between the driven wheels. ***Figure 13.12*** indicates that the limited slip value of the simulation model was set too low. In reality, the wheel speed difference between the inner and outer driven wheels is always smaller than that calculated by the model.
- Determining if the suspension stiffness parameters are correct can be confirmed by observing the roll and pitch movements ***(Figure 13.13)***.

Once the model is validated against the recorded track data, it can be used to explore the influence of certain vehicle parameters on lap times. This enables the engineer to establish a basic setup prior to arriving at the track.

***Figure 13.9** The speed, RPM, and gear ratio for the simulated and real lap. The lower graph is the time difference between the two laps.*

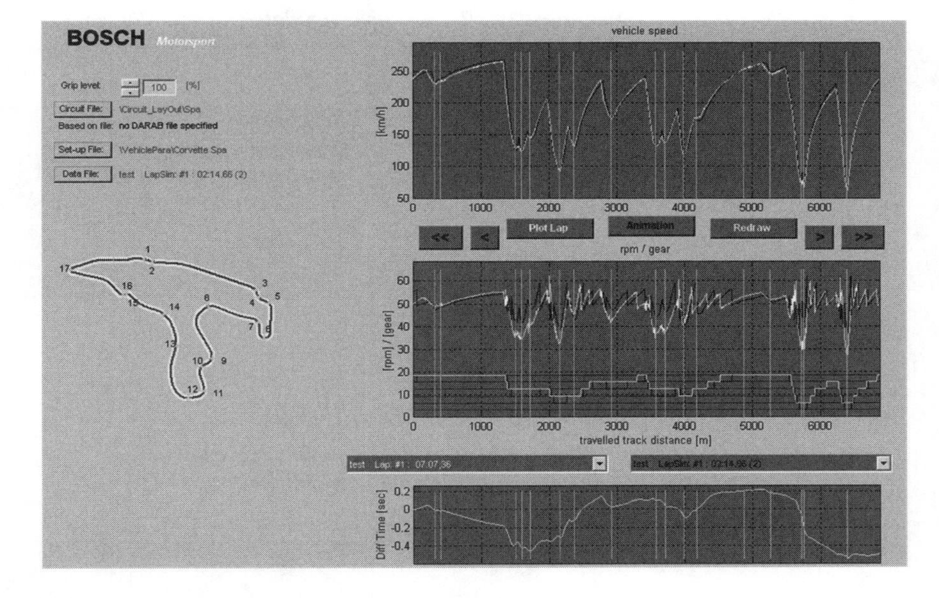

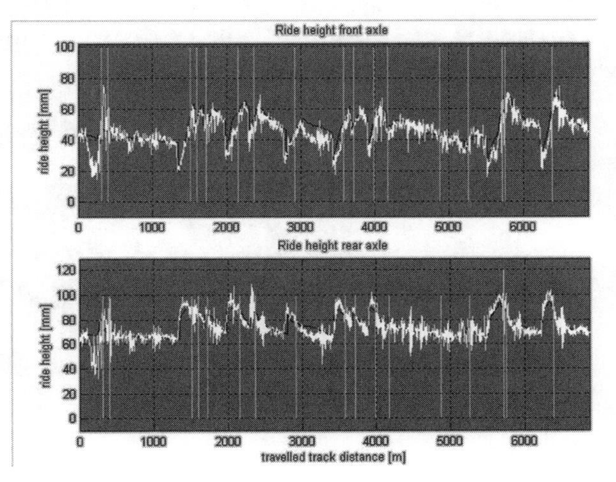

***Figure 13.10** A front and rear ride height comparison between a real car and the simulation model. The white trace is the recorded data; the dark trace is the simulated result.*

Figure 13.11 *Aerodynamic forces calculated from the recorded data and by the simulation model*

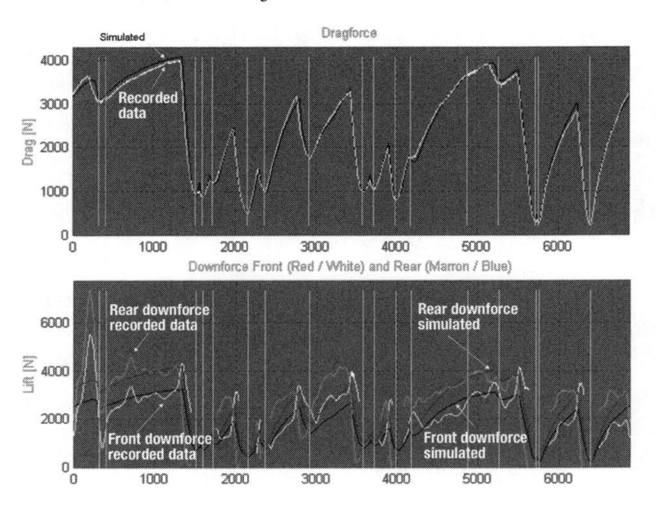

Figure 13.12 *Real and simulated differential work. The limited slip parameter of the model is too low.*

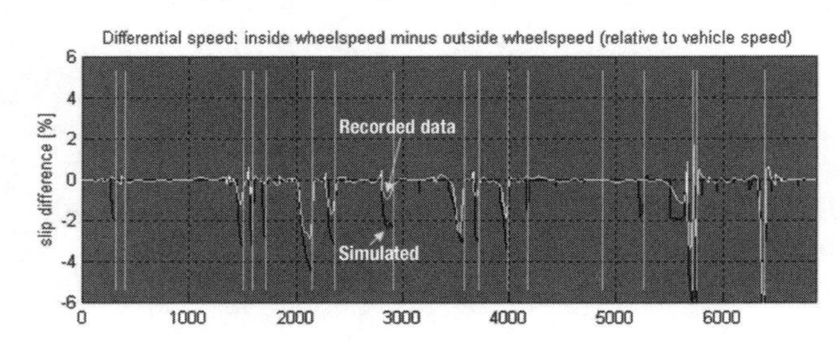

Figure 13.13 *Roll and pitch angle comparison between recorded data and simulation*

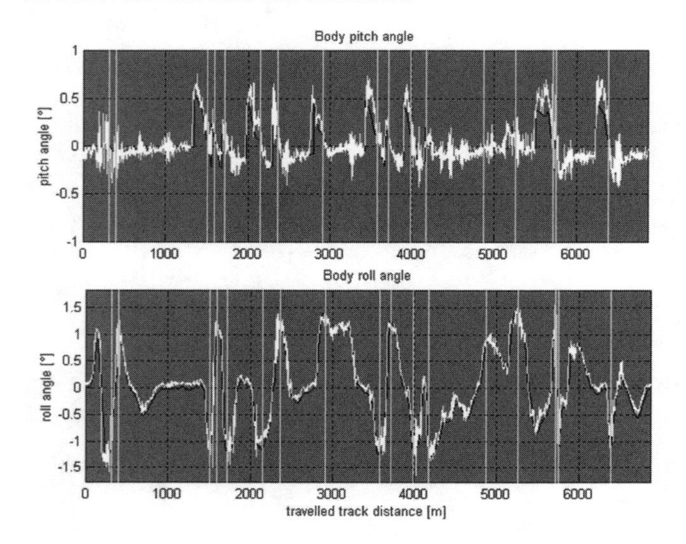

CHAPTER 14

USING THE DATA ACQUISITION SYSTEM FOR DEVELOPING A RACE STRATEGY

On race day, the engineer should be armed with enough knowledge to predetermine a race strategy. This knowledge also allows the engineer to be flexibile when circumstances change during the race. Fuel consumption, tire wear, and driver consistency should be investigated to obtain the necessary knowledge. This chapter discusses methods for measuring these parameters.

Fuel Consumption

During the practice sessions preceding a race, the engineer wants to address the following questions concerning the vehicle's fuel economy.

1. How much fuel is consumed per lap?

In racing, fuel consumption commonly is not expressed in liters/100 km, as with road-going vehicles. On a racetrack, it is important to know what the consumption per lap will be.

2. How many laps can be completed on a tank?

This number determines the amount of fuel that should be in the tank to complete the race. When refueling is necessary during the race, this figure determines the minimum amount of stops that must be made and establishes the pitstop window.

3. What is the weight penalty resulting from fuel load? What influence does this have on lap time?

The performance potential of the vehicle changes as the fuel load changes during the race. A lighter load helps decrease lap times. As the fuel level drops, the car's center of gravity shifts in height. If the car does not have a centrally located fuel cell, it also changes longitudinally. This modifies the vehicle's balance.

Fuel consumption can be measured by the data acquisition system in several ways. The simplest solution is measuring the tank level. The quality of the sensor mounting in the fuel tank determines how useful this signal is.

Another way is to assess the fuel flow between the tank and the engine, both on the supply and return pipe. The difference between the two is the amount of fuel burned by the engine. Care should be taken with this measurement because the temperature of the fuel supplied to the engine may differ from that of the fuel returning to the tank in the return pipe. This modifies the fuel density; temperature compensation in the flow measurement is necessary to obtain an accurate result.

Most modern motorsport ECUs have a programmed algorithm for calculating fuel usage. The ECU software uses an injection table to determine injector timing for every possible engine RPM and load. Every time an injector opens, the fuel passing through it equals the injector opening time multiplied by the flow rate of the injector. The sum of the amount of fuel for all injectors during a certain time period is the fuel consumed.

A variable containing total injector opening time as a function of elapsed time is stored primarily by the ECU, leaving it to the user to scale this variable to the liters used. ***Figure 14.1*** illustrates an example of this in the MoTeC Dash Manager Software.

From this, the following variables can be calculated using the analysis software:

- amount of fuel used,
- amount of fuel left in the tank,
- amount of fuel used per lap, and
- laps remaining on fuel tank.

Figure 14.1 *MoTeC Dash Manager fuel prediction calculation*

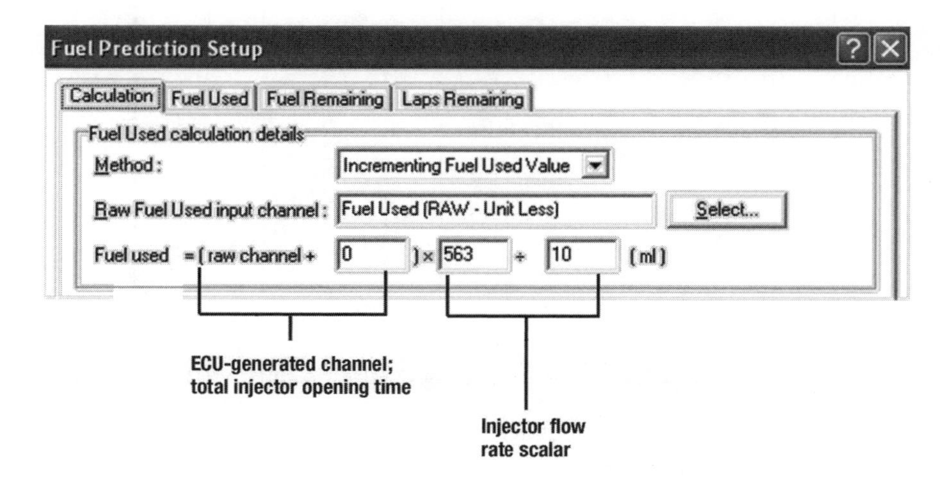

To guarantee sufficient accuracy, the configuration of this calculation must be checked against actual fuel consumption. The best way to do this is manually measuring the weight of the fuel put into the tank before driving and the weight of the fuel that is left over. This compensates for differences in temperature.

Figure 14.2 was taken from a GT car during a free practice session at Zhuhai International Circuit. The chart illustrates the last outing of this car. The fuel tank level is plotted against time. The amount of fuel used per lap is shown at the bottom of the screenshot.

The average fuel consumption (in liters) per lap can be calculated from the data by subtracting the remaining tank level at the end of this outing from that measured at the beginning and dividing it by the number of covered laps ***(Equation 14.1)***.

$$\frac{60.69\ \text{L} - 30.67\ \text{L}}{12\ \text{laps}} = 2.50\ \text{L/lap} \qquad \textit{(Eq. 14.1)}$$

This figure includes the fuel consumption during the in and out laps. From a strategic perspective, knowing the specific fuel consumption during these two laps may be necessary; this incorporates the amount of distance covered at a controlled speed in the pit lane. However, it does not provide a 100% accurate average for the normal racing laps.

Figure 14.2 ***Fuel tank level during a free practice outing around Zhuhai***

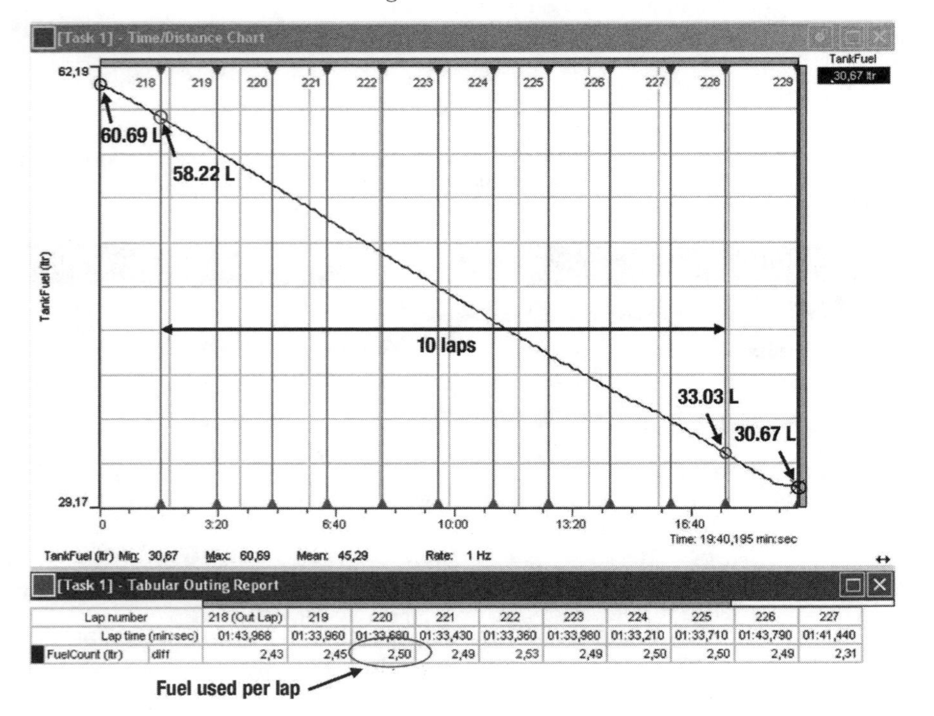

Lap number		218 (Out Lap)	219	220	221	222	223	224	225	226	227
Lap time (min:sec)		01:43,968	01:33,960	01:33,680	01:33,430	01:33,360	01:33,980	01:33,210	01:33,710	01:43,790	01:41,440
FuelCount (ltr)	diff	2,43	2,45	2,50	2,49	2,53	2,49	2,50	2,50	2,49	2,31

When the in and out laps (which are not representative of fuel consumption during normal racing laps) are ignored, the average fuel used per lap becomes ***Equation 14.2***.

$$\frac{58.22\ \text{L} - 33.03\ \text{L}}{10\ \text{laps}} = 2.52\ \text{L/lap} \qquad \textit{(Eq. 14.2)}$$

In this outing the amount of fuel used during the in and out laps is the following:

in lap 2.43 L

out lap 2.36 L

To evaluate the accuracy of this measurement, the following method should be applied continuously. Taking into account the entire practice session, the following fuel tank levels were recorded:

fuel tank level before start of session 96.40 L

fuel tank level at end of session 30.67 L

The logged data indicated 65.73 L were used over a distance of 27 laps (including all in and out laps), which results in an average of 2.43 L consumed per lap.

In reality, 52.5 kg of fuel was put into the car before and during the practice session. The fuel temperature at that moment (i.e., before it was put into the tank) was 29°C. The fuel specification sheet shows a density of 0.75 kg/L at 15°C. Fuel volume and temperature are correlated using ***Equation 14.3***.

$$\frac{\Delta V}{V} = \left(\frac{1}{1 - a\dfrac{(T-15)}{\rho_{15}}} \right) - 1 \qquad \textit{(Eq. 14.3)}$$

with V = fuel volume (L)
T = fuel temperature (°C)
ρ_{15} = fuel density at 15°C (kg/L)
a = constant = 0.0008

This means the density of the fuel put into the car is as shown in ***Equation 14.4***. In other words, 76.06 L of fuel was put in the fuel tank. After the session, 4.15 kg of fuel (at a temperature of 39°C) was removed from the tank. At this temperature the fuel density is 0.731 kg/L, resulting in a volume of 5.68 L. Therefore, 65.38 L were used, compared to 65.73 L calculated by the data logging system. This indicates a relative error of 0.53%.

The accuracy of the data in this example means it is feasible to investigate the effect of various parameters on fuel consumption (e.g., shift RPM, throttle blipping during downshifts, and throttle application). In addition, the car weight can be estimated at any given time to analyze changes in the vehicle balance as the fuel load decreases.

The fuel tank level can be displayed for the driver on the dashboard and the reading communicated to the engineer by radio. When telemetry is available, the level can be observed in the pits at any time.

Tire Wear and Driver Consistency

Tire wear is the second important factor for developing a race strategy. The performance of the tires over the race distance is not constant. The overall grip level decreases. If tire wear is greater on one axle, the car's balance changes as well.

To measure the effect of tire drop-off, record longer runs during the practice sessions preceding the race. Grip and balance then can be investigated over the lifetime of a tire set. The following questions should be answered to obtain information on the lifetime of a set of tires.

- From the time they are new, how many laps are completed before the performance of the tires peak?
- In which lap are the highest grip levels recorded?
- From the performance peak, what is the drop-off in grip as a function of the lap number?
- What is the realistic maximum number of laps this tire set is going to last?
- In which direction is the balance of the car developing (oversteer or understeer)? Which axle produces the highest tire wear?
- What is the average possible lap time over the life of the tire set?

Figures 14.3 and ***14.4*** illustrate the performance of three drivers during a 3-hour GT race on the Dubai Motodrom. The car in question was a 600-hp rear-wheel drive vehicle. The first illustration represents the lap times achieved by the three drivers. The race was split into three stints, each beginning with a full fuel tank and a new set of tires made of the same compound. The end of the second stint was interrupted by a safety car situation. The average lap times for each driver are given in ***Table 14.1***. In and out laps and the laps completed under the safety car were not taken into account. The first driver is the fastest, with the slowest of the three tackling the middle stint. Assuming similar conditions, the second driver loses approximately 1.4 sec to the quickest driver, while the third driver limits this difference to 0.5 sec.

$$\frac{\rho_{15}}{\left(1+\frac{\Delta V}{V}\right)} = \frac{0.75}{\left(\frac{1}{1-0.0008\frac{(29-15)}{0.75}}\right)} = \frac{0.75}{1.015} = 0.739 \text{ kg/l} \qquad (Eq.\ 14.4)$$

Figure 14.3 *Lap time graph of a 3-hour GT race in Dubai*

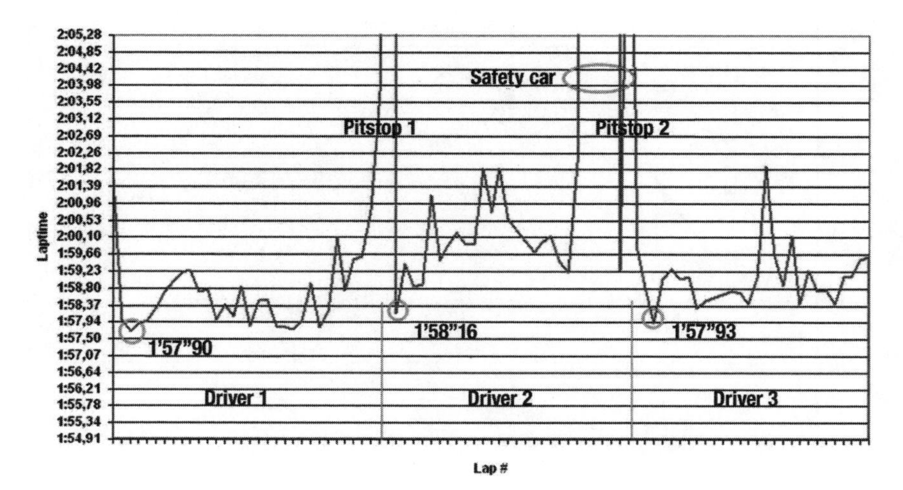

Table 14.1 Average race lap time and average understeer angle per driver

	Average Lap Time	Average Understeer Angle
Driver 1	1'58"56	0.85 deg
Driver 2	1'59"99	0.97 deg
Driver 3	1'59"05	0.81 deg

During the first stint, the driver complains about a diabolic understeer for the first 30% of the covered laps. The situation improves once the driver adjusts his driving style to the changing balance. This is evident in Figure 14.3. During the first 10 laps of the race, there is a drop-off from the fastest lap (which was achieved in the third race lap) of 1.3 sec. After that, the times come down and the driver begins to record some high 57s again. During the last quarter of the stint (after approximately 24 laps), tire wear becomes noticeable as the lap times increase again.

The second stint indicates a similar pattern. The difference here is that the stint generally is slower and it takes until the middle of the stint before the tire set reaches its second life. The driver has more difficulty adapting to the changing situation. The third driver has the least difficulty coping with the understeering characteristic.

The first driver's comments are confirmed in Figure 14.4. Here the average understeer angle (see Chapter 7) per lap is plotted. The higher the understeer angle, the more understeer the vehicle is developing. In Table 14.1, the average understeer angle for each driver's stint is calculated, illustrating that the second driver deals with the highest degree of understeer (or is inducing the most understeer through a specific driving style). This is confirmed in the graph.

All three drivers are confronted with increasing understeer for the first 10–15 laps in their stint. After that, the balance begins to develop into increasing oversteer, probably assisted by the decreasing fuel load on the rear axle. The third driver has the most consistent balance over a complete stint, but eventually the third driver's average lap time is 0.5 sec slower than that of the first driver.

This example perfectly illustrates how drivers must adapt to changing situations. Driver consistency is the third important factor in a race strategy. A driver can be faster than the competition in a qualifying lap, but over a race distance a driver might have a greater degradation in lap times. Of course, the physical condition of the driver plays a major role in this. Problems with fitness and concentration often are indicated in the data as driver error. Gear-shifting mistakes, changes in throttle blipping, earlier-than-normal braking points, and other abnormalities can indicate a fatigued driver. See Chapter 12 for more details on driver analysis.

Figure 14.4 Average understeer angle per lap

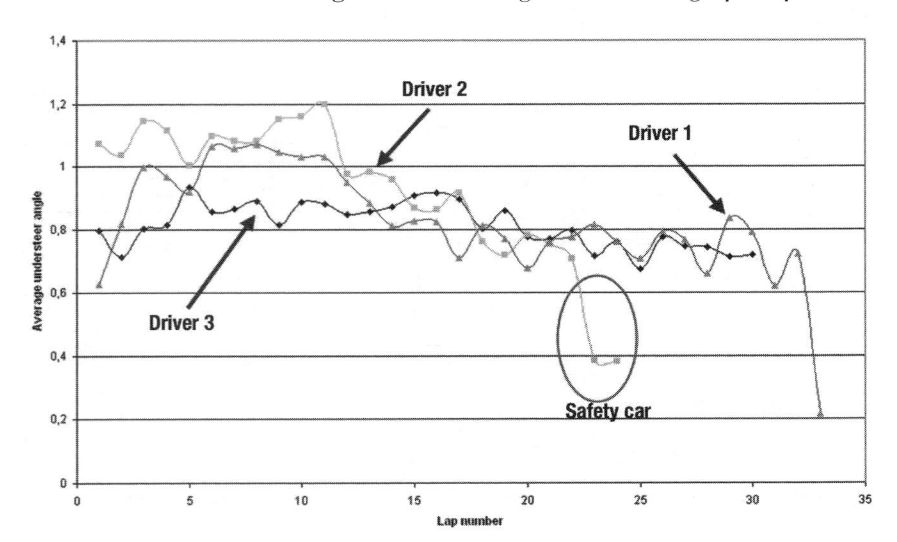

CHAPTER 15

INTRODUCTION TO MEASUREMENT

The previous chapters investigated how numerous sensor signals are analyzed to provide information about vehicle and driver performance. These signals always are taken for granted. However, when a physical phenomenon must be measured, it is necessary to understand how this is done. In addition, to draw the right conclusions from the data, the measurement must be evaluated to meet the necessary requirements. This chapter discusses the basics of sensor technology and metrology to arm the reader with the necessary knowledge to select and apply the correct sensors for obtaining measurements.

Analog-Digital Conversion: Accuracy Implications

The data acquisition system is an extensive measurement tool with a memory. Like any other measurement tool, it has limitations regarding precision. The memory can store only data converted to digital form. This conversion takes a finite amount of time; as this process takes place, a change of magnitude in the signal goes undetected. One is forced to approximate a continuous signal by a succession of sampled points.

The digitizing process converts the analog values to a stream of data bits with values of zero or one. The resolution of the signal (the smallest change in signal that the system can detect) is limited to the voltage that corresponds to one bit of variation.

Consider an 8-bit device that stores each data point as an integer with a value of 1 to 2^8 or 256 (actually stored as 0 to 255). This means the data point is stored with a resolution of 1/256 of the full scale or approximately 0.4%. Take a look at a linear potentiometer measuring suspension travel. The measurement range of this sensor is given in ***Table 15.1***.

One mm of suspension travel creates a variation in the sensor signal of 0.05 V. The data recorder measures a signal in the range of 0 to 5 V. This means that it can only detect a change in the signal greater than 5/256 or 0.019 V. The smallest suspension travel that can be measured therefore is 0.019/0.05 or 0.38 mm.

In ***Table 15.2,*** the resolution for 8-, 10-, and 12-bit systems is given. The regeneration of a signal is a succession of sampled points. Therefore, the preservation of the original signal depends on the number of sampled points per unit of time—the sampling frequency. Errors in signal recording can arise from a sampling frequency that is too low, a phenomenon known as *aliasing*.

The continuous line in ***Figure 15.1*** represents the signal to measure, in this case a simple sine wave. The squares are the samples stored by the data logger. Everything happening between these points is ignored. The graph shows that the sampling frequency used was a bit lower than the frequency of the original sine wave. The signal produced by the data logger is a sine wave with a much lower frequency, an alias.

To avoid aliasing, the Nyquist-Shannon sampling theorem states that the frequency at which a signal is sampled must be greater than twice the

Table 15.1. *Output voltage for a typical linear potentiometer (with a range of 100 mm)*

Lin. dist.	Output voltage
0 mm	5 V
100 mm	0 V

Table 15.2 *Resolution for 8-, 10-, and 12-bit data recorders*

	8-bit	10-bit	12-bit
Resolution (%)	0.39	0.10	0.02

Figure 15.1 *A too-slow sampling rate results in a false signal representation called* aliasing

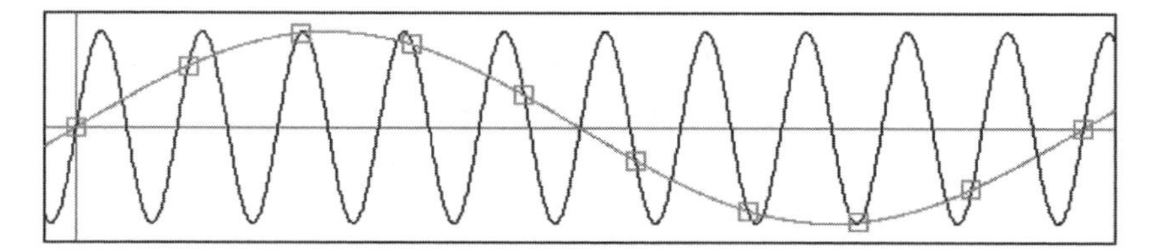

highest frequency encountered in the signal. This means that to digitize a sine wave with a frequency of 20 Hz, the sampling frequency should be more than 40 Hz.

A sampling frequency that is too low can cause the data acquisition system to miss valuable events. Most systems allow sampling frequencies to be set according to specific needs. High sampling frequencies increase accuracy but require more memory, thus decreasing the available logging time.

The following sampling frequencies can be used as guidelines:

fluid temperatures and pressures	**1–5 Hz**
chassis and driver activity	**20 Hz**
suspension motion	**100–200 Hz**

Sensor signals also are affected by noise resulting from vibration or electromagnetic fields from the ignition system. Noise is added to the measured signal when it exceeds the resolution of the analog-digital (A-D) conversion. Prevention of parasite signals is aided significantly by adequate wiring. Of course, the sensor itself also poses a question as to the accuracy of the measurement.

The required accuracy of a data acquisition system should be chosen carefully. Again, it comes down to the specific needs of the user. It is not necessary to log water temperature with an accuracy of 0.01 °C; this does not make one any smarter. For shock absorber motion, more accuracy is always better. Measuring the longitudinal *g*-force does not require a high sampling frequency, except when gearshift times must be derived from this signal. Think about what needs measuring and decide on the required accuracy.

Sensor Selection and Application

A sensor is a device that outputs an electrical signal in reaction to a physical phenomenon. For any imaginable measurement, a sensor probably exists. To justify the application of a given sensor, the requirements of the measurement must be determined first. Therefore, the following questions require answers:

- *What requires measuring?*
  The answer to this question often falls into one of the following categories: temperature, pressure, flow, displacement or position, velocity, acceleration, and force.
- *What is the expected measurement interval?*
  The minimum and maximum estimated measurement values need to be inside the sensor's range.
- *To what type of environmental circumstances will the sensor be exposed?*
  Because environmental effects can introduce errors into the measurement, this should be carefully considered when selecting a sensor. Temperature, pressure, and vibration can influence the output signal of the sensor. In addition, the mounting of the sensor and contamination by fluid, dirt, and similar contaminants can have an effect.
- *What kind of accuracy is required?*
  The highest possible accuracy for a sensor always is wanted, but there are economical factors to consider. In addition, using a sensor within a greater degree of accuracy than the data logger can record is not necessary. Signal conditioning and A-D conversion can cause inaccuracy in a highly accurate sensor signal.
- *What is the available budget?*
  Sensors come in different shapes and sizes, but there are also price differences between various sensors. A more expensive sensor usually has more functions than a cheaper one and scores better on the four points mentioned previously.

To evaluate these questions, one must understand the performance characteristics of the sensor mentioned on the sensor data sheet. It is absolutely necessary to grasp precisely what the data on the sensor data sheet means to appropriately evaluate a sensor. The important sensor properties are covered using the Bosch AM 600 accelerometer (http://www.bosch-motorsport.com) as an example.

1. Transfer Function

This determines the relationship between physical input and electrical output. This often is illustrated in the data sheet as a graph showing this relationship. For the AM 600, the linear function shown in ***Equation 15.1*** applies.

$$V(Acc) = 2.5\ V + \left(Acc \cdot 440\ \frac{mV}{G} \right) \quad \text{(Eq. 15.1)}$$

with V(Acc) = sensor output voltage at the measured acceleration value

Acc = the acceleration value measured by the sensor

2. Sensitivity

This is the ratio between a small change in the electrical signal to a small change in physical input (i.e., the derivative of the transfer function). This means that for the AM 600, the sensitivity equals 440 mV/G.

3. Offset

This is the value of the electrical signal at zero input. The AM 600 measures 2.5 V at 0 G.

4. Measurement Range

This is the range of input signal that can be converted into an electrical signal. Signals outside the measurement range result in unacceptable inaccuracies. The measurement range of the Bosch AM 600 equals ±4.5 G.

These first four sensor properties allow the creation of a graphical representation of electrical output versus physical input. ***Figure 15.2*** illustrates the transfer function calculated over the complete measurement range.

5. Uncertainty

Also called *tolerance*, uncertainty is the largest expected error between the actual and ideal output signal. On sensor data sheets, this usually is referred to as one of the following:

- fraction of the full-scale output,
- fraction of the reading, and
- fraction of the sensor's sensitivity.

The acceleration sensor in this example has a quoted tolerance of ±3% of the sensor's sensitivity. Assuming that the sensor is exposed to an acceleration of 2.5 G, the ideal output voltage is ***Equation 15.2***.

$$V(2.5\ G) = 2.5\ V + \left(2.5\ G \cdot 440\ \frac{mV}{G} \right) = 3.6\ V \quad \text{(Eq. 15.2)}$$

With a stated tolerance of ±3% of sensitivity, the range in which the sensitivity can vary is from 426.8 to 453.2 mV. This means that the output voltage can vary between 3.567 to 3.633 V. For a real acceleration of 2.5 G, the value measured by the sensor can therefore range from 2.425 to 2.575 G.

6. Nonlinearity

This is the maximum deviation from a linear transfer function over the specified measurement range. For the AM 600, a nonlinearity of ±2% of sensitivity is stated, which is within the tolerance of the sensor.

7. Hysteresis

This is the variation of the output value when the input value is cycled up and down. The AM 600 data sheet does not mention hysteresis.

8. Noise

All sensors produce some noise output, which must be considered in addition to the other electronic elements in the measurement system.

9. Resolution

The sensor's resolution is the smallest detectable physical input factor. A sensor's resolution often is limited by the noise it produces.

The best possible sensor choice can still provide bad data if not properly applied. A sensor responds to its total environment, and therefore everything in this environment must be taken into account. This not only includes the external influences acting on the sensor but also the complete measurement system. Therefore, connectors, cables, power supply, signal conditioners, and logging unit all must work together.

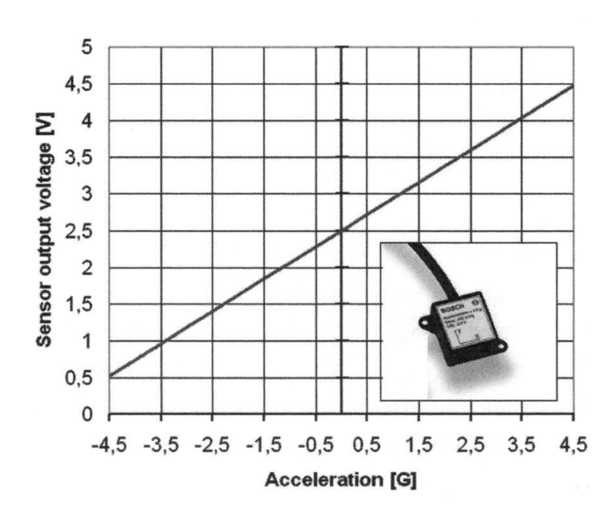

Figure 15.2
Bosch AM 600 transfer function

Measurement Uncertainty

Any measurement is valid only if accompanied by a consideration of the errors involved. Before taking a measurement, identify significant sources of error and eliminate them if possible. After the measurement is taken, maintain an impression of the probable remaining measurement error. Every time a sensor measurement is repeated, the results obtained will vary. Each measurement deviates by a certain amount from the true value for the following reasons:

- inadequacies in the measurement object,
- inadequacies in the measuring instruments,
- inadequacies in the measurement method,
- environmental influences,
- influences by the person performing the measurements, and
- changes over time.

A quantity has a true value that one tries to calculate through measurement. However, recognizing that no measuring instrument is perfect and outside influences never can be eliminated completely, the best that can be achieved is an estimate of the true value. The difference between the measured value and the true value is called the error of the measurement, or accuracy. Because the true value of a quantity never can be determined, it is also impossible to know the exact accuracy of a measurement. However, estimating the effect that various errors have on the measured value is possible. When this effect is estimated, an uncertainty can be attached to the measured value, which indicates a range of values within which the true value is expected to lie.

The total error in a measurement is comprised of the following components:

- large errors,
- systematic errors, and
- random errors.

Large errors occur as a result of an improper measurement method, circuit errors, incorrect sensor application, and logging errors. They cannot be corrected for and only can be eliminated if detected before performing the measurement. When significant errors occur, the measurement is rendered obsolete; it serves no purpose incorporating them in any error quantification.

Systematic error, sometimes called *statistical bias*, is caused by deficiencies in the measured object, in the measurement method, and in the measuring instruments.

Statistical bias can be eliminated or reduced by calibrating the relevant instrument. Correcting the measurements to the results obtained with a reference instrument also reduces bias. For instance, an often-observed practice by race teams is comparing the output of tire pressure gauges to those used by the tire manufacturers to obtain the same measurement results. The difference between the two readings is the gauge's statistical bias.

Random errors occur because of factors beyond the engineer's control. Examples of these factors include the following:

- environment (e.g., temperature, humidity, pressure, presence of magnetic fields, radiation),
- aging of the measured object, and
- aging of the measurement instruments.

An example of a random error often occurs in the least significant digit in digital balances. Three measurements of a single object might be 0.567 g, 0.566 g, and 0.568 g.

Random error can be estimated statistically by attaching an uncertainty to the measurement result.[17] Random uncertainties generally follow a normal distribution, which basically means that small random deviations from the average measured value are much more probable than large ones.

In a series of measurements, if *n* single values $x_1, x_2, \ldots x_n$ are measured under comparable conditions, the average μ of these *n* values usually is considered an estimate of the true value. In a normal distribution, approximately 68% of all measured values fall within $\pm 1 \cdot \sigma$, and 95% fall within $\pm 1 \cdot \sigma$ (where σ is the standard deviation of the data). This is illustrated in ***Figure 15.3***. The standard deviation determines the scatter in the data.

Do not assume that μ is equal to the true value. When the series of measurements mentioned earlier is repeated, another result for μ might be obtained. The question that needs to be answered is, *What uncertainty can be assigned to the average value of the data?* For a normal distribution, the standard deviation of the average is defined by ***Equation 15.3***.

$$\text{Standard Deviation of the Average} = \frac{\sigma}{\sqrt{n}} \quad (Eq.\ 15.3)$$

with σ = the standard deviation of the data
n = the number of data values

The standard deviation of the average, or in this context often called *standard uncertainty*, expresses the uncertainty of a measurement ***(Equation 15.4)***.

$$\text{Measurement result} = \mu \pm \frac{\sigma}{\sqrt{n}} \quad (Eq.\ 15.4)$$

In practical terms, this means that there is a 68% probability that the true value lies within

$$\pm \frac{\sigma}{\sqrt{n}}$$

of the average measured value and a 95% probability that it lies within twice this distance from the average.

As an example, the uncertainty of a measurement of a vehicle's roll angle ratio (see Chapter 8) is evaluated. The roll ratio is calculated by dividing the rear roll angle by the front roll angle. These inputs are calculated in turn from the signals measured by suspension position sensors.

For this measurement, the data of a complete lap is evaluated, which resulted in 10,421 samples for the front and rear roll angles. For each sample, the roll ratio is calculated by dividing the rear roll angle by the front roll angle (assuming a linear relationship between the front and rear roll angle as ***Figure 15.4*** suggests).

Following the theories explained earlier, the measured value of the roll ratio for the complete lap (the coefficient of direction of the straight line in Figure 15.4) is the average of all 10,421 roll ratio samples ***(Figure 15.5)***.

The calculated average in this example equals 1.12. The standard deviation of the data is 3.35, which is a measure of the scatter of the data samples. This is quite high. Two main issues significantly influence the standard deviation of the data samples:

- When the car hits a curb, data also is incorporated into the calculation. This temporarily results in very large roll ratios.
- Data scatter around the y-axis in Figure 15.4 (where the front roll angle is zero) in theory causes the roll ratio to approach infinity. In reality, because of a finite sampling rate, it induces very high values for the roll ratio into the data.

The standard error is calculated using Equation 15.3 and equals 0.03. Therefore, it can be stated (with a confidence level of 68%) that the true value for the roll angle ratio is located within the following limits.

lower confidence limit 1.12 – 0.03 = 1.09
upper confidence limit 1.12 + 0.03 = 1.15

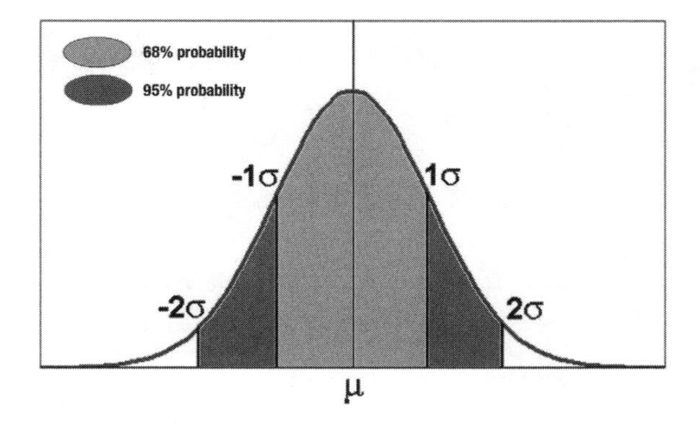

Figure 15.3 *A normal distribution and its probability limits*

Figure 15.4 *The relationship between the front and rear roll angle*

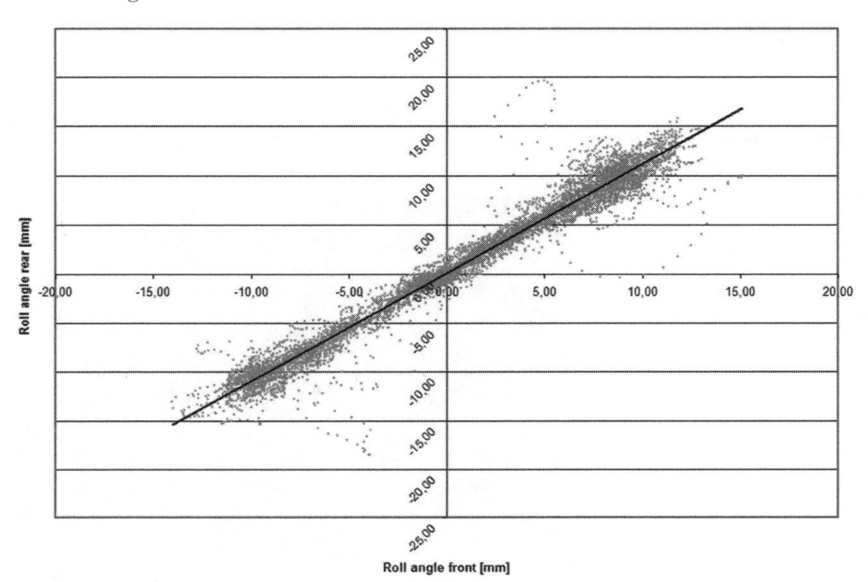

Or, with a 95% confidence level the true value is within these limits:

lower confidence limit **1.12 – (2 • 0.03) = 1.06**

upper confidence limit **1.12 + (2 • 0.03) = 1.18**

The measurement result now is expressed as follows:

roll angle ratio = 1.12 ± 0.03 for a confidence level of 68%

roll angle ratio = 1.12 ± 0.06 for a confidence level of 95%

Temperature Sensors

Most methods for measuring temperature rely on measuring some physical property of a metal that varies with temperature. In some cases, it is possible to determine temperature by measuring a target's thermal radiation. However, a temperature sensor's output always consists of an output voltage that corresponds to a temperature change. There are two basic types of temperature measurement:

Figure 15.5 Roll ratio uncertainty calculation

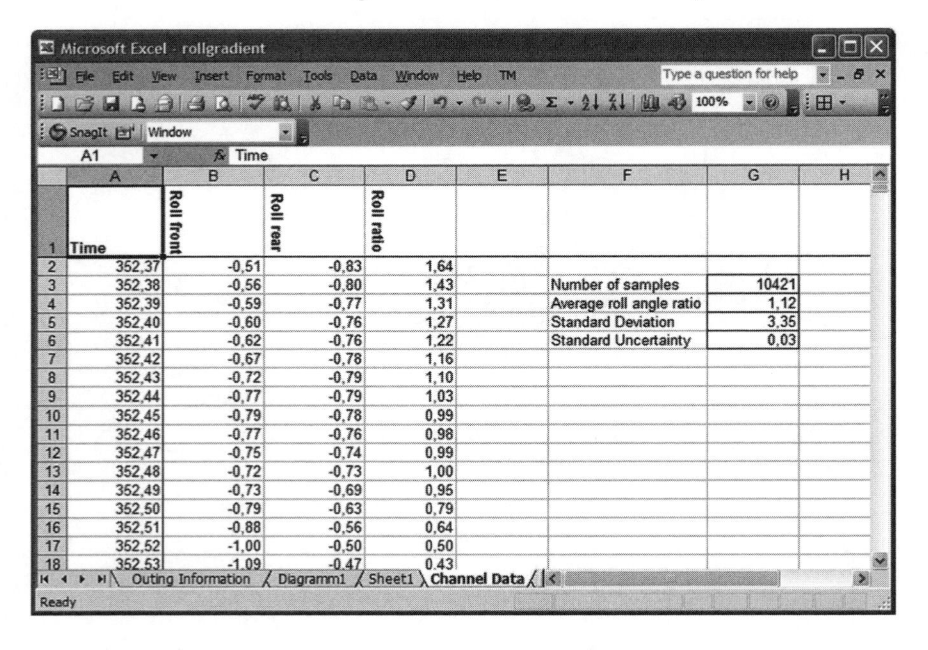

	A	B	C	D	E	F	G	H
1	Time	Roll front	Roll rear	Roll ratio				
2	352,37	-0,51	-0,83	1,64				
3	352,38	-0,56	-0,80	1,43		Number of samples	10421	
4	352,39	-0,59	-0,77	1,31		Average roll angle ratio	1,12	
5	352,40	-0,60	-0,76	1,27		Standard Deviation	3,35	
6	352,41	-0,62	-0,76	1,22		Standard Uncertainty	0,03	
7	352,42	-0,67	-0,78	1,16				
8	352,43	-0,72	-0,79	1,10				
9	352,44	-0,77	-0,79	1,03				
10	352,45	-0,79	-0,78	0,99				
11	352,46	-0,77	-0,76	0,98				
12	352,47	-0,75	-0,74	0,99				
13	352,48	-0,72	-0,73	1,00				
14	352,49	-0,73	-0,69	0,95				
15	352,50	-0,79	-0,63	0,79				
16	352,51	-0,88	-0,56	0,64				
17	352,52	-1,00	-0,50	0,50				
18	352,53	-1,09	-0,47	0,43				

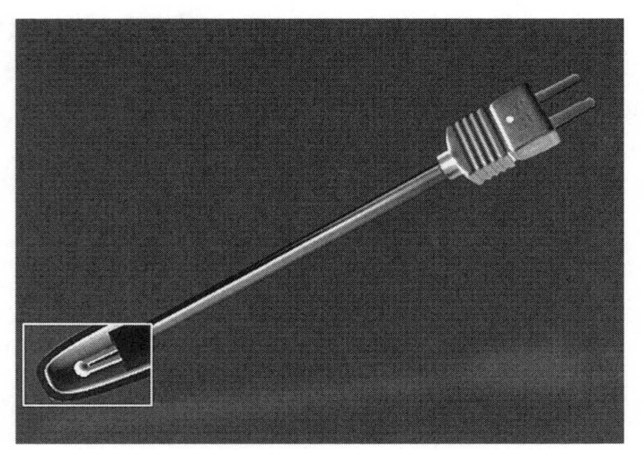

Figure 15.6 Thermocouple measurement junction

1. Contact Temperature Measurement

This requires the sensor to be in direct contact with the medium to be measured. Oil, water, and air temperature fall into this category. For these applications, thermocouples, thermistors, or resistive temperature devices (RTDs) commonly are used.

2. Noncontact Temperature Measurement

This is used for logging tire or brake disc temperatures with infrared (IR) sensors.

Thermocouple Temperature Sensors

A thermocouple temperature sensor consists of two wires of different materials welded together into a junction, called the measurement junction. At the other end of the signal wires is another junction, called the reference junction. A change in temperature within the measurement junction generates a current in the wires proportional to the temperature change. Temperature at the measurement junction then can be determined from the type of thermocouple used, the magnitude of the millivolt potential, and the temperature of the reference junction ***(Figure 15.6)***.

The big advantage of a thermocouple temperature measurement is the potential measurement range. Thermocouples may be rated from –270 to 1800 °C. They are also very reliable under vibration and shock because of their simple design. The disadvantage of using this sensor type is that it requires special extension wires and reference junction compensation.

Thermocouples are available in different combinations of metals or calibrations. The four most popular configurations are named J, K, E, and T. Each calibration has a different measurement range, although the maximum temperature varies with the diameter of the wire used in the thermocouple. ***Figure 15.7*** shows the upper temperature limit of the four common thermocouple calibrations for different wire sizes.[18]

Thermocouple probes are available in three junction types: grounded, ungrounded, or exposed

(Figure 15.8). With a grounded junction, the thermocouple wires are attached to the inside of the probe wall. In an ungrounded junction, the wires are detached from the probe wall. Therefore, the response time of an ungrounded junction is slower than that of a grounded junction. In an exposed junction, the thermocouple wires protrude from the probe wall and are in direct contact with their environment. This type of junction has the fastest response time but is limited in use to dry, noncorrosive, and nonpressurized environments. Thermocouples with exposed junctions often are used for air temperature measurement.

Thermistors

Thermistors ***(Figure 15.9)*** change their electrical resistance in relation to their temperature. They are typically composed of two metal oxides encapsulated in glass or epoxy. Thermistors are available in two types:

- positive temperature coefficient (PTC), where resistance increases with a rise in temperature, or
- negative temperature coefficient (NTC), where resistance decreases as temperature rises.

The change in resistance of thermistors is generally quite large, resulting in high sensor sensitivity, but the measuring range is smaller than that of thermocouples. The relationship between temperature and resistance is not a linear one, but with external circuitry it can be made virtually linear. Thermistors are one of the most accurate types of temperature sensors.

Resistive Temperature Devices

RTDs work on the same principle as a thermistor. A change in electrical resistance is used to measure temperature ***(Figure 15.10)***. The sensing element consists of a wire coil or deposited film of pure metal, whose resistance has been documented at various temperatures. Common materials used in resistors are platinum (the most popular and accurate), nickel, or copper.

An RTD can have a similar measurement range as a thermocouple, with the sensitivity advantage of a thermistor. However, their construction makes them unsuitable for measurements in high-vibration environments.

Infrared

IR technology is not a new phenomenon. It has been used in research and industrial applications for decades, but lately innovations have developed for noncontact IR sensors on racecar applications. Especially popular are tire temperature ***(Figure 15.11)*** and brake disc temperature measurements.

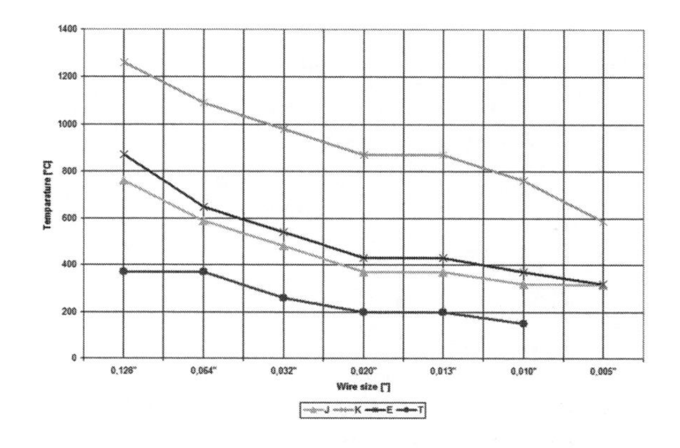

Figure 15.7 *Upper temperature limit in degrees Celsius of protected bare wire thermocouples versus wire diameter*

Figure 15.8 *Common thermocouple junctions*

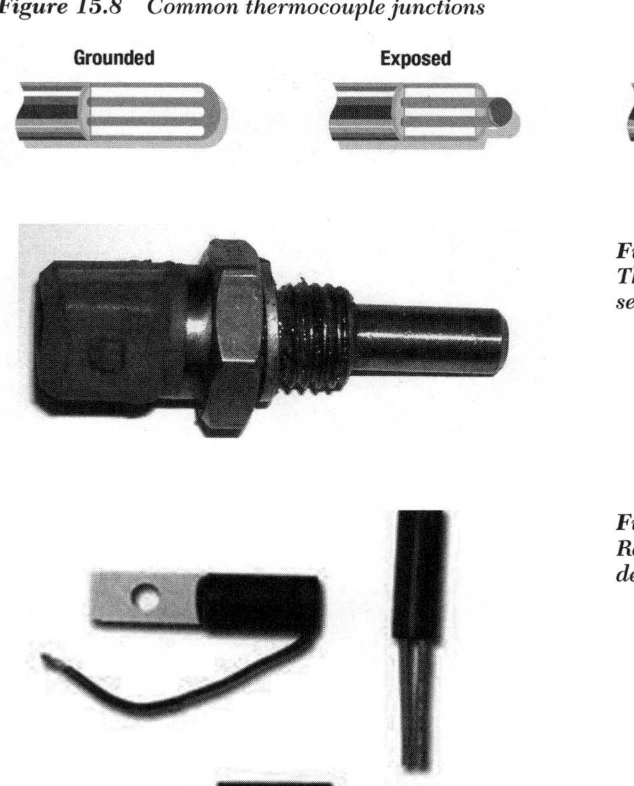

Figure 15.9 *Thermistor temperature sensor*

Figure 15.10 *Resistive temperature devices*

The following are the advantages of IR temperature measurement:

- fast response times (in the ms range),
- temperature measurement on moving targets,
- measuring physically inaccessible objects,
- high measurement range, and
- no heat distortion (no heat energy lost from the target).

Nevertheless, keep in mind the following disadvantages of IR measurements:

- The target must remain optically visible to the IR sensor. Dirt or dust can cause inaccuracies.
- Only surface temperatures can be measured.
- Emissivity of the surface material must be taken into account.

Every form of matter with a temperature above zero emits IR radiation proportional to its temperature. This is called *characteristic radiation*, and the spectrum of this radiation ranges from 0.7 to 1000 μm wavelength ***(Figure 15.12)***.

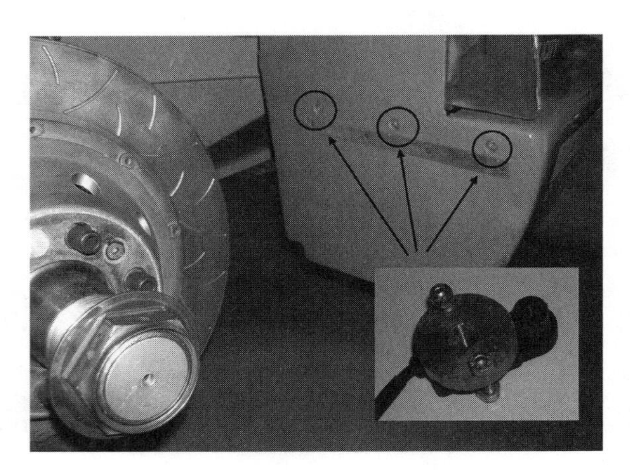

Figure 15.11 *Infrared tire temperature sensors (Courtesy of GLPK-Carsport)*

Figure 15.12 *IR radiation spectrum*[19]

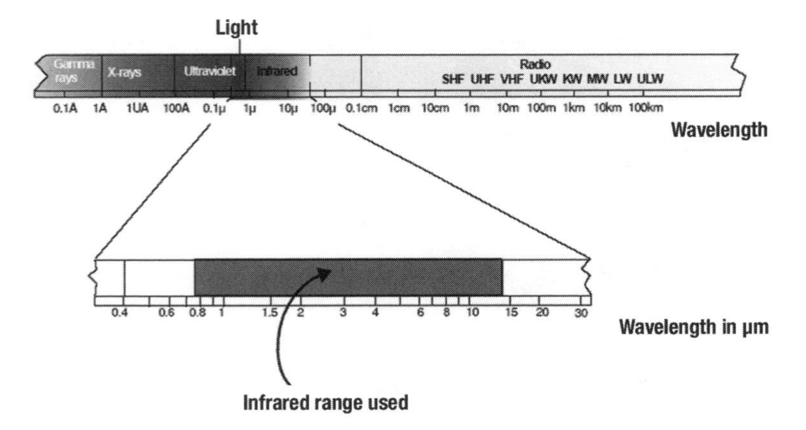

The functioning of IR sensors is complicated by the fact that surfaces other than so-called *black-bodies* emit less radiation at the same temperature. The relationship between the real emissive power and that of an ideal black-body is called emissivity (ε) and can have a maximum of one (for ideal black-bodies) and a minimum of zero. Bodies with an emissivity less than one are called *gray bodies*. Bodies where emissivity also depends on temperature and wavelength are called *nongray bodies*.

Emissivity of the measured surface should be known to determine the correct wavelength in which the IR sensor needs to operate.

The IR sensor core converts the received radiation energy into an electrical signal. This core falls into one of two categories: quantum detectors or thermal detectors. Quantum detectors (or photodiodes) interact directly with the impacting photons, resulting in electron pairs and ultimately an electrical signal. Thermal detectors change their temperature according to the impacting radiation. As with a thermocouple, this temperature change creates voltage.

Pressure Sensors

Pressure sensor applications on racecars include engine oil, brake line, coolant pressure, fuel, manifold air, and aerodynamic. Most pressure-sensing devices in automotive applications rely on piezoresistive semiconductor technology. A piezoresistive pressure sensor is essentially a strain gage. It contains a sensing element made up of a silicon chip with a thin silicon diaphragm and three or four piezoresistors ***(Figure 15.13)***. The piezoresistance of the semiconductor refers to the change in resistance by strain on the diaphragm, compared to a reference pressure. The resistor values change proportionally to the amount of pressure applied to the diaphragm. The thickness of the diaphragm determines the pressure range of the sensor.

Depending on the reference pressure, piezoresistive sensors can be divided into the following categories:

- absolute pressure sensor in which the reference pressure is vacuum ***(Figure 15.14);***

- differential pressure sensor which has two ports for measuring two different pressures ***(Figure 15.15);*** and
- gauge pressure sensor, which is a differential pressure measurement with atmospheric pressure as a reference ***(Figure 15.16)***.

Displacement Sensors

Displacement sensors are generally divided into two broad categories: linear motion and angular motion. The well-known linear potentiometer ***(Figure 15.17)*** and the string-potentiometer fall into the first category, while rotary potentiometers ***(Figure 15.18)*** measure angular motion.

In racecars, linear and rotary potentiometers measure throttle position, gear position, suspension movement, steering angle, hydraulic level, and clutch or brake pedal position. They all work according to the principle illustrated in ***Figure 15.19***. A potentiometer transforms a linear or rotary motion into a change in resistance. It is basically a voltage divider.

The string potentiometer is a suitable sensor when size and mounting restrictions eliminate other choices. It can be used to measure multiaxis movements. The retractable cable allows for flexible mounting and can be routed around obstacles using pulleys and flexible guides.

Acceleration Sensors

Accelerometers have two main applications on racecars. The most popular is measuring the acceleration acting on the vehicle in lateral, longitudinal, or vertical direction. Second, they often measure vibration on various vehicle components (i.e., engine knock or track surface profiling by upright vibration measurements).

Based on their operation, accelerometers belong to either the capacitive or piezoelectric category.

Capacitive Accelerometers

Capacitive acceleration sensors measure a change in electrical capacitance, proportional to the acceleration acting on the sensor. Their operating principle is illustrated in ***Figure 15.20***. A diaphragm with a known spring rate and mass is sandwiched between two fixed electrode plates.

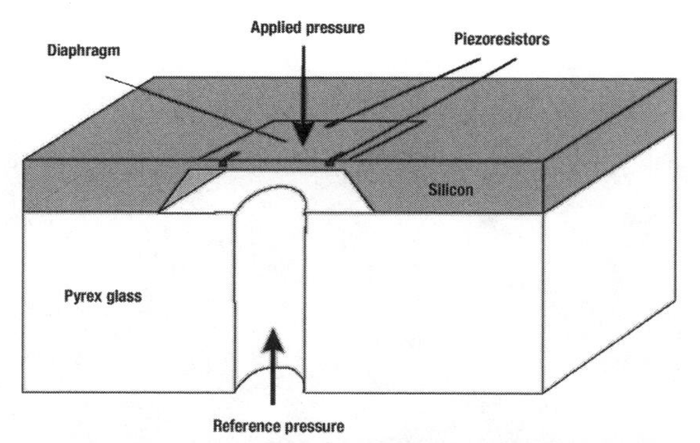

Figure 15.13
Piezoresistive pressure sensing element

Figure 15.14
Absolute pressure sensor (fluid pressure)

Figure 15.15
Differential pressure sensor (for aerodynamic applications)

Figure 15.16
Gauge pressure sensor

With these plates, the diaphragm forms two capacitors. As acceleration acts on the sensor, the diaphragm spring-mass experiences a force resulting in deflection. This deflection causes the distance between the spring-mass and the electrodes to vary, effectively changing the capacitor gaps.

Capacitive accelerometers generally can measure smaller acceleration levels, making them more suitable for vehicle inertial measurements.

Figure 15.17 Linear potentiometer measuring rear suspension movement on a Dodge Viper GTS-R (Courtesy of GLPK-Carsport)

Figure 15.18 Rotary potentiometer

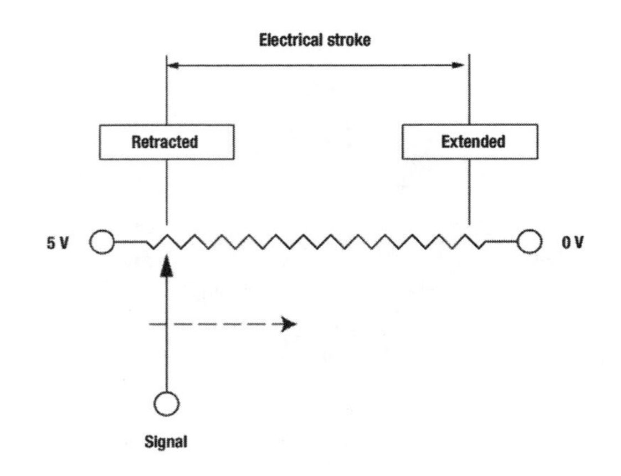

Figure 15.19 How a potentiometer works

Piezoelectric Accelerometers

Piezoelectric accelerometers ***(Figure 15.21)*** use a piezoelectric material as a sensing element, which can output an electrical signal proportional to the stress applied to it. Most piezoelectric acceleration sensors are made of quartz crystal, piezoelectric ceramics, or tourmaline or lithium niobate. The piezoelectric elements in the sensor act as a spring, which is connected to the seismic masses. When acceleration acts on the sensor base, a force is created on the piezoelectric elements proportional to the applied acceleration and the size of the seismic mass (Newton's law of motion). Therefore, the more mass or acceleration there is, the higher the applied force and the more electrical output from the crystal.

An important characteristic to keep in mind when selecting an acceleration sensor is the useful frequency range of a piezoelectric sensor, which is determined by its resonant frequency. This frequency can be estimated by ***Equation 15.5,*** where k is the spring rate of the piezoelectric element and m the size of the seismic mass.

$$\omega = \sqrt{\frac{k}{m}} \qquad \text{(Eq. 15.5)}$$

A typical frequency response of a piezoelectric accelerometer is depicted in ***Figure 15.22***.

Because of their wide dynamic measurement interval and frequency range and the fact that they can be made very small, piezoelectric accelerometers are found in applications where shock and vibration need mapping to understand the dynamic behavior of the object. Knock sensors and upright acceleration are the most common applications for racecars.

Speed Sensors

Measuring the speed of a rotating shaft finds multiple applications on a racecar, of which engine RPM and wheel speed are the most common. Shaft speeds usually are measured using a Hall effect sensor. If electric current flows through a conductor placed in a magnetic field, this field forces electrons to one side of the conductor resulting in a voltage potential. This phenomenon is known as the Hall effect, after the scientist who discovered it in 1879.

A Hall effect sensor used with a ferrous trigger gear placed on the shaft (of which the speed needs to be determined) measures the variation in magnetic field between a magnet and the passing gear-teeth. The signal from the sensor then is converted into a digital block signal by external circuitry. This is illustrated in ***Figure 15.23***.

To detect the trigger gear, it is necessary to provide a source of magnetic energy. Therefore, most Hall effect sensors incorporate a permanent magnet with its axis of magnetization pointing toward the gear-teeth surface. When a tooth passes in front of the sensor, the flux density between the ferrous surface and the sensor face increases. When a valley passes before the sensor face, the flux density decreases.

Because Hall effect sensors pick up the presence of a magnetic field, they essentially are immune to dust, oil, and other contaminants found on automotive components. The trigger for this kind of sensor is not necessarily a gear; other objects such as bolt heads or other metal profiles can be used.

Strain Gages

When external forces are applied to an object, stress and strain result. Stress is the object's internal resisting forces. Strain is the deformation of the object that takes place because of these internal forces. Strain typically is measured by strain gages ***(Figure 15.24),*** which are designed to convert mechanical motion into an electrical signal. They rely on the fact that metallic conductors subjected to mechanical strain exhibit a change in electrical resistance.

Racecar applications include suspension and steering loads, wheel forces, driveshaft torque, chassis loads, and ignition cut load cells. Strain

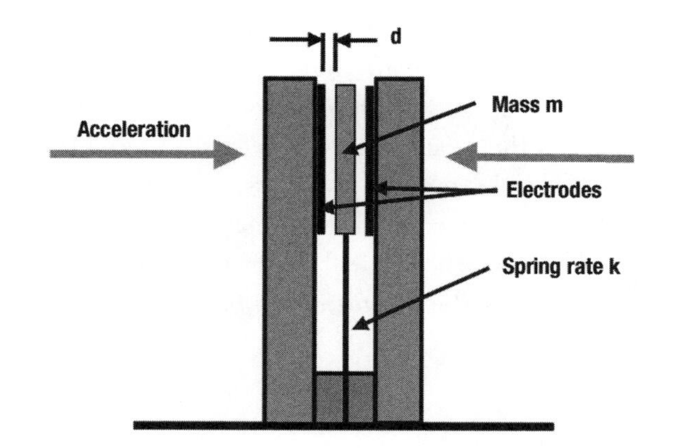

Figure 15.20
Capacitive accelerometer working principle

Figure 15.21
Piezoelectric accelerometer

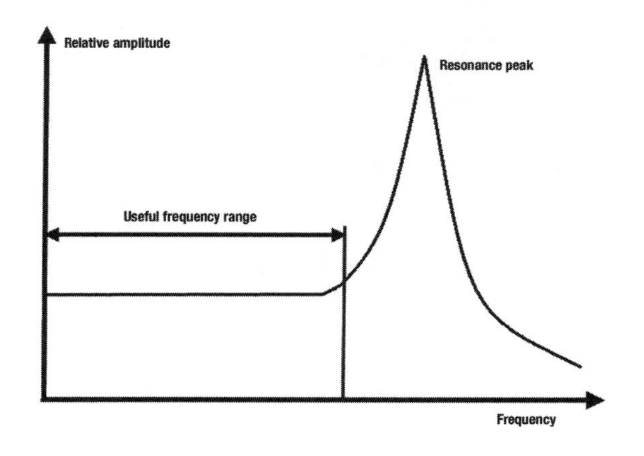

Figure 15.22
Typical frequency response of a piezoelectric accelerometer

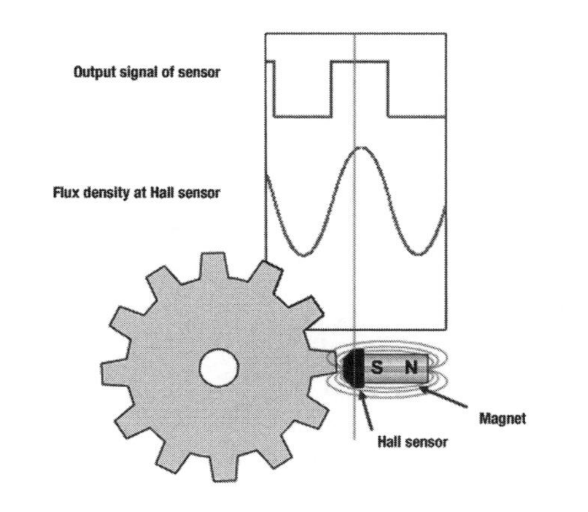

Figure 15.23
Hall effect geartooth sensor

gages also are used often as an integral part in (piezoresistive) pressure transducers.

The operating principle of a strain gage is based on the relationship between strain and the resistance of electrical conductors. Strain is defined as the ratio between total deformation of the original length and the original length ***(Figure 15.25)*** and is expressed in ***Equation 15.6***.

$$\varepsilon = \frac{\Delta L}{L} \qquad (Eq.\ 15.6)$$

Any electrical conductor changes its resistance with mechanical stress. This relationship is expressed with the gauge factor (GF) ***(Equation 15.7)***.

$$GF = \frac{\Delta R / R_0}{\Delta L / L} = \frac{\Delta R / R_0}{\varepsilon} \qquad (Eq.\ 15.7)$$

In this equation, R_0 is the resistance of the electrical conductor when no mechanical stress is applied. ΔR is the difference in resistance from R_0 when the conductor experiences a strain equal to ε.

Figure 15.24
Examples of typical strain gage

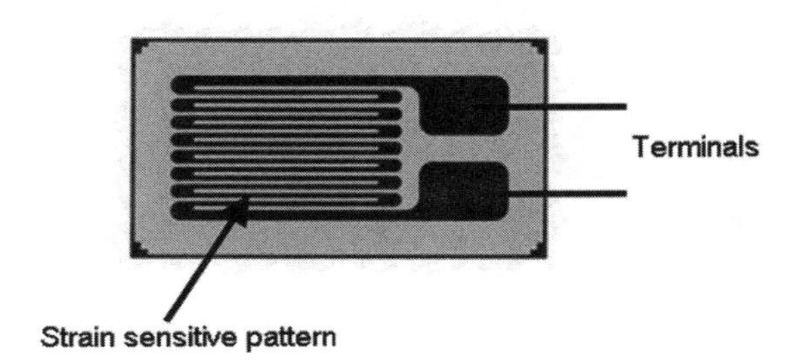

Figure 15.25 *Definition of strain*

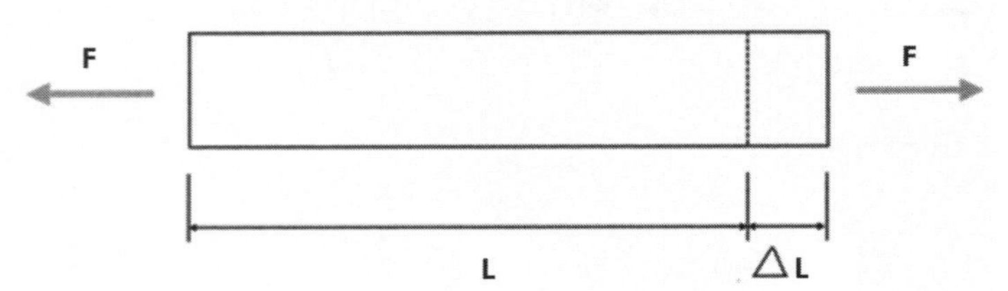

Figure 15.26
Strain gage measurement system using a Wheatstone bridge

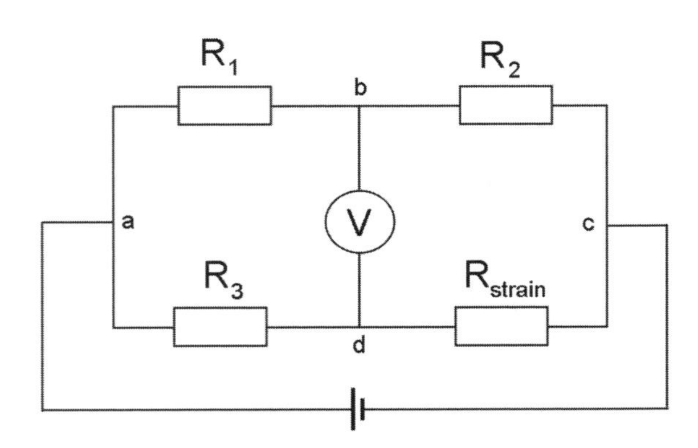

The strains measured with strain gages are normally very small. Consequently, the changes in electrical resistance also are very small. The strain gage must be included in a measurement system that can precisely determine this change in resistance. To measure relative changes in resistance around the order of 10^{-4} to 10^{-2} Ω/Ω, the strain gage should be integrated into a Wheatstone bridge ***(Figure 15.26)***.

In Figure 15.26, R_1, R_2, and R_3 are equal and R_{strain} is equal to this as well in an unstressed condition. A voltage (V_{in}) is applied between points *a* and *c*. As long as the strain gage does not experience a change in resistance, the output between points *b* and *d* exhibits no potential difference. However, when strain is applied, R_{strain} changes to a value unequal to R_1, R_2, and R_3. The bridge becomes unbalanced and an output voltage (V_{out}) exists between points *b* and *d*. V_{out} can be expressed with ***Equation 15.8***.

$$V_{out} = V_{in} \cdot \left(\frac{R_3}{R_3 + R_{strain}} - \frac{R_1}{R_1 + R_2} \right) \qquad (Eq.\ 15.8)$$

V_{out} typically fluctuates between zero and a couple of millivolts, so the sensor signal must be amplified before it can be directed into the datalogger.

Strain-sensing materials change their structure at higher temperatures. Temperature also can alter the properties of the base material to which the strain gage is attached. This means that the gauge factor of the strain gage can change with varying temperature. Therefore, the manufacturer always should include temperature sensitivity data on the sensor's data sheet. If the temperature changes while strain is being measured, this should be compensated for.

Once measured, strain must be converted into an absolute value for the mechanical stress experienced by the object in question. This calculation can be done using Hooke's law, which applies to the elastic deformation range of linear elastic materials. In its simplest form, Hooke's law can be expressed as ***Equation 15.9***.

$$\sigma = \varepsilon \cdot E \qquad \text{(Eq. 15.9)}$$

with σ = material stress
ε = strain
E = material's elasticity modulus

This version of Hooke's law only applies to uniaxial stress states (i.e., tension and compression bars). Multiaxial stress states require extended versions of Hooke's law.[20]

Pitot Tube

Aerodynamic measurements (downforce, drag, and pressure distributions) often use a pitot tube to determine dynamic pressure (see Chapter 11).

As illustrated in ***Figure 15.27,*** a pitot tube consists of two concentric tubes each with an inlet port. The outer tube measures the static air pressure, while the inner one is exposed to the sum of static and dynamic pressure. The tubes then are connected to a differential pressure sensor that outputs the dynamic air pressure to the datalogger. To obtain good results, the pitot tube must be aligned with the flow velocity. Any misalignment should not exceed ±5 deg.

Oxygen Sensors

Oxygen sensors (or what automotive applications often refer to as lambda sensors) measure the volume of oxygen remaining in the engine's exhaust gas after combustion. With these sensors fitted to the exhaust manifold, the engineer and/or ECU can determine if the fuel-air mixture going into the engine is too rich (i.e., too much fuel) or too lean (i.e., too little fuel).

The sensor element consists of a ceramic cylinder plated inside and outside with porous platinum electrodes. It measures the difference in oxygen between the exhaust gas and the external air by generating a change in resistance proportional to the difference between the two.

Because lambda sensors ***(Figure 15.28)*** only work effectively when heated to approximately 300°C, most have heating elements incorporated in the ceramic to quickly raise the sensor tip to the correct temperature when the exhaust is still cold.

GPS

In previous chapters, the use of a GPS signal to measure location, distance, and speed was mentioned in numerous examples. GPS refers to a group of U.S. Department of Defense (DoD) satellites that constantly circle the Earth, making two complete orbits around the planet every 24 hours. These satellites transmit very low-power radio signals allowing a GPS receiver to determine its precise location on Earth.

GPS consists of three segments: space (satellites), control (ground stations), and user (GPS receiver).

The space segment is the heart of the system. It consists of at least twenty-four satellites, of which twenty-one are active. They orbit approximately 12 km above the Earth's surface and are arranged so that a GPS receiver always can receive at least four satellites at any given time. Civilian GPS

Figure 15.27 *Working principle of a pitot tube*

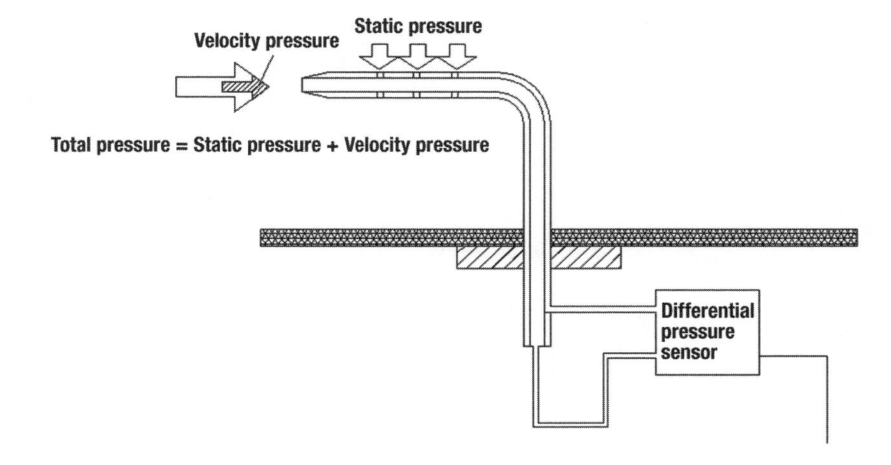

Figure 15.28
Lambda sensor

receivers transmit a low-power radio signal at a frequency of 1575.42 MHz on the UHF band. This signal passes through clouds, glass, and plastic, but it does not penetrate solid objects such as trees and buildings.

Each satellite transmits a unique code, allowing a GPS receiver to identify the satellite. The main purpose of these signals is to calculate the travel time from the satellite to the receiver, also called *time of arrival.* The arrival time multiplied by the speed of light gives the distance between the satellite and the receiver. In addition to the satellite identification signal, a navigation message containing satellite orbital and clock information is sent to the GPS receiver.

The control segment, consisting of five Earth-based ground stations, tracks the GPS satellites and provides them with corrected orbital and time information.

The user segment is the GPS receiver. To establish its location, the receiver must know the exact location of the satellite in space and the distance to it. The navigational data transmitted by the satellites contain two types of information. The *almanac* data contains the approximate locations of the satellites. This data is transmitted continuously and stored in the memory of the GPS receiver so it knows the orbit in which every satellite is supposed to be. The almanac data is updated periodically as the satellites orbit.

The ground stations keep track of satellite orbit, altitude, location, and speed and send corrected data, called the *ephemeris* data, back to the satellites. Ephemeris data is valid for 4–6 hours.

The combination of almanac and ephemeris data tells the GPS receiver the exact location of the satellite in space. The distance between the receiver and respective satellite is calculated from the arrival time. The coded identification signal is a so-called "pseudo-random" signal because it looks like a noise signal. The GPS receiver generates the same signal and tries to match it to the satellite's signal. The receiver then compares the two to determine how much it needs to delay its signal to match the satellite signal. A radio wave travels at the speed of light ($2.99 \cdot 10^8$ m/s); multiplied with the delay time, the distance between receiver and satellite is determined.

The receiver needs at least four visible satellites to determine a three-dimensional location (longitude, latitude, and altitude) on Earth. The more satellites the receiver can see, the better the accuracy.

Most modern GPS receivers are parallel multicircuit receivers, with each circuit devoted to one particular satellite. In this way, strong locks can be maintained on all satellites at all times, even in difficult conditions such as blockage from trees, buildings, and other solid objects.

GPS signals can be logged by the racecar data logging system to measure the position, speed, and heading of the vehicle. Standalone GPS position measurements are accurate to within 2–3 m. When corrections are applied using the data from lateral and longitudinal accelerometers, accuracy improves to 1–2 m.[21] GPS signals can be degraded by one or more of the following factors.

- Ionosphere or troposphere delay: As it passes through the atmosphere, the satellite signal slows down. The GPS uses a built-in model that partially corrects this type of error.
- Signal multipath: This occurs when the GPS signal is reflected off objects such as tall buildings before it reaches the receiver. This increases the signal travel time.
- Receiver clock error
- Ephemeris error: This is the inaccuracy of the satellite's reported location.
- Number of satellites visible
- Intentional degradation of the GPS signal: To prevent military adversaries using the highly accurate GPS signal the U.S. DoD intentionally degraded the signal prior to May 2000 (selective availability).

By combining the data from two separate GPS antennas located at a fixed distance from each other on a vehicle, it is even possible to measure vehicle yaw and pitch. This system can be used to determine the vehicle's slip angle.[22, 23]

Laser Distance Sensors

Laser distance sensors measure a racecar's ride height or ground clearance. This sensor operates on the principle of triangulation. A laser emitter projects a beam onto an object ***(Figure 15.29)***. The reflection of this beam passes through a lens that focuses the beam onto a receiving photodiode element. A change in distance between sensor and target changes the angle of the reflected beam, thereby changing the location of the beam on the receiving element.

This receiving element is coupled to a microcontroller, which calculates the distance to the target from the reflected beam's location on the receiver and outputs a voltage proportional to the target distance.

Through the use of a microcontroller, a high linearity and accuracy is achieved. The signal can be filtered with user-determined rates to smooth the sensor's output signal. This is particularly useful for ride height measurements to filter out the roughness of the asphalt. Also, the emissivities of different target materials can be accounted for by the microcontroller.

Resolution and accuracy depends on the distance between the sensor and object. An object within close proximity to the sensor creates a significant difference in the angle between the emitted and reflected beam given a small change in distance. When the target is located further from the sensor, a small change in distance results in a small difference in angle. Therefore, the highest resolution is achieved with laser distance sensors with a relatively small measurement range.

Figure 15.29 *Principle of operation of a laser distance sensor*

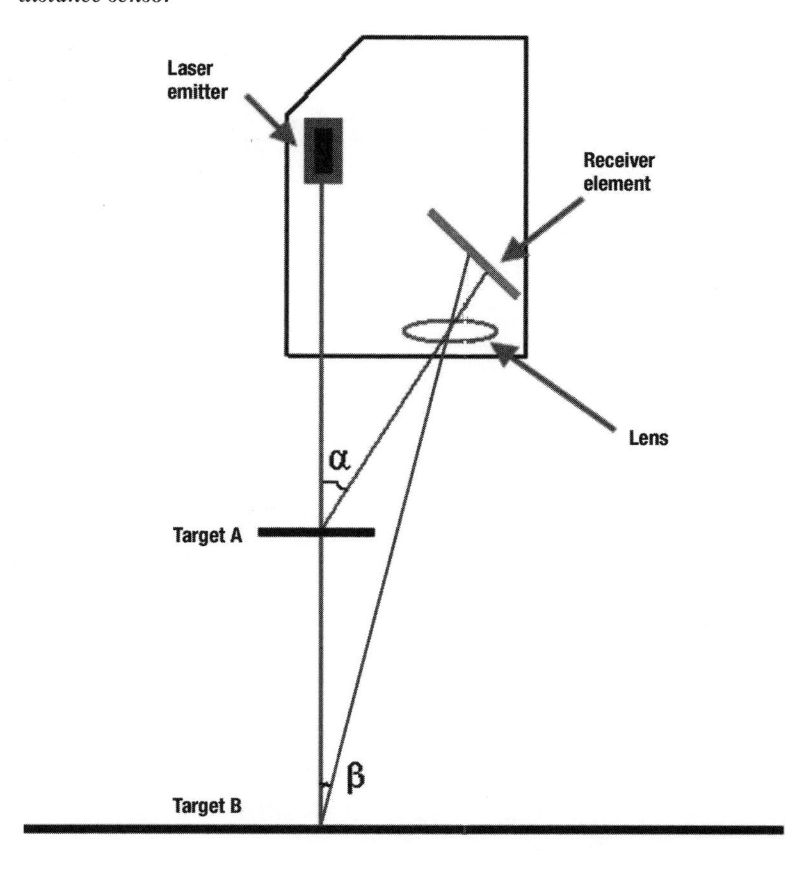

LIST OF SYMBOLS

This book contains a number of equations that use a variety of English letters and Greek symbols. The following list will help you better understand the equations.

English Letters

A	Frontal vehicle surface
A	Skewness of a normal distribution
a	Acceleration
a_{yaw}	Yaw acceleration
a_{hub}	Hub vertical acceleration
$a_{suspension}$	Suspension acceleration
C_D	Aerodynamic drag coefficient
C_L	Aerodynamic lift coefficient
C	Damping coefficient
C_H	Damping constant for heave
C_P	Damping constant for pitch
C_R	Damping constant for roll
C_X	Damping constant for warp
d	Distance, lap distance
D	Aerodynamic drag force
E	Elasticity modulus
F	Force
F_N	Normal force
F_{mass}	Acceleration force
F_{spring}	Spring force
F_{shock}	Shock absorber force
$F_{rolling}$	Rolling resistance force
F_{res}	Resisting force
F_{aero}	Aerodynamic force
F_{CP}	Tire contact patch force
$F_{Suspension}$	Suspension force
$FRF_{CP\text{-}body}$	Frequency response function tire contact patch—body
$FRF_{CP\text{-}hub}$	Frequency response function tire contact patch—hub
g	Gravitational acceleration ($g = 9.81\ m/s^2$)
G_{lat}	Lateral acceleration
G_{long}	Longitudinal acceleration
G_{vert}	Vertical acceleration
$G_{combined}$	Combined acceleration
GF	Gage factor
G_2	Kurtosis of a distribution
h	Height
h_{roll}	Distance between vehicle center of gravity and roll axis
h_{RCf}	Height front roll center from ground
h_{RCr}	Height rear roll center from ground
h_{CoG}	Height center of gravity from ground
h_F	Height front unsprung weight center of gravity from ground
h_R	Height rear unsprung weight center of gravity from ground
i	Gear ratio
i_{total}	Total gear ratio
k	Spring rate
K	Spring rate
$K_{rolltot}$	Total roll stiffness
K_{rollf}	Front roll stiffness
K_{rollr}	Rear roll stiffness
$K_{rollSPRINGS}$	Spring roll stiffness
$K_{rollfSPRINGS}$	Front spring roll stiffness
$K_{rollrSPRINGS}$	Rear spring roll stiffness
$K_{rollARB}$	Roll stiffness antiroll bar
$K_{rollfARB}$	Roll stiffness front antiroll bar
$K_{rollrARB}$	Roll stiffness rear antiroll bar
K_H	Vehicle heave spring rate
K_P	Vehicle pitch spring rate
K_R	Vehicle roll spring rate
K_X	Vehicle warp spring rate
K_{total}	Total spring rate
K_{spring}	Spring rate
K_{tire}	Tire spring rate
L	Aerodynamic lift force
L	Length
m	Mass
m_{wheel}	Wheel mass
M_R	Suspension motion ratio
MR_F	Front suspension motion ratio
MR_R	Rear suspension motion ratio

MR_{RollR}	Rear antiroll bar motion ratio
M_{roll}	Roll moment
M	Total vehicle mass
$M_{1/4}$	One-quarter body mass
M	Translational mass
M_r	Equivalent rotational mass
M_R	Wheel equivalent mass for roll
M_P	Wheel equivalent mass for pitch
M_f	Mass factor
n	Amount of samples
n	Amount of moles
n_{engine}	Engine RPM
$n_{driveshaft}$	Driveshaft RPM
p	Pressure
P_a	Pressure of dry air
P_w	Pressure of water vapor
P_{engine}	Driven wheel power
PG	Pitch gradient
q	Dynamic pressure
q	Roll stiffness distribution factor
R	Corner radius
R	Electrical resistance
R	Gas constant
R_a	Gas constant of dry air
R_w	Gas constant of water vapor
R_x	Tire rolling resistance coefficient
$r_{rolling}$	Dynamic tire radius
RG	Roll gradient
RG_F	Front roll gradient
RG_R	Rear roll gradient
SR	Slip ratio
SR_F	Front spring rate
SR_R	Rear spring rate
$SR_{chassis}$	Chassis torsion spring rate
t	Time
T	Temperature
T	Track width
T_F	Front track width
T_R	Rear track width
T_{wheel}	Wheel torque
T_{mass}	Acceleration torque
v	Shock absorber velocity
V	Volume
V	Speed
V_{slip}	Slip velocity
V_0	Free rolling velocity
V_{out}	Output voltage
V_{in}	Input voltage
W	Vehicle weight
$WSPD$	Wheel speed
WB	Wheelbase
W_s	Sprung weight
W_{sF}	Sprung weight on front axle
W_{sR}	Sprung weight on rear axle
W_{uF}	Front unsprung weight
W_{uR}	Rear unsprung weight
WR_f	Front wheel rate
WR_r	Rear wheel rate
$WR_{SPRINGF}$	Wheel rate of front springs
$WR_{SPRINGR}$	Wheel rate of rear springs
WR_{ROLLF}	Wheel rate of front antiroll bar
WR_{ROLLR}	Wheel rate of rear antiroll bar
W_{f1}	Front wheel weight measured with car on level surface
W_{f2}	Front wheel weight measured with raised rear axle
W_{LF}	Left-front corner weight
W_{RF}	Right-front corner weight
W_{LR}	Left-rear corner weight
W_{RR}	Right-rear corner weight
x_{wheel}	Wheel movement
x_{LF}	Left-front wheel movement
x_{RF}	Right-front wheel movement
x_{LR}	Left-rear wheel movement
x_{RR}	Right-rear wheel movement
$x_{suspension}$	Suspension movement
$x_{suspensionLF}$	Left-front suspension movement
$x_{suspensionRF}$	Right-front suspension movement
$x_{suspensionLR}$	Left-rear suspension movement
$x_{suspensionRR}$	Right-rear suspension movement
x_H	Heave movement
x_P	Pitch movement
x_R	Roll movement
x_X	Warp movement
x_{CP}	Tire contact patch movement
x_{hub}	Hub movement

Greek Symbols

α	Banking angle
α_{roll}	Roll angle
α_{rollF}	Front roll angle
α_{rollR}	Rear roll angle
$\alpha_{rolltires}$	Tire roll angle
$\alpha_{rolltiresF}$	Front tire roll angle
$\alpha_{rolltiresR}$	Rear tire roll angle
$\alpha_{torsion}$	Chassis torsion angle
β_{pitch}	Pitch angle
δ	Steered angle
δ_{u}	Understeer angle
δ_{SW}	Steering wheel angle
δ_{Acker}	Ackermann steering angle
ε	Emissivity
ε	Strain
ζ	Roll angle ratio
θ	Track slope angle
μ	Average
μ	Friction coefficient
$\mu_{1/2}$	Median
ρ	Density of air
ρ_{15}	Fuel density at 15°C
σ	Standard deviation
σ	Material stress
σ^2	Variance
ω	Frequency

REFERENCES

1. Mitchell, William C., "A Method For Data Alignment," SAE paper No. 983087, SAE International, Warrendale, PA, 1998.
2. SAE International, *Vehicle Dynamics Terminology,* SAE Standard J670, SAE International, Warrendale, PA, 1976.
3. Metz, Gregory L. and Metz, Daniel L., "Deriving Wheel HP and Torque from Accelerometer Data," SAE Paper No. 2000-01-3544, SAE International, Warrendale, PA, 2000.
4. Milliken, William F. and Milliken, Douglas L., *Race Car Vehicle Dynamics,* SAE International, Warrendale, PA, 1995.
5. NHRA, "Introduction To Drag Racing," www.nhra.com
6. Fey, Buddy, *Data Power—Using Racecar Data Acquisition,* Towery Publishing, London, UK, 1993.
7. Davison, Paul, "Check Your Brake Balance Consistency," *Optimum G Newsletter,* Vol. 1, Issue 2, 2005 p. 4.
8. Van Valkenburg, Paul, *Race Car Engineering and Mechanics,* HP Books, Tucson, AZ, 1976.
9. Giaraffa, Matt, "Springs & Dampers Part 2," *Optimum G Newsletter,* Vol. 1, Issue 9, 2005 p. 4.
10. Rouelle, Claude, "Race Car Vehicle Dynamic and Data Acquisition," Seminar binder, Race Car Dynamics and Data Acquisition Training Seminar, Denver, CO, 2001.
11. Fontdecaba, Josep I.B., "Integral Suspension System for Motor Vehicles Based on Passive Components," SAE Paper No. 2002-01-3105, SAE International, Warrendale, PA, 2002.
12. SAE International, *Road Load Measurement and Dynamometer Simulation Using Coastdown Techniques,* SAE Standard J1263, SAE International, Warrendale, PA, 1996.
13. Pi Research, "Aerodynamics Application Note," Issue 1.0, Pi Research, Cambridge, UK, 1998.
14. Editors of The American Heritage Dictionaries, *The American Heritage Dictionary of the English Language,* Fourth Edition, Houghton Mifflin, Boston, MA 2006.
15. Hakewill, James, "Laptime Simulation," www.jameshakewill.com.
16. Rutten, Chris van, "Accessible Simulations," *Racetech Magazine,* 2003 p. 66.
17. Taylor, Barry N. and Kuyatt, Chris E., "Guidelines for Evaluating and Expressing the Uncertainty of NIST Measurement Results," NIST Technical Note 1297, National Institute of Standards and Technology, Gaithersburg, MD, 1994.
18. Omega Engineering Inc., "Omega Engineering Technical Reference," www.omega.com.
19. Gruner, Klaus-Dieter, *Principles of Non-contact Temperature Measurement,* Raytek GmbH, Berlin, Germany, 2003.

20. Hoffmann, Karl, *An Introduction to Measurements Using Strain Gages,* Hottinger Baldwin Messtechnik GmbH, Darmstadt, Germany, 1989.

21. Durant, Andrew J. and Hill, Martin J., *Technical Assessment of the DL2s GPS Data Quality,* Race Technology, Ltd., Nottingham, UK, 2005.

22. Race Technology Ltd., "Race Technology 2006/7 Catalogue," Race Technology Ltd., Nottingham, UK, 2006.

23. Bevly, David M., Daily, Robert and Travis, William, "Estimation of Critical Tire Parameters Using GPS Based Sideslip Measurements," SAE Paper 06ADSC-026, SAE International, Warrendale, PA, 2005.

BIBLIOGRAPHY

Smith, Carroll, *Tune To Win—The art and science of race car development and tuning,* Aero Publishers Inc., Fallbrook, CA, 1978.

Smith, Carroll, *Drive to Win,* Carroll Smith Consulting, Palos Verdes Estates, CA, 1996.

Puhn, Fred, *How to Make Your Car Handle,* HP Books, New York, 1981.

Searle, John, "Equations for Speed, Time and Distance for Vehicles Under Maximum Acceleration," SAE Paper No. 1999-01-0078, SAE International, Warrendale, PA, 1999.

McBeath, Simon, *Competition Car Data Logging—A Practical Handbook,* Haynes Publishing, Newbury Park, CA, 2002.

Hakewill, James, "Suspension Position Measurement," www.jameshakewill.com.

Dixon, John C., *The Shock Absorber Handbook,* SAE International, Warrendale, PA, 1999.

Haney, Paul, *The Racing and High Performance Tire,* SAE International & TV Motorsports, Warrendale, PA, 2003.

Glimmerveen, John H., *Hands-on Race Car Engineer,* SAE International, Warrendale, PA, 2004.

Leuschen, Jason and Cooper, Kevin R., "Effect of Ambient Conditions on the Measured Top Speed of a Winston Cup Car," SAE Paper No. 2004-01-3507, SAE International, Warrendale, PA, 2004.

Lopez, Carl, *Going Faster—Mastering the Art of Race Driving,* Robert Bentley Inc., Cambridge, MA, 1997.

Vaduri, Sunder and Law, Harry E., "Development of an Expert System for the Analysis of Track Test Data," SAE Paper No. 2000-01-1628, SAE International, Warrendale, PA, 2000.

Martin, B.T. and Law, Harry E., "Development of an Expert System for Race Car Driver & Chassis Diagnostics," SAE Paper No. 2002-01-1574, SAE International, Warrendale, PA, 2002.

Kirkup, Les, *Calculating and Expressing Uncertainty in Measurement,* University of Technology Sydney, New South Wales, Australia, 2002.

Pi Research, "Inertial Sensors Application Note," Issue 1.0, Pi Research, Cambridge, UK, 1997.

Wilson, John S., *Sensor Technology Handbook,* Elsevier, Oxford, UK, 2005.

INDEX

NOTE: Page references followed by* f *refer to figures and* t *refer to tables.

ABOUT THE AUTHOR

The author (pictured left) in the company of Pedro Lamy at the Zolder 24 Hours in 2004.

With an educational background in automotive engineering, Jörge Segers has been involved with racing disciplines such as GT and sportscar racing, single seaters, and touring cars since 1998. He started with an apprenticeship at GLPK Carsport, a Belgian team active in international GT racing. Mr. Segers became the team manager only three years later.

After finishing his studies, he was employed at BPR Competition Engineering as track engineer in the International Sports Racing Series and later as development manager at Eurotech Racing. At Eurotech, he was responsible for the GT racing activities of British sportscar manufacturer Marcos Cars.

In 2001, Mr. Segers became the youngest team manager ever in an FIA organized championship. At GLPK Carsport, he is still responsible for the team's activities and the FIA GT Championship. Subsequently, he has been working for other teams such as Henrik Roos Motorsports (FIA GT), Racing for Holland (Le Mans 24 Hours), and Carsport Modena.

His interest in technical writing was triggered when he was asked to write part of the manual for a computer racing game for Simbin Development. A special interest in data acquisition and racecar performance optimization led him to write this book.